Springer Tracts in Civil Engineering

Springer Tracts in Civil Engineering (STCE) publishes the latest developments in Civil Engineering - quickly, informally and in top quality. The series scope includes monographs, professional books, graduate textbooks and edited volumes, as well as outstanding PhD theses. Its goal is to cover all the main branches of civil engineering, both theoretical and applied, including:

- Construction and Structural Mechanics
- Building Materials
- Concrete, Steel and Timber Structures
- Geotechnical Engineering
- Earthquake Engineering
- Coastal Engineering; Ocean and Offshore Engineering
- Hydraulics, Hydrology and Water Resources Engineering
- Environmental Engineering and Sustainability
- Structural Health and Monitoring
- Surveying and Geographical Information Systems
- Heating, Ventilation and Air Conditioning (HVAC)
- Transportation and Traffic
- Risk Analysis
- Safety and Security

Indexed by Scopus

To submit a proposal or request further information, please contact:
Pierpaolo Riva at Pierpaolo.Riva@springer.com (Europe and Americas)
Wayne Hu at wayne.hu@springer.com (China)

Gerhard Fink · Robert Jockwer ·
José Manuel Cabrero
Editors

Holistic Design of Taller Timber Buildings

Springer

Editors
Gerhard Fink
Department of Civil Engineering
Aalto University
Espoo, Finland

Robert Jockwer
Inst. of Steel and Timber Construction
TU Dresden
Dresden, Germany

José Manuel Cabrero
School of Architecture
University of Navarra
Pamplona, Spain

ISSN 2366-259X ISSN 2366-2603 (electronic)
Springer Tracts in Civil Engineering
ISBN 978-3-032-02097-0 ISBN 978-3-032-02098-7 (eBook)
https://doi.org/10.1007/978-3-032-02098-7

This Springer imprint is published by the registered company Springer Nature Switzerland AG
The registered company address is: Gewerbestrasse 11, 6330 Cham, Switzerland

Preface

This book is the final publication of the COST Action CA 20139, holistic design of taller timber buildings, that was carried out between 2021–2025 in an attempt to create a common understanding of the challenges in the holistic design of taller timber buildings.

Buildings, by nature, are complex systems that function optimally when all their subsystems, with their specific roles and functions, operate in harmony. This is particularly relevant for taller timber buildings, which introduce a layer of complexity beyond their counterparts in conventional construction materials such as concrete, steel, or masonry.

Timber, as a natural and renewable material, presents a unique set of challenges. When designed and executed correctly, it can endure for centuries, but it requires particular attention to detail due to its bio-based nature. Unlike other materials, timber is less for-giving when mistakes are made regarding durability or excessive and unbeneficial loading. Therefore, it is essential to approach the design process in a comprehensive and parallel manner, addressing all demands on the building simultaneously rather than in isolation.

The book provides in depth knowledge and insights into the complexities of the designing, planning, assessing, and construction of multi-storey timber buildings. The intrinsic need for a holistic and interdisciplinary approach in these activities is highlighted.

The book aims at going beyond traditional design codes, which often focus on isolated aspects of building design. It integrates a spectrum of engineering topics alongside architects, builders, and other key stakeholders, such as investors, municipalities, and policymakers. By fostering an interdisciplinary dialogue, potential issues can be identified early in the design process, enabling proactive solutions before challenges arise in practice.

Structured in five distinct parts, the book addresses a wide range of contemporary challenges in multistorey timber buildings. After an introduction and overview of the COST Action aims and activities and the mapping of the challenges related to the design of taller timber buildings, the general planning of taller timber buildings with regard to the different structural and service demands is discussed. This is followed by the topic of performance and assessment of taller timber buildings from the construction phase throughout the entire service life and beyond. The particular exposures and hazard, that taller timber buildings are exposed to, are analyzed in the 4th chapter and a framework towards multi-hazard design of taller timber buildings is developed. The end-of-life scenarios and the aspects and potentials for circularity are discussed in the final chapter of the book. Finally, also short summaries of related articles of the special issue of the COST Action (Wood Material Science & Engineering, Volume 20, Issue 4 (2025)) are provided delivering the basis for further research focused literature.

Ultimately, this book provides timber construction professionals, building investors, and users with valuable guidelines to create safer, more robust, and more comfortable multi-storey timber buildings. By optimizing the use of raw materials and integrating diverse expertise, it aims to contribute to the future of timber construction in an ever-evolving built environment.

This book consists of individual chapters, for which the respective authors are solely responsible. We would like to thank all the authors, contributors, and reviewers of this book for the interesting discussions, their great work, and for the punctual delivery!

This book is based upon the work from COST Action HELEN-Holistic design of taller timber buildings CA20139, supported by COST (European Cooperation in Science and Technology).

The financial support is greatly acknowledged!

Gerhard Fink
Robert Jockwer
José Manuel Cabrero

COST Acknowledgments

This book is based upon the work from COST Action HELEN-Holistic design of taller timber buildings CA20139, supported by COST (European Cooperation in Science and Technology).

COST (European Cooperation in Science and Technology) is a funding agency for research and innovation networks. Our actions help connect research initiatives across Europe and enable scientists to grow their ideas by sharing them with their peers. This boosts their research, career, and innovation.

https://www.cost.eu/

EUROPEAN COOPERATION IN SCIENCE & TECHNOLOGY

Funded by the European Union

Contents

Introduction

COST Action HELEN – Holistic Design of Taller Timber Buildings

Gerhard Fink[1](✉), Robert Jockwer[2], Iztok Šušteršič[3], José Manuel Cabrero[4], Mislav Stepinac[5], Eleni Toumpanaki[6], Pedro Palma[7], Chiara Bedon[8], Daniele Casagrande[9], and Steffen Franke[10]

[1] Aalto University, Espoo, Finland
gerhard.fink@aalto.fi
[2] TUD Dresden University of Technology, Dresden, Germany
robert.jockwer@tu-dresden.de
[3] LynxCraft, Ljubljana, Slovenia
iztok@lynx-craft.com
[4] Universidad de Navarra, Pamplona, Spain
jcabrero@unav.es
[5] Faculty of Civil Engineering, University of Zagreb, Zagreb, Croatia
mstepinac@grad.hr
[6] University of Bristol, Bristol, UK
eleni.toumpanaki@bristol.ac.uk
[7] Empa – Swiss Federal Laboratories for Material Science and Technology, Dubendorf, Switzerland
pedro.palma@empa.ch
[8] University of Trieste, Trieste, Italy
chiara.bedon@dia.units.it
[9] University of Trento, Trento, Italy
daniele.casagrande@unitn.it
[10] ETH Zürich, Zurich, Switzerland
steffen.franke@ibk.baug.ethz.ch

Abstract. The construction sector is responsible for a large amount of greenhouse gas emissions, energy and raw materials consumption and waste production. Thus, a shift towards sustainable construction methods is essential. Engineered timber, a leading material for sustainable construction, has advanced to a point where it can now be efficiently used for taller buildings. Unfortunately, designing taller buildings made from timber is more demanding compared to their concrete and steel counterparts. A successful design and construction of taller timber buildings requires intensive collaboration between different fields. COST Action HELEN aims to support the transformation from research on isolated topics to a more integrated, interdisciplinary approach. This shift is essential for the safe and economic design, construction, maintenance, and recycling of taller timber buildings.

Keywords: Taller timber buildings · Holistic design · Robustness · Reuse and repairability · Deformations and Vibrations · Accidental Load Situations · Sustainability · Durability

G. Fink et al. (Eds.): *Holistic Design of Taller Timber Buildings*, SPRTRCIENG, pp. 3–15, 2025.
https://doi.org/10.1007/978-3-032-02098-7_1

This book chapter contains content from other documents and contributions created from the authors [1–3], where you can also find more detailed information about COST Action HELEN.

1 Introduction

With the worldwide construction sector being responsible for a significant amount of the carbon dioxide emissions, the world's energy use and waste production, a shift towards sustainable and renewable construction is crucial. Engineered timber will play an essential role in this transition, as it has evolved to a stage that enables the construction of taller buildings that are commonly built from concrete or steel. The number and height of multi-storey timber buildings has seen significant growth over the past decade. Each year, these boundaries are being pushed further. As of 2020, the tallest pure timber multi-storey apartment building stands at 18 storeys (85 m), while the tallest timber-concrete hybrid reaches 24 storeys (84 m). Buildings up to 10 storeys are now considered midrise. Contemporary multi-storey timber buildings are increasingly seen as a long-term sustainable solution, particularly in urban areas, where they offer a greener alternative to concrete and steel constructions while also improving living quality and fostering healthier environments.

The current design process for taller timber buildings is, in principle, not different to other buildings. Typically, architects create the concept, followed by various engineering disciplines working on their respective areas. Structural engineers are responsible for dimensioning building elements and connections, mechanical engineers design heating, ventilation, and plumbing systems, fire engineers ensure fire safety measures are in place, and acousticians help architects to select the right components to minimize sound transmission. Nevertheless, designing taller buildings made from timber is more demanding than their concrete and steel counterparts. Whereas different designers (architects, structural, fire engineers etc.) of concrete buildings can work almost independently, the design of taller timber buildings should be performed with intensive collaboration among the design team members. Designing multi-storey timber buildings has always involved highly specialized engineering teams with in-depth knowledge of the unique challenges and demands these buildings present.

Despite the construction of several midrise and some taller timber buildings, the understanding of how to design such structures remains far behind that of concrete and steel buildings. Taller timber buildings present additional challenges that are yet to be fully addressed. Over the past 15 years, research in timber engineering has intensified, and this knowledge is gradually being integrated into design codes and handbooks. North American codes are being updated more rapidly than European ones, with the 2021 US International Building Code permitting timber buildings up to 18 storeys tall. In contrast, European building directives vary by country, and the revised Eurocode for general structural design is not expected to be available before the end of 2025. Despite these efforts, most global research on multi-storey timber buildings has focused on specific aspects such as connections, vibrations, acoustics, fire safety, and durability, rather than considering the design in a holistic manner. However, an integrated approach is crucial for designing taller timber buildings, and is the primary challenge addressed by this COST Action.

2 Aims and Objectives

COST Action CA20139 – Holistic Design of Taller Timber Buildings (HELEN) is a research network that started on 12/10/2021 and will run until 11/10/2025 [2] and is funded by the European Cooperation in Science and Technology (COST). The main goal of HELEN is to support the transformation of building construction research from isolated topics to a more integrated, interdisciplinary approach. This shift is essential for the safe design, construction, maintenance, and recycling of taller timber buildings.

A key challenge in the design of timber buildings lies in the inherent design complexity of timber buildings due to the material's nature and properties. Without the collaboration of experts from different fields, significant conflicts can arise, affecting both the load-bearing capacity and serviceability of the building. An example is the acoustic insulation methods used in timber buildings that often contradict the design strategies against wind or earthquake loads. Whereas acoustic requirements might result in decoupling building elements in order to reduce the noise transfer, wind and seismic forces demand elements to be tightly connected to resist lateral forces.

HELEN aims to promote international collaboration and interest in developing a shared understanding and common guidelines for the holistic design of taller timber buildings. This is achieved through the exchange of technical and scientific expertise from the diverse research profiles within the network, as well as leveraging their research facilities. Cooperation within this network enables coordinated research efforts to accomplish the objectives (Fig. 1).

Fig. 1 Illustration of the interaction of different building design fields and their inherent collisions [1, 2]

3 Working Groups

HELEN is structured in four working groups (WGs). It was acknowledged early on that overlap existed between the WGs due to the multifaceted, timely and challenging research questions to be addressed and collaboration between WGs was also prioritised (see also [1, 3]).

3.1 WG 1 – Robust Design for Adoption, Reuse and Repair

The Working Group 1 (WG1) deals with aspects related to robustness, adaptability, design for disassembly and reuse, and repairability. Given the broad range and interdisciplinary nature of the topics assigned to WG1, it has members with different backgrounds in both engineering and architecture, as well as in research and industry. After the first years, WG1 has been reorganised into two sub-groups (SG), one dealing with robustness and another dealing with design for extended service life.

The SG "Robustness" deals with the topics of resistance to disproportionate damages, including structural and non-structural robustness and resistance to progressive collapse. The SG has worked on developing a framework for the design of timber buildings against disproportionate collapse, which includes identifying all stakeholders and their interests and responsibilities. Case studies of structural design for increased robustness have been analysed and the most important strategies have been identified. The ongoing research projects and exchanges with structural engineers involved in designing timber buildings against progressive collapses have allowed the SG to identify some key issues: guidance to adjust the necessary measures to increase resistance to disproportionate collapse to the risk category of the building; simplified structural analysis models for alternative load-paths (ALPs), e.g., with dynamic amplification factors; behaviour of connections under large deformations, e.g. catenary action; connections as fuse elements in segmentation strategies, which has similarities with capacity design for earthquake resistance; and "power storeys" for vertical segmentation in taller timber buildings.

The SG "Design for extended service" results from the merger of previous SGs "Adaptability", "Design for disassembly and re-use" and it dealt with topics related to changes in the functional use of buildings, how the design of tall timber buildings can account for adaptability-related requirements, and their impact on other requirements (e.g., robustness, acoustics, durability). The SG has focused on the evaluation of the demountability of timber connections and on identifying solutions that facilitate disassembly of timber buildings. Design for disassembly is not only important for increasing the reuse and recyclability potential, but primarily for repairs in case of local damage. Damages and repair of modern timber buildings have also been addressed, since it is a critical aspect for insurance purposes.

3.2 WG 2 – Deformation and Vibration

WG2 focused on aspects and design issues primarily related to deformations and vibrations, in the framework of taller timber structures. WG2 took a primary advantage from the interaction of scientist and professional engineers (i.e., representative of research institutions, universities and industrial partners) that have different experiences and technical skills on these themes. To optimize the impact of networking activities, the internal organization of WG2 was established into two SGs: "Deformations" and "Vibrations".

Talking about the SG "deformations" in tall timber structures, the attention was spent on a multitude of aspects and issues that have major effects in research and industrial applications, and are often fairly addressed by existing standards and regulations. As a matter of fact, deformations in timber structures are primarily associated to joints and connections. There are however no doubts about the complexity and variability of possible technological solutions in the field of joints and connections for timber structures. Also, the type of load, the boundary conditions and the assessment of their mechanical performance suggests the need of a robust background in support of optimal and safe mechanical design of these systems. The SG "vibrations" is implicitly related to deformations and corresponding gaps in engineering knowledge / design tasks. Starting from the assumption that vibration itself is a rather general definition and can cover a multitude of practical / technical aspects in the framework of timber structures, WG2 members actively contributed to the elaboration of a State-of-the-Art document in which most of engineering terms and problems could be first defined in their context. So far, do we implicitly talk about vibrations in floors or partition walls for timber structures? And which kind of design action should be primarily addressed in terms of vibration serviceability, for the specific solutions in use in tall timber structures? But indeed, how can we monitor and control, or possibly minimize and mitigate the effect of vibrations in typical load-bearing components for tall timber structures? There are no doubts, first of all, about the inter-correlation of vibrations and deformations, which again suggest an intrinsic mutual interaction of load-bearing components for tall timber structures and the final user/the design actions.

In most of WG2 elaborations, one major gap emerged from various sub-tasks, that is the need of standardized operational procedures and guidelines which could be efficiently applied to any type of building component. This general consideration implicitly recalls the complexity of the topics, as they are strictly interconnected to the effects of different design actions (see WG3).

3.3 WG 3 – Accidental Load Situations

WG3 activities were aimed at exploring optimized approaches for the holistic design of taller timber buildings under accidental load situations. Specifically, accidental load scenarios due to earthquake, fire and blast were considered acting either as separate actions (single hazard) or as simultaneous or cumulative events (multi-hazard).

The WG3 was composed of four different SGs, dealing with "seismic", "fire", "blast" and "multi-hazard accidental load". Each subgroup focused on a deep examination of the state-of-the-art for each load situation in order to identify the current limitations that can be met in the design phases of mid-to-high rise timber buildings. Moreover, the potential interactions and collisions among different accidental load scenarios were analysed. A survey was circulated among WG3 members to assess the urgency of filling gaps in knowledge in different design situations as well as evaluating the collisions between seismic and blast, seismic and fire and fire and blast loads.

For seismic load situations, the need of developing innovative high-performance connections can be considered as a priority in the field. Most of proprietary connections are characterized by values of resistance primarily suitable for low-to-midrise buildings but further investigations are still needed to explore the ductility performance and the brittle failure modes of customized anchors (e.g. hold-down) in taller buildings.

Three main research topics were identified as primary to overcome the current limitations in taller timber buildings under fire loads, namely the contribution of timber element to both external and internal fire spread, the structural robustness and the timber's contribution to fire development. Regarding the design for blast loads, research and technological development in the near future should be mainly aimed at investigating the redistribution of internal loads after an element loss and ensuring appropriate redundancy.

The design of timber connections characterized by adequate ductility, the adoption of proper capacity-design approaches and the design of shearwalls for both in-plane (seismic) and out-of-plane (blast) load are example of strong interactions for accidental load design situations involving seismic and blast. Regarding fire and blast, further investigations should be conducted to investigate how the structures behave under fire occurring after blast (fire following blast). The damages of structural components and of the protection elements caused by blast need to be carefully assessed to determine the residual fire resistance of structural components. Similarly, the damages of structural connections and non-structural elements caused by strong earthquakes may significantly influence the capacity of structure under fire load (fire following earthquakes).

Many of these topics have been largely discussed withing WG3 meeting and summarized in some of the chapter of this book with the aim of addressing new research advancements.

3.4 WG 4 – Sustainability and Durability

Timber constructions have gained the (rightful) reputation of being a sustainable building option. On the other hand, they also raise questions regarding their durability. They are more susceptible to damage, either induced by moisture or insects, as well as design mistakes due to their complexity. They are also less forgiving when it comes to construction mistakes, possibly leading to premature failure of their building components. WG4 looks into the issues dealing with taller timber buildings' environmental footprint and their longevity based on the design details, all assessed through the interdisciplinary prism of the consortium's experts. The results of this Working Group's work are in close correlation to WG1, where the initial design assumptions are considered. As in other WGs, work in this group is also country dependent as, apart from local legislation, local climate properties are also of great importance. The possibility to build safely and effectively in areas with heavy rain and snow differs greatly from drier places. This, in turn, influences the construction technologies, which, in turn, affect the building erection price, which makes the timber alternatives to concrete or steel more or less viable. For Europe, which strives for an increase in sustainable timber construction, this opens a discussion on state subsidies for timber construction in order to make them more attractive to investors. The interdisciplinary consortium, also including Life Cycle Costing (LCC) and Social Life Cycle Assessment (S-LCA) experts, is able to provide answers to such questions. WG4 is mainly divided into two SGs: "Sustainability" and "Durability".

4 Activities and Outcomes

4.1 Network and Participation

The COST Action HELEN has successfully built a global network of researchers, educators, and practitioners from across Europe and beyond. The network has been continually growing during the last years and has currently more than 410 members from more than 40 countries. The network is well-distributed between the individual working groups, with more than 150 participants in each of them. More than half of the WG members are early-career investigators. About one-third of the WG members are from inclusive target countries, which ensures knowledge distribution among various COST members. HELEN targets a high female participation, and could achieve 31%. In the second year of HELEN, a Diversity, Equity, and Inclusion (DEI) Coordinator was appointed to oversee and implement initiatives aimed at promoting diversity and inclusivity within the COST Action, providing equitable opportunities among members and facilitating an open network of knowledge among a diverse community of researchers, professionals, and manufacturers, that reflects the wide range of skills required for the design, construction, and research of taller timber buildings. To promote collaboration in such a diverse group, a tailored session was designed in the fourth working group meeting in Hasselt, Belgium, in May 2024. The session aimed at understanding different social identities (e.g., age, gender, religion etc.), individual experiences and personality traits, and creating a safe environment for open dialogue and effective communication, fostering productive collaboration within and between working groups. Key aspects highlighted from this session are: (a) members may perceive themselves at a disadvantage, if they come from a country with no forestry industry or without seismic activity, as this can limit their background knowledge in timber engineering and earthquake engineering; (b) there are still steps to be done in the engineering sector towards more inclusive career opportunities, and (c) increased work/study mobility within Europe facilitates broadening horizons with exposure to different cultures and knowledge hubs. The HELEN activities facilitate cross-pollination and knowledge transfer between practitioners and researchers irrespective of their backgrounds, providing equitable opportunities to expand their knowledge (e.g., through Short-Term Scientific Missions), and showcase their work (e.g., through WG meetings), supporting effective collaboration.

4.2 Workshops and Conferences, and Training Schools

HELEN organizes conferences, working group meetings, management committee sessions, and specialized events (see also Fig. 2). These gatherings facilitate the transfer of knowledge from basic research to practical industrial applications, with contributions from industry experts and selected keynote speakers. Face-to-face meetings are considered key elements for engaging more actively those involved in research, education, and practice. Particular focus is placed on Inclusiveness Target Countries (ITCs) and Early Career Investigators (ECIs), with attention to gender equality. Synergies between institutions, as well as collaboration with other COST Actions, help extend the reach.

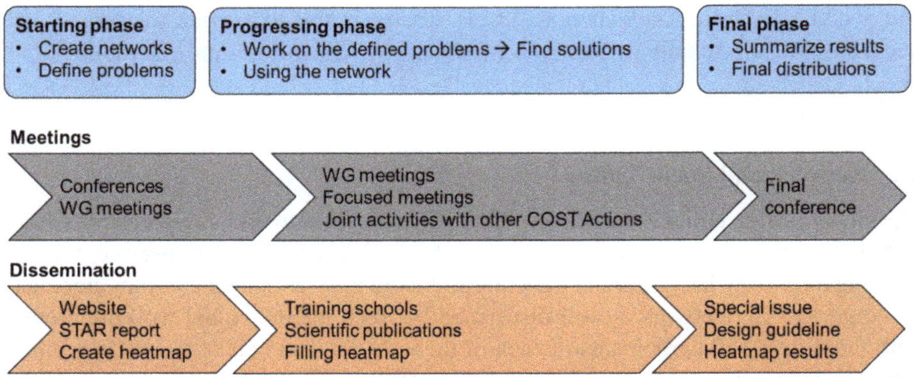

Fig. 2 Overview of aims and activities during the course of the COST Action HELEN

4.3 Training Schools

During the course of HELEN three training schools have been organized on targeted topics related to taller timber buildings. These training schools are carried out by members and for members of the COST Action. Through a holistic concept, it provides learning opportunities for members at all career levels.

The first training school, held in Cortaccia, Italy, covered a broad range of topics around timber connections [4], from the basic theories towards innovation and developments in structural modelling. The lectures included the evolution of timber connections, with emphasis on dowel-type connections in line with the new Eurocode 5. Special connection aspects, such as 3D connectors and reinforcement techniques, were explored, along with the use of Cross-Laminated Timber (CLT) and screws in modern timber construction.

The second 3-day training school, held in Zagreb, Croatia, focused on Sustainability of Taller Timber Buildings focused on pathways to achieve zero-emission buildings, exploring decarbonization strategies across different materials: steel, concrete, and timber. A significant portion was dedicated to sustainability in timber buildings, emphasizing sustainable forest management practices. The training further examined the concept of whole life carbon, introducing participants to the distinction between embodied and operational carbon. Through lectures and hands-on sessions, students gained familiarity with related standards and the practical application of life cycle assessment tools. Special attention was given to the modelling and validation of LCA results, with discussions on sensitivity and uncertainty analysis to strengthen the reliability of environmental performance evaluations.

The third training school, held in Zurich, Switzerland, focused on the topic "Case studies in Taller Timber Buildings" [5]. During the training school, experts from various fields who have been involved in the design of tall timber buildings presented their projects and shared their expertise with the participants.

4.4 Short-Term Scientific Missions and Conference Grants

Short-Term Scientific Missions (STSMs) are a valuable tool for promoting collaboration within HELEN and are especially encouraged for ECIs. These missions foster synergies between institutions, accelerate student learning and knowledge transfer, and offer both academia and industry access to emerging talent and fresh perspectives. So far the Action supported 41 STSMs. Documentation from these missions is available on the HELEN website, and all STSM reports will be compiled and published in a dedicated "Book of STSMs". This publication will provide an overview of each mobility, including the problem statement, a concise summary of the research activity, and details about the grantee and host institution. The submission, evaluation, and approval process for STSM proposals followed DEI principles and considered ITC, ECI, and gender balance. The outcomes of STSMs are regularly presented and discussed in plenary sessions during workshops. In total, researchers from 20 different countries participated, either as grantees who travelled abroad or as hosts who welcomed visiting scholars. This wide geographical distribution reflects the strong collaborative network fostered within the Action. Furthermore, HELEN has funded so far 6 researchers from ITC to participate in international conferences to present their research results.

4.5 Dissemination

HELEN provides a network framework for individual researchers, which has resulted in several national and international research proposals and research initiatives. Beside individual contributions and initiatives three joint contributions have been realised or are currently under development.

The key achievement of the first year was the creation of the state-of-the-art report titled "Holistic Design of Taller Timber Buildings (HELEN)" [6–9]. This report explores various aspects of the subject, including both the drivers and barriers. It is available on the HELEN website, where it can be downloaded as a complete document or in four separate sections: (1) Design for robustness, adaptability, disassembly, reuse, and repairability of taller timber buildings; (2) Design of taller timber buildings to prevent deformations and vibrations; (3) Design of taller timber buildings to withstand accidental loads; (4) Sustainability and durability of taller timber buildings.

Furthermore, HELEN published a special issue in the journal *Wood Material Science and Engineering* [10]. The special issue includes a variety of research topics that have been carried out during the Action: The topics covered in the special issue are related to timber connections, timber-concrete composites, reinforced timber elements, fire safety, and durability. Particular attention has been given to the reuse of structural timber as well as the potential of disassembling wood construction.

The final book publication of the COST Action HELEN, which you are currently reading, provides in-depth knowledge and insights into the complexities of the designing, planning, assessing and construction of multi-storey timber buildings. The intrinsic need for a holistic and interdisciplinary approach in these activities is highlighted. Buildings, by nature, are complex systems that function optimally when all their subsystems, with their specific roles and functions, operate in harmony. This is particularly relevant for taller timber buildings, which introduce a layer of complexity beyond their counterparts in conventional construction materials such as concrete, steel, or masonry.

5 Heatmap

The HELEN COST Action aims to support a paradigm of building construction research, shifting R&D from isolated topics to an integrated interdisciplinary approach, which is critically necessary to safely design and build as well as correctly maintain and recycle taller timber buildings. In order to achieve this aim, potential conflicts between the different areas of expertise need to be identified. Analysis of such data are assumed to be used to direct policy and shape research directions.

Identifying and addressing potential design conflicts, is the cornerstone to effectively contribute to improved and safer timber buildings and, hence, transition to a decarbonised built environment. During the MC meeting in 2020, a list of topics with potential conflicts was drafted and further developed in the subsequent meetings. Eight main areas, each with several sub-categories, resulting in a total number of 42 categories, have been identified. A compilation of the criteria is presented in Table 1. Along these criteria and topics, experts from the COST Action have identified and commented on the different conflict risks throughout the duration of the action.

From the responses, possible conflicts can be mapped and visualised. The results are illustrated in Fig. 3, where the number on both axis represents the conflict categories as given in Table 1 and the colour represents the "average conflict" from the participants who filled out the specific field, with '3' representing the highest degree of conflict. It can be seen that for the construction and the use phases significantly higher conflicts were identified.

The following 10 categories are those with the highest number of identified conflicts (all of them had 15 or more conflicts): Timber structures, Timber connections, Fire behaviour of timber, Seismic response, Wind response, Maintenance, Repair, Fire spread modelling, Moisture, Architectural design. The category Timber Structures may have been interpreted too broadly (akin to Timber Engineering) and thus showed conflicts across nearly all other categories. The remaining conflict areas are explored in more detail below. From these matrices, the conflict of different topics in the different processes during the entire lifetime of taller timber buildings from design, construction, use, to recycling can be identified.

For connections, challenges are identified for all stages of the life cycle of the building, but particularly regarding waste management (#33). Proper detailing is important to ensure maintainability but this can be compromised during construction due to issues like moisture ingress. Training and education is needed in order to impact and enhance the quality of timber structures for their entire lifetime.

Both the fire behaviour and the fire spread modelling are considered to be in large conflict with architectural design (#23–27). Experts also noted that certain materials used to improve sound and acoustic performance often perform poorly in fire scenarios.

In terms of seismic and wind response, conflicts were observed with acoustic performance (#18) and floor vibrations (#9). A key challenge arises from the need to increase the building's mass to reduce wind response. This, however, leads to higher inertial loads during seismic events.

During the construction and use phases, notable tensions exist between structural reliability (#7) and robustness (#8) on one hand, and maintenance (#10) and repair (#12)

Table 1 Summary of the conflicts categories

Timber engineering	
1. Timber structures 2. Timber connections 3. Fire behaviour of timber 4. Environmental influences 5. Seismic response 6. Wind response 7. Structural reliability	8. Robustness 9. Floor vibration 10. Maintenance 11. Repair 12. Monitoring 13. Disassembly 14. Duration of load
Computational modelling	
15. Fire spread modelling 16. Seismic modelling	17. General finite element modelling
Building physics	
18. Acoustics 19. Volatile organic compounds 20. Indoor air quality	21. Thermal behaviour 22. Moisture
Architecture	
23. Architectural design 24. Room design 25. Facades	26. Urban planning 27. Roofs
Construction management	
28. Factory management 29. Construction site management 30. Logistics	31. Industrialisation 32. Prefabrication 33. Waste management
Material science	
34. Material production (EWP) 35. Adhesive	36. Coatings 37. Wood modification
Human health	
38. Restorative design	39. Ergonomics
LCA	
40. Life cycle analysis 41. Life cycle cost	42. Social life cycle analysis

on the other hand. Currently, there is a lack of clear guidance on how robustness relates to reuse and recycling.

Maintenance and repair are especially critical in the construction and operational phases, yet they face challenges related to computational modelling (#15–17) and architecture (#23–27). Experts emphasize the need to first develop effective repair strategies for timber buildings to enable future reuse and recycling.

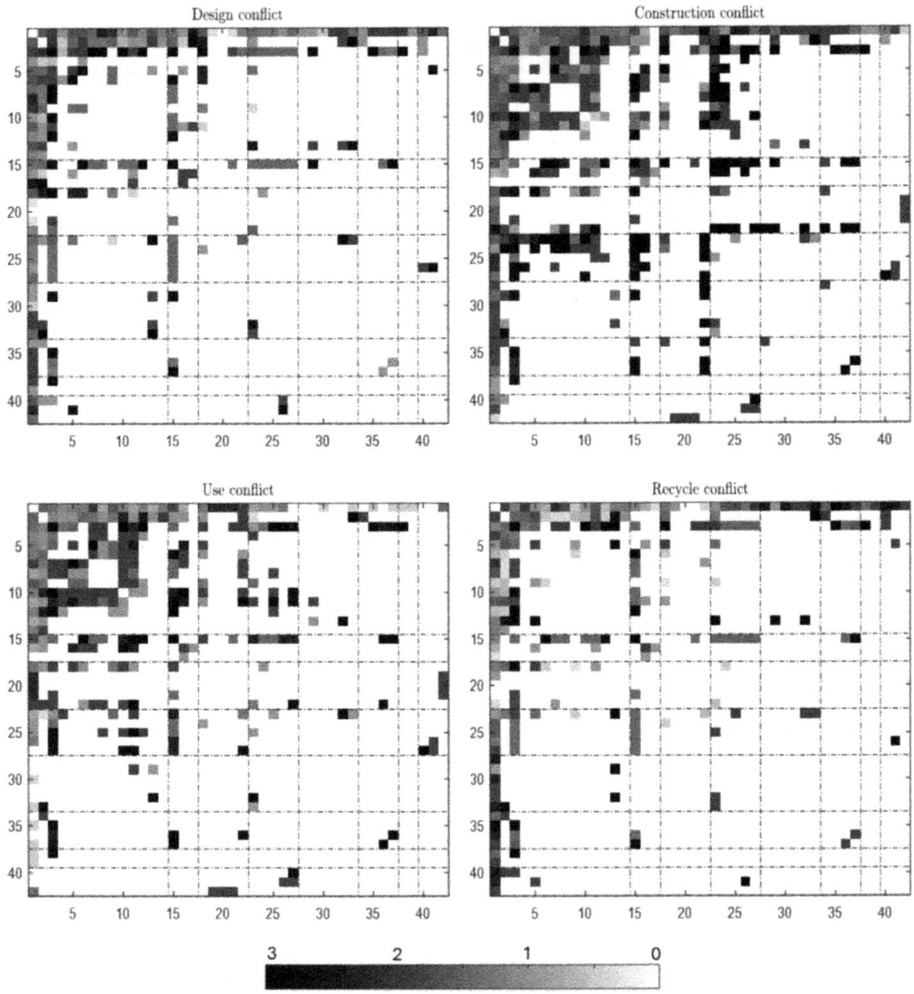

Fig. 3 Identified collision for different construction stages depending on the conflict categories. The colour represents the "average conflict" from the participants who filled out the specific field.

Finally, in timber construction, architectural design must not be viewed solely as an aesthetic discipline but rather as a performance-oriented tool, that takes account of the unique demands and opportunities of the material.

Acknowledgments. This publication is based upon work from COST Action CA20139 - Holistic design of taller timber buildings (HELEN), supported by COST (European Cooperation in Science and Technology).

For further information please consult also the COST website https://www.cost.eu/actions/CA20139/ and the Actions website: https://cahelen.eu/. The authors thank all members of the COST Action CA20139 HELEN, for the discussions, presentations, and contributions. Special thanks also go to Martina Sciomenta and Shady Attia.

References

1. Fink, G., Jockwer, R., Šušteršič, I., Stepinac, M., Palma, P., Bedon, C., Casagrande, D., Franke, S., D'Arenzo, G., Brandon, D., et al.: Holistic design of taller timber buildings–cost action helen (CA20139). In: Proceedings of the World Conference on Timber Engineering, Oslo, Norway, pp. 1001–1008 (19–22 June 2023)
2. COST: Memorandum of Understanding for the Implementation of the COST Action "Holistic Design of Taller Timber Buildings" (HELEN) CA20139; European Coopera-tion in Science and Technology (COST): Brussels, Belgium (2021) Available online: https://www.cost.eu/actions/CA20139/. Accessed 30 March 2025
3. Jockwer, J., Fink, G., Cabrero, J.M., Stepinac, M., Toumpanaki, E., Palma, P., Bedon, C., Casagrande, D., Franke, S.: Addressing the challenges in the holistic design of taller timber buildings – COST Action HELEN. In: World Conference on Timber Engineering, Brisbane, Australia (2025)
4. Brandon, D., Cabrero, J.M., Casagrande, D., D'Arenzo, G., Franke, S., Grossi, P., Izzi, M., Jockwer, R., Ottenhaus, L., Palma, P., Polastri, A., Viau, C.: Training school on holistic design of timber connections. In: Casagrande, D. (ed.) Cortaccia, Italy: COST Action CA 20139 Holistic Design of Taller Timber Buildings (HELEN) (2023)
5. Palma, P., Felicita, M., Franke, S.: Training school "case studies in taller timber buildings" (under preparation). In: COST Action CA 20139 Holistic design of Taller Timber Buildings (HELEN), Zurich (2025)
6. Palma, P., Fink, G. (eds.): Design for robustness, adaptability, disassembly and reuse, and repairability of taller timber buildings: a state of the art report. In: COST Action CA 20139 Holistic Design of Taller Timber Buildings (HELEN) (2022)
7. Bedon, C., Reynolds, T., Aloisio, A. (eds.): Design of taller timber buildings against deformations and vibrations: a state-of-the-art review. In: COST Action CA 20139 Holistic Design of Taller Timber Buildings (HELEN) (2022)
8. Casagrande, D., Brandon, D., D'Arenzo, G., Viau, C. (eds.): Design of taller timber buildings subjected to accidental loads: a state-of-the-art review. In: COST Action CA 20139 Holistic Design of Taller Timber Buildings (HELEN) (2022)
9. Franke, S. (ed.): Sustainability and durability of taller timber buildings: a state-of-the-art report. In: COST Action CA 20139 Holistic design of taller timber buildings (HELEN) (2022)
10. Fink, G., Jockwer, R., Cabrero, J.M.: COST Action CA20139 – Holistic design of taller timber buildings (HELEN). Wood Mater. Sci. Eng. **20**(4), 753–754 (2025). https://doi.org/10.1080/17480272.2025.2519550

Planning of Taller Timber Buildings - ULS and SLS design

Vertical Deformations in Tall Timber Buildings: Sources and In-Field Estimations

Martina Sciomenta[1(✉)] and Chiara Bedon[2]

[1] Department of Civil, Architectural and Building and Environmental Engineering, University of L'Aquila, L'Aquila, Italy
martina.sciomenta@univaq.it

[2] Department of Engineering and Architecture, University of Trieste, Trieste, Italy

Abstract. Tall timber building 'structural components may undergo severe vertical deformations due to the action of different loads/load combinations during their lifetime. These deformations, basically unusual or negligible for low and mid-rise buildings 'components, could harshly affect the overall stability of TTBs; therefore, specific estimation methods and mitigation intervention must be established. A deeper understanding of the short- and long-term deformation behavior of large span and highly loaded components is crucial for improving future design possibilities for TTBs, particularly concerning serviceability during both construction and operational phases. In the following paragraphs, an overview of the current analytical formulations and experimental methods for the estimation of vertical deformations is provided. Firstly, since for practitioners it is essential to adopt clear formulation, the provision included in the main standards are discussed. Secondly, the discussion concerns the experimental methodology for the assessment of in-situ vertical deformations.

Keywords: Vertical deformations · Axial shortening · Axial shrinkage · Connection settlements · Brittle failures

1 Introduction

In the context of developing high-rise structures, the vertical carriers play a predominant mechanical role (Fig. 1). Differential shortening in high-rise building structures is a well-documented phenomenon in the literature [1–3]. It arises from two primary factors: (i) variation in axial loading/moisture conditions within structural elements, and (ii) differences in the design of adjacent vertical load bearing elements. Due to these differences, the reaction of the adjacent vertical structural elements expressed as length change can be different [4]. At the serviceability limit state, the cumulative compressive deformation generates vertical displacements that can no longer be ignored or neglected [5].

In the following, an overview of deformation methods provided by Standards is discussed, with the aim to highlight the current situation and the gap of knowledge: Then, novel proposals and experimental campaigns carried out in situ are summarized.

G. Fink et al. (Eds.): *Holistic Design of Taller Timber Buildings*, SPRTRCIENG, pp. 19–27, 2025.
https://doi.org/10.1007/978-3-032-02098-7_2

Fig. 1. Example of timber structure. Figure reproduced from [6] under the terms and conditions of a CC-BY license agreement.

2 Source of Vertical Deformations

According to [7] the main sources causing vertical deformations in tall mass timber structures can include: axial shortening including creep, shrinkage on beams, columns and panels, crushing perpendicular to the grain on beams and panels, connection tolerances and joint settlements.

2.1 Axial Shortening

The main source of short-term axial shortening on columns are gravity loads acting on them. A variable that can affect column axial shortening is the fire-resistance rating (FRR) of the column, which can require a cross-sectional area oversizing for fire purposes. The considerations for shortening of mass timber bearing walls are similar to those for mass timber columns, although the magnitude of anticipated vertical movement will likely be lower. Due to the continuous nature of these systems and their uniformly supported loads, vertical movements would be expected to be low. For systems such as CLT, vertical movement is typically restrained by horizontally oriented laminations, making for a more dimensionally stable wall panel [7]. An additional contribution to columns 'axial shortening is due to creep effect under long-term loading conditions. Three primary components contribute to the total creep deflection of a structural timber element. These three components are: (i) time-dependent or viscoelastic creep which mostly depends on the stress-level, the temperature and moisture, (ii) mechano-sorptive creep due to moisture changes and (iii) pseudo-creep that is attributed to swelling and shrinkage of the timber.

2.2 Axial Shrinkage

Another source of deformation comes from the shrinkage of timber. This change of dimensions is due to the hygroscopic behavior of timber which absorbs and releases moisture from the surrounding environment.

Due to the cellular structure of wood, shrinkage in timber elements occurs primarily perpendicular to grain, meaning that shrinkage effects on a timber member are much more significant in its width and depth than its length. This means that the main deformation due to shrinkage would affect beams or floor joist which will shrink in their cross-section dimensions while their length will remain unchanged. In multi-story buildings, wood shrinkage is therefore concentrated at the wall plates, floor and roof joists, and rim boards. The total shrinkage in conventional framed buildings can be determined by summing the estimated shrinkage of horizontal lumber members in walls and floors, such as wall plates and floor joists [8].

The American Softwood Lumber Standard (PS 20) offers a general guideline for calculating shrinkage in most softwood species. It specifies that for every 4% point reduction in moisture content below the fiber saturation point, there is a corresponding 1% shrinkage. This corresponds to a shrinkage coefficient of 0.0025.

Nevertheless, the shrinkage S, could also be calculated as Eq. (1):

$$S = D \cdot M \cdot C \tag{1}$$

Being D one of the dimensions of the cross section, M the change in moisture content expressed as percentage and C the shrinkage coefficient.

In wood-frame buildings of three or more stories, cumulative shrinkage can be significant and have an impact on the function and performance of finishes, openings, mechanical/electrical/plumbing (MEP) systems, and structural connections.

Since the column-to-beam zone is one of the most critical for vertical movement due to shrinkage in the beam cross section, [7] provided an alternative solution to the "traditional" platform-frame detail in which the upper column stands on the floor panel supported by the beam. The suggested choice would imply to notch the lower column to create a shelf to host the beam or to use a steel device to connect the upper and lower column and so eliminate the cross-grain shrinkage zone.

2.3 Crushing Perpendicular-to-the-Grain

Due to the reduced strength of timber members in the direction perpendicular to the grain, they are susceptible to isolated crushing. This must be accounted when designing structural components, especially given the potential vertical movements that may arise from the aforementioned factors. Certain configurations, such as steel-on-timber bearing conditions where timber is loaded perpendicular to the grain, increase the likelihood of crushing. This configuration is typical of TTBs with post and beams and diagonals as lateral load resisting system, in particular the joint made of slotted-in steel plates that connect diagonals to the continues beams generates on it pressure in the direction perpendicular to the grain which can cause crushing. This issue can partially be solved by using steel plates on top of and underneath the beam at the location of the connection to spread the force over a larger area [9].

2.4 Joint Settlement

According to the National House Building Council (NHBC) [10] provisions, joints should be detailed to accommodate the expected amount of shrinkage or expansion safely and provide an additional allowance for the residual thickness of any compressible filler materials after movement has occurred. Nevertheless, quantifying the joints vertical movement in the design phase is challenging firstly due to the complexity to estimate the connections 'stiffness' and then to their interaction with the surrounding structural and non-structural components. Due to the nature of timber and other construction materials (i.e. concrete) it is never actually possible to create fully fixed end conditions on timber members which compose the systems and most often connections are semi-rigid [11].

Additionally, although mass timber is often fabricated with exceptionally tight tolerances for overall size, as well as size and locations of holes, notches and any other alterations, and can be constructed to within as little as ±1.58 mm. of the specified dimensions, this tolerance cannot be overall settled since the attachment of mass timber elements to other materials could require larger tolerances [12].

3 Standard Formulations and Main Gap

3.1 US Codes

Council (ICC) [13] provides at Chapter 35 a list of referenced standards, which represent consensus on how a material, product or assembly is to be designed, manufactured, tested or installed to achieve a specified level of performance.

For timber the aforementioned standards are the National Design Specification (NDS) for Wood Construction published the American Wood Council (AWC) [14] and the product standards for cross laminated timber (ANSI/APA PRG 320) [15] and glulam (ANSI 190.1) [16]. Additionally, regardless of the framing type, IBC Section 2304.3.3 requires that designs for buildings over three stories consider the fact that wood shrinks as it dries. It stipulates that shrinkage in a wood building not have adverse effects on systems such as roof drainage, electrical, mechanical, or other equipment.

None of the current standards explicitly address vertical movements in TTBs or consider the effects of creep and long-term behavior in compressed elements. For instance, Chap. 3.5.1 and Annex F of the NDS only account for the additional deformation caused by long-term loading in the deflection formulation for bent members. In such cases, the total deflection $\varDelta_{\mathrm{T}}$ is calculated according to Eq. (2):

$$\varDelta_{\mathrm{T}} = K_{\mathrm{cr}} \cdot \varDelta_{\mathrm{LT}} + \varDelta_{\mathrm{ST}} \tag{2}$$

Being K_{cr} the time dependent deformation creep factor, $\varDelta_{\mathrm{LT}}$ the immediate deflection due to the long-term component of the design load and $\varDelta_{\mathrm{ST}}$ the deflection due to the short-term or normal component of the design load.

3.2 Eurocode 5

Eurocode 5 integrates the effect of creep on the long-term behavior of wood structures only in the case of the calculation of the deflection of bending elements and in the long-term stiffness of fasteners. This consideration has been reflected by an amplification of the deflections through a coefficient k_{def} modulated by the duration of the given loading and service class [7].

3.3 Gap of Knowledge

The aforementioned codes and standards lack explicit guidance on certain aspects of vertical movement in tall timber structures. For instance, they do not provide detailed procedures for calculating shrinkage in wood members or the creep factor for axial shortening in columns. Consequently, engineers must rely on supplementary analyses and assumptions to address these critical design considerations [7, 8].

Concerning the choice of the creep coefficient, most of the design engineers use creep coefficients determined for bending also for the calculation of the axial shortening.

4 Experimental Tests

The availability of long-term experimental data is a concern for the industry but there has been a significant effort to utilise finite element modelling to predict this behaviour over longer periods of time. Timber is a challenging material to model numerically due to its natural variability in properties and anisotropic behaviour; however, in recent years, timber has been modelled successfully under long-term loading situations thereby increasing the reliability and safety of structural timber design. The recent advances in the modelling of the creep behaviour of timber by [17–19] have resulted in validated fully coupled three-dimensional moisture-displacement models that can be used to predict the long-term behaviour of timber elements under varying climates and relative humidity conditions. Experimental monitoring of the long-term behaviour of engineered wood products is a costly and time-consuming process and the benefits of a validated model will provide a powerful tool to aid the further development of such engineered wood products in the future. This is coupled with a significant effort around the world to instrument and monitor some of the many demonstrator structures using CLT and novel mass timber solutions [20, 21]. Other studies monitoring elements such as CLT shear walls and Timber Concrete Composite (TCC) beams are also being successfully utilised in a number of structures as novel solutions to examine the deformation performance and the vibration criteria. Similar numerical studies of this technology were analysed in [22, 23] with modelling efforts proving successful. Fragiacomo & Ceccotti [22] developed a validated model based on two long-term experimental tests in outdoor conditions. Despite some uncertainties in environmental conditions and material properties, a good fit between experimental and numerical results was obtained. Binder et al. [23] examined and compared TCC and CLT panels and demonstrated similar creep behaviour after a 50-year design life based on a numerical study. Baas et al. [24] and Riggio et al. [21] reported on the structural health monitoring data collected during the construction of Peavy Hall, a mass-timber university building which focused on creating a comprehensive building performance dataset, including static and hygrothermal data collected during the phases of building construction. These data were used to validate a proposed methodological approach to structural health monitoring for mass-timber buildings under construction, potentially encompassing aspects related to creep behaviour through the monitoring of wood moisture content, displacements, and tension loss in structural components [24]. This study provides insights into the hygrothermal performance of mass timber structures during construction, which is a critical aspect affecting the long-term structural behaviour, including creep and future data will aid the design of taller timber buildings, however, there still remains a gap in the knowledge related to the long-term creep performance of timber buildings.

Jockwer et al. [25] examined the long-term behaviour of timber columns in Arbo, a tall timber building with RC core and timber skeleton in Switzerland consisting of 15 floors. Fibre optic sensors were used to measure the strain in the building components over a length covering 8 storeys from the floor on the ground level first floor to the ceiling of the eighth floor (Fig. 2). In Fig. 2 the measurement numbers represent the height of the timber structure during the construction stage: measurement no.1 is associated with 2 floors already realized, measurement no.2 is associated with 4 floors realized,

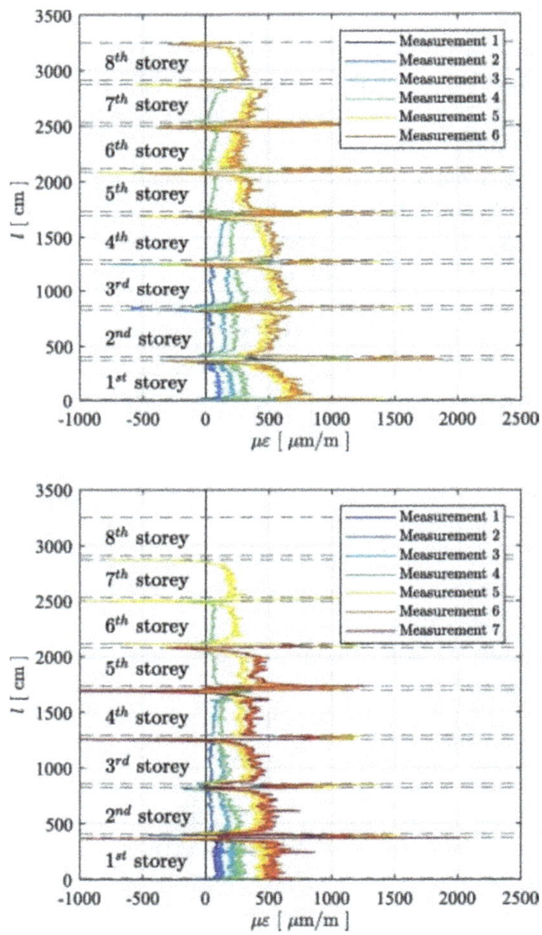

Fig. 2. Experimental strain from the test carried out by [25]. Strain development along the column positions "wall" (top) and "corner" (bottom). Measurement data is expressed in micro-strain $(10^{-4}\%)$ and interval length between the individual measurement points is 1.02 cm. Positive measured values correspond to a compression of the elements. Figure reproduced under the terms and conditions of a CC-BY license agreement.

measurement no.3 is associated with 7 floors already realized, measurement no.4 is associated with 9 floors realized and measurement from 5 to 7 were taken with the entire number of floors equal to 15.

Their research demonstrated the increased deformation due to the addition of load during the construction phase within the spruce glued laminated timber columns and beech LVL columns. The deformation of the columns was compared to model calculations to compare the deformation of individual timber elements and the total deformation experienced. Furthermore, their study indicated that approximately 25% and 8% of the total deformation after 420 days of monitoring can be attributed to viscoelastic and mechano-sportive creep, respectively and comparison with current structural design codes show that the conservative results are provided for highly stressed timber columns.

5 Conclusions

Tall timber buildings are subject to significant vertical deformations due to several factors such axial shortening, shrinkage, crushing, and joint settlement. Current design codes offer limited guidance on these long-term effects, particularly for axial components. Experimental studies and advanced numerical models are helping to fill this gap by improving understanding and prediction of deformation behavior. To support the safe growth of tall timber construction, future standards must better integrate these insights for more accurate and reliable design.

References

1. Zhao, X., Wang, L.: Vertical shortening prediction for super-tall buildings considering enclosure effect and coupling effect. Struct. Des. Tall Spec. Build. **29**(2) (2020). https://doi.org/10.1002/tal.1685
2. Zou, D., Liu, T., Teng, J., Du, C., Li, B.: Influence of creep and drying shrinkage of reinforced concrete shear walls on the axial shortening of high-rise buildings. Constr. Build. Mater. **55**, 46–56 (2014). https://doi.org/10.1016/j.conbuildmat.2014.01.034
3. Wu, J., Pang, C., Dong, B.: Dual compensation method for vertical shortening and differential shortening control in high-rise buildings. J. Build. Eng. **57** (2022). https://doi.org/10.1016/j.jobe.2022.104967
4. Matar, S.S., Faschan, W.J.: A structural engineer's approach to differential vertical shortening in tall buildings. Int. J. High-Rise Build. **6**(1), 73–82 (2017). https://doi.org/10.21022/ijhrb.2017.6.1.73
5. Dubois, F., Dopeux, J., Pop, O., Metrope, M.: Long-term creep behavior of timber columns: experimental and numerical protocols. Eng. Struct. **275**, 115283 (2023). https://doi.org/10.1016/j.engstruct.2022.115283
6. Iqbal, A.: Developments in tall wood and hybrid buildings and environmental impacts. Sustain. **13**(21) (2021). https://doi.org/10.3390/su132111881
7. McLain, R., Racine, J., Lumpkin, B.: Differential Material Movement Lv Tall Mass Timber Structures. In: 13th World Conference on Timber Engineering, WCTE 2023, vol. 5, pp. 2930–2939 (2023). https://doi.org/10.52202/069179-0383
8. R. McLain and D. Steimle, 'Accommodating Shrinkage in Multi-Story Wood-Frame Structures', 2017

9. De Jong, R.: Tall Timber Buildings. Eindhoven University of Technology (2017)
10. National House-Building Counci: 'NHBC Standards 2025. NHBC Ltd. (2025). Available: http://nhbccampaigns.co.uk/landingpages/techzone/previous_versions/2010/Part6/section9/default.htm#I3
11. S. I: Joints in timber structures: State-of-the-art knowledge in North America. In: International Conference on Control of Semi-rigid Behaviour of Civil En-gineering Structural Connections (1998)
12. Kinder, E., Kingsley, G.: Mass Timber Connections Index:Optimal Connection Considerations. Wood Products Council (2021). Available: https://www.woodworks.org/wp-content/uploads/mass-timber-connections-paper-woodworks.pdf
13. I. C. Council: International Building Code. (2025)
14. American Wood Council: National Design Specification (NDS) for Wood Construction (2024)
15. ANSI and APA: ANSI/APA PRG 320-2025: Standard for Performance-Rated Cross-Laminated Timber (2025)
16. ANSI and APA: Product Standard for Structural Glued Laminated Timber: ANSI A190.1' (2022)
17. O'Ceallaigh, C., Sikora, K., McPolin, D., Harte, A.M.: Modelling the hygro-mechanical creep behaviour of FRP reinforced timber elements. Constr. Build. Mater. **259** (2020). https://doi.org/10.1016/j.conbuildmat.2020.119899
18. Hanhijärvi, A., Mackenzie-Helnwein, P.: Computational analysis of quality reduction during drying of lumber due to irrecoverable deformation. I: Orthotropic Viscoelastic-mechanosorptive-plastic material model for the transverse plane of wood. J. Eng. Mech. **129**(9), 996–1005 (2003). https://doi.org/10.1061/(asce)0733-9399(2003)129:9(996)
19. Fortino, S., Mirianon, F., Toratti, T.: A 3D moisture-stress FEM analysis for time dependent problems in timber structures. Mech. Time-Depend. Mater. **13**(4), 333–356 (2009). https://doi.org/10.1007/s11043-009-9103-z
20. Riggio, M., Dilmaghani, M.: Structural health monitoring of timber buildings: a literature survey. Build. Res. Inf. **48**(8), 817–837 (2020). https://doi.org/10.1080/09613218.2019.1681253
21. Riggio, M., Schmidt, E., Mustapha, G.: Moisture monitoring data of mass timber elements during prolonged construction exposure: the case of the forest science complex (Peavy Hall) at Oregon State University. Front. Built Environ. **5**, 98 (2019). https://doi.org/10.3389/fbuil.2019.00098
22. Fragiacomo, M., Ceccotti, A.: Long-term behavior of timber–concrete composite beams. I: finite element modeling and validation. J. Struct. Eng. **132**(1), 13–22 (2006). https://doi.org/10.1061/(asce)0733-9445(2006)132:1(13)
23. Binder, E., Derkowski, W., Bader, T.K.: Development of creep deformations during service life: a comparison of CLT and TCC floor constructions. Buildings. **12**(2), 239 (2022). https://doi.org/10.3390/buildings12020239
24. Baas, E., Riggio, M., Barbosa, A.R., Mugabo, I., Baas, E.J., Schmidt, E.L.: Living lab at peavy hall: structural health monitoring of mass timber buildings. In: Shatis (2019). Available: https://www.researchgate.net/publication/338612314
25. Jockwer, R., Grönquist, P., Frangi, A.: Long-term deformation behaviour of timber columns: monitoring of a tall timber building in Switzerland. Eng. Struct. **234**, 111855 (2021). https://doi.org/10.1016/j.engstruct.2021.111855

Design of Timber Columns of Tall Timber Buildings with Consideration of the Long-Term Behaviour

Janusch Töpler[1,2](✉) and Jörg Schänzlin[1]

[1] Institute for Timber Design, Biberach University of Applied Sciences, Biberach, Germany
toepler@hochschule-bc.de

[2] Department of Civil Engineering, Aalto University, Espoo, Finland

Abstract. Timber columns of tall timber buildings need to be designed for combined axial compression and bending under consideration of in-plane buckling and the material-specific long-term behaviour. This contribution provides an overview of the load-bearing behaviour of such columns and their design according to EN 1995-1-1. The background, assumptions, and limitations of different design concepts are discussed. The aim is to give a comprehensive overview of the load-bearing behaviour and design of timber columns.

Keywords: Stability · columns · flexural buckling · long-term behaviour · design · EN 1995-1-1

1 Introduction

With the increasing number of storeys in timber structures, the loads and dimensions of timber columns are advancing into new dimensions, and high-performance engineered wood products are employed; see → Fig. 1. The high proportion of permanent loads in such columns additionally puts a focus on the influence of long-term behaviour on the load-bearing capacity. Therefore, timber columns of tall timber buildings should not only be designed for in-plane buckling under short-term loads, but the influence of long-term behaviour should also be considered.

This contribution provides an overview of the stability behaviour of timber columns, see → Sect. 2, describes different design methods according to EN 1995-1-1, see → Sect. 3, and discusses the advantages and limitations of the different design methods, see → Sect. 4. The objective is to provide a comprehensive overview of the load-bearing behaviour and the design of timber columns.

2 Stability Behaviour

Whether stability phenomena appear for columns depends on whether the unavoidable imperfections and deformations lead to significant additional internal forces. For columns, stability behaviour can occur for axial compressive forces $N_{\_x,c}$ combined with

G. Fink et al. (Eds.): *Holistic Design of Taller Timber Buildings*, SPRTRCIENG, pp. 28–35, 2025.
https://doi.org/10.1007/978-3-032-02098-7_3

Fig. 1 Larch GL columns of the Schönbuchturm, Herrenberg, Germany.

bow or sway imperfections or for combined N_x,c and bending, i.e. $N_{x,c}$-$M_{y,1}$-interaction (subscript 1 indicates geometrically linear analyses, i.e. according to first order theory). Additional deformations w_2 occur in the same plane as the imperfections or deformations w_1 due to bending $M_{y,1}$ (subscript 2 indicates geometrically nonlinear analyses, e.g. according to second order theory). The stability behaviour of columns is therefore named *in-plane buckling* or *flexural buckling*.

The stability behaviour is characterised by a nonlinear increase in deformations and internal forces with increasing load, see → Fig. 2. The load-bearing capacity is governed either by a member failure due to exceeding the strength (C1), by the maximum of the load-deformation curve (C2), or by a deformation limit criterion (C3). The common load-deformation behaviour of timber columns is represented by the solid line in → Fig. 2 with C2, see, e.g. Theiler [15]. The load drop after the peak C2 is caused by the reduced member stiffness due to compressive plasticising in grain direction. For very slender columns, the load-deformation behaviour with C3, and for high bending moments $M_{y,1}$, the load-deformation behaviour with C1 can occur, see, e.g. Buchanan et al. [4]. Phenomenologically, the nonlinear stability behaviour can be subdivided into the *geometrically nonlinear load-bearing behaviour* and the *materially nonlinear load-bearing behaviour*.

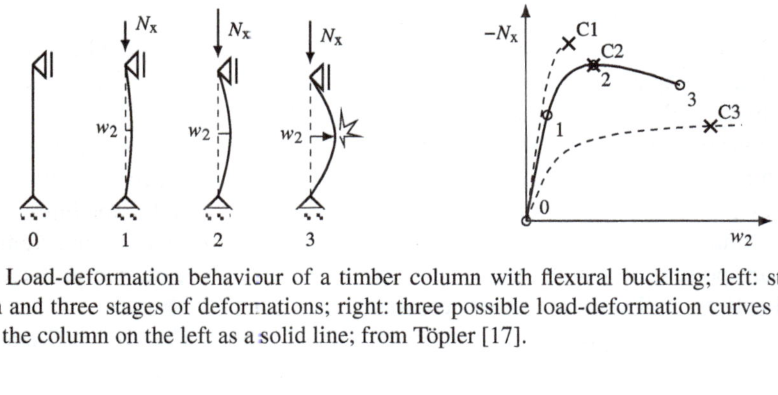

Fig. 2 Load-deformation behaviour of a timber column with flexural buckling; left: structural system and three stages of deformations; right: three possible load-deformation curves with the one of the column on the left as a solid line; from Töpler [17].

With increasing slenderness, the flexural buckling load-bearing capacity of columns decreases, see → Fig. 3. This curve of the reduction in load-bearing capacity due to flexural buckling plotted over the slenderness is referred to as *buckling curve*. The flexural buckling behaviour of timber columns is influenced by the member geometry, the structural system, the actions, the bow imperfections, the sway imperfections, and the material properties, i.e. the Modulus of Elasticity (MoE), the compressive strength, the compressive plasticising, and the tensile strength in grain direction. For timber products with low shear stiffness and strength, e.g. CLT, the shear stiffness and strength can be relevant, see Narcy et al. [9].

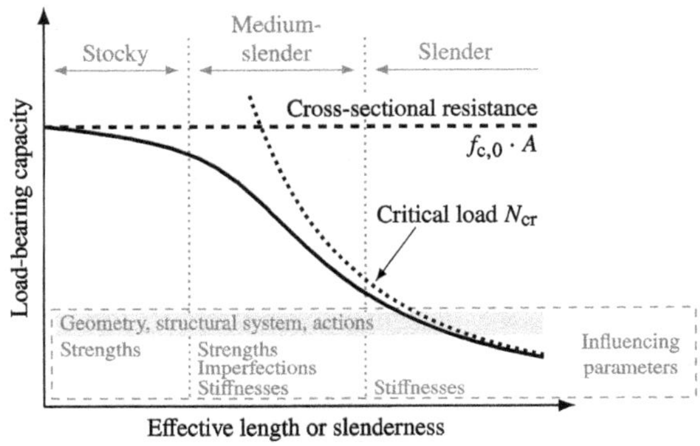

Fig. 3 Load-bearing capacity of timber columns (solid line) plotted over the effective length or slenderness; with critical load and cross-sectional resistance; illustration of slenderness ranges and associated influencing parameters; from Töpler [17].

The column failure mode depends on the slenderness as described by Buchanan et al. [4] and Theiler [15]. While compressive failure occurs for short, *stocky* columns, stability failure occurs for long, *slender* columns. For the former, the load-bearing capacity depends on the compressive strength in grain direction $f_{c,0}$ and for the latter, the load-bearing capacity approaches the critical load N_{cr} and depends on the stiffness, see → Fig. 3. For *medium-slender* columns, a transition between both failure modes takes place, and imperfections additionally influence the results.

Besides the short-term behaviour, timber members exhibit additional deformations over time, i.e. creep deformations, and decreasing strengths, i.e. duration of load effect / creep strength. While the influence of decreasing strengths is obvious, that of creep deformations depends on column slenderness and stress level, see Blaß [2] and Rautenstrauch and Becker [11]. They reported no influence of creep deformations for low slenderness, as no significant nonlinear lateral deformations occur. For medium slenderness, they found the highest influence, as nonlinear lateral deformations and significant stresses are present. For high slenderness, as the stresses decrease due to decreasing critical loads, the influence of creep decreases.

The flexural buckling behaviour of softwood Solid Timber (ST) and Glued Laminated timber (GL) columns has been comprehensively experimentally investigated by, e.g. Buchanan et al. [4], Theiler [15], and Zahn [20]. A variety of models were developed by, e.g. Blaß [2], Hörsting [7], and [15, 20]. But only a few studies were conducted on other European wood species or products, e.g. on beech GL columns by Ehrhart [5] and on beech Laminated Veneer Lumber (LVL) columns by Töpler and Kuhlmann [16], which demonstrated a decrease in load-bearing capacity due to increased compressive plasticising.

3 Design According to EN 1995-1-1

3.1 General

The design methods in EN 1995-1-1:2004 and FprEN 1995-1-1:2025 were derived based on the described extensive experimental, analytical, and numerical investigations. Only an ultimate limit state (ULS) design is required for timber columns by the Eurocodes and no serviceability limit state (SLS) design.

Timber columns can be verified with *unmagnified* internal forces, i.e. *geometrically linear* calculations (T1O), or with *magnified* internal forces, i.e. *geometrically nonlinear* calculations (T2O), see → Sects. 3.2 and 3.3. Materially nonlinear calculations are not intended by the standards, but some effects of plasticising are included in the design equations. Internal forces (and deformations) can be determined with tables, analytical equations, or numerical (FE) models. Tables and analytical equations allow for a straightforward calculation of standard cases, whereas FE models can be applied universally for stability problems.

From EN 1995-1-1:2004 to FprEN 1995-1-1:2025, only a few adjustments on the design methods for timber columns were made: separate bow imperfections for ST and GL were defined; equations for a geometrically nonlinear calculation of internal forces and deformations were provided; and an equation for calculation of the fitting factor β_c of the k_c-method was given. Subsequently, the design of timber columns according to FprEN 1995-1-1:2025 is discussed.

3.2 Geometrically Nonlinear Analysis

Timber columns should be designed for axial compression and bending according to FprEN 1995-1-1:2025 8.1.10:

$$\left(\frac{N_{x,c,d}}{Af_{c,0,d}}\right)^p + \frac{M_{y,2,d}}{W_y f_{m,y,d}} + k_{red}\frac{M_{z,2,d}}{W_z f_{m,z,d}} \leq 1 \tag{1}$$

With $p = 2$ for rectangular cross-sections and 1 for other cross-sections and CLT. p is a factor accounting for the positive influence of compressive plasticising on the cross-sectional resistance and was derived by Blaß [2], Buchanan et al. [4], and Zahn [20] for rectangular softwood members. With $k_{red} = 0.7$ for rectangular cross-sections and 1.0 for other cross-sections. k_{red} considers the size effect on the bending strength at biaxial bending and was derived by [4] and van der Put [19] for softwood with Weibull's weakest link theory. $M_{y,2,d}$ and $M_{z,2,d}$ are the design bending moments from geometrically nonlinear analyses. They can be calculated according to FprEN 1995-1-1:2025 Annex B.4.3, literature, e.g. Petersen [10], or with FE models. In FprEN 1995-1-1:2025 Annex

B.4.3, and most other analytical equations from literature, the stiffness reduction due to compressive plasticising in grain direction is neglected. For softwood this is acceptable, but for other wood products like beech GL and LVL this should be questioned, see Ehrhart [5] and Töpler and Kuhlmann [16].

3.3 Geometrically Linear Analysis (k$_c$-Method)

For a relative slenderness ratio of $\lambda_{c,z,rel} \leq 0.3$, flexural buckling may be neglected, and Eq. (1) with $M_{y/z,1,d}$ may be utilised. For $\lambda_{c,z,rel} > 0.3$, timber columns should be designed for axial compression and bending according to FprEN 1995-1-1:2025 8.2.2.2:

$$\frac{N_{x,c,d}}{k_{c,y} A f_{c,0,d}} + \frac{M_{y,1,d}}{W_y f_{m,y,d}} + k_{red} \frac{M_{z,1,d}}{W_z f_{m,z,d}} \leq 1 \tag{2}$$

$$k_{c,y} = \frac{1}{\phi_{c,y} + \sqrt{\phi_{c,y}^2 - \lambda_{c,y,rel}^2}} \tag{3}$$

$$\phi_{c,y} = 0.5\left(1 + \beta_{c,y}\left(\lambda_{c,y,rel} - 0.3\right) + \lambda_{c,y,rel}^2\right) \tag{4}$$

Equations (3) and (4) for calculating k_c are the exact solution of the differential equations for flexural buckling, Euler buckling case 2, pure axial compressive forces, and linear-elastic material behaviour, see Taras [14], except for the term ($\lambda_{c,y,rel}$ – 0.3), which represents the slenderness ratio of the tests for determining the compressive strength $f_{c,0,k}$, see Brüninghoff and Klapp [3]. The fitting factor $\beta_c = 0.1$ for GL and LVL and 0.2 for ST accounts for the ratio of the MoE to the compressive strength to the bending strength ($E_0{:}f_{c,0}{:}f_m$), the geometrical and structural imperfections, and the plasticising. It was calibrated by means of curve fitting to the results of numerical calculations on softwood columns with scattering input values and realistic stress-strain behaviour by Blaß [2]. As discussed in → Sect. 3.2, the transfer of β_c to other wood products without an adaptation considering the negative effect of stiffness reduction due to compressive plasticising should be questioned.

3.4 Effective Length, Critical Buckling Load

The critical buckling load N_{cr} and the effective flexural buckling length $L_{c,ef}$ for its calculation are influenced by the structural system, the column dimensions, and the MoE in grain direction, or, in summary, the stiffness, see → Fig. 3. In some cases, e.g. CLT, the shear stiffness should be additionally considered, see, e.g. Narcy et al. [9]. $L_{c,ef}$ can be determined according to FprEN 1995-1-1:2025 Annex B.3.2.2, literature, e.g. Petersen [10], or with numerical eigenvalue analyses.

3.5 Imperfections

Imperfections are considered in design with geometrically nonlinear analyses explicitly in the calculation of internal forces and in design with geometrically linear analyses, i.e. the k_c-method, implicitly in the fitting factor β_c.

FprEN 1995-1-1:2025 7.4.1 recommends equivalent bow imperfections of $e = L/400$ and $L/1000$ for the design of ST and GL columns, which cover the effects of geometrical and structural imperfections and are based on the investigations by Blaß [2] on softwood ST and GL columns. Effects of (un)planned load-eccentricities are not included in these values. Additionally, FprEN 1995-1-1:2025 7.4.1 recommends sway imperfections , which need to be considered when designing (partly) clamped columns. The basis of the harmonised values of sway imperfections across different materials in the Eurocodes are the measurements of Lindner and Giezelt [8]. Sway imperfections of (partly) clamped columns can be considered in the k_c-method by modifying $\beta_c = \beta_{c,sway} = 100\ \phi/2$, see Schänzlin et al. [13].

3.6 Long-Term Behaviour

The DoL effect is considered in FprEN 1995-1-1:2025 by modifying the strengths $f_{c,0,d}$ and $f_{m,d}$ in Eqs. (1) and (2) with k_{mod}.

The creep deformations should be considered according to FprEN 1995-1-1:2025 7.3.4 if the design value of the permanent and quasi-permanent action is more than 70% of the total action. In this case, creep may be taken into account by reducing the stiffness by the factor $1/(1 + k_{def})$. In line with EN 1995-1-1:2004, k_{def} may be reduced to $k_{def,eff} = k_{def}\ E_{d,qper}\ /\ E_{d,char}$ to account for the ratio of effects of actions with quasi-permanent combination of actions $E_{d,qper}$ to effects of actions with characteristic combination of actions $E_{d,char}$. Alternatively, creep may be considered by increasing the imperfections according to FprEN 1995-1-1:2025 Annex B.4.6 or, in FE analyses, by applying creep strains in line with rheological models from literature, e.g. Schänzlin [12] or Toratti [18]. The background on the 70% limit criterion and the stiffness reduction was discussed by Hartnack and Rautenstrauch [6], who proposed limits of 50% for SC 2 and 75% for SC 3. The imperfection approach is based on the investigations by Abeysekera et al. [1].

4 Discussion

Major advantages of the flexural buckling design with geometrically nonlinear internal forces compared to the k_c-method are: the actual internal forces and deformations are known; the nonlinear effect of $M_{y,1}$ is considered; and the results are continuous over no buckling, flexural buckling, and lateral torsional buckling. An advantage of the k_c-method is the possibility of superposition of load cases. While in analytical calculations according to FprEN 1995-1-1:2025, the influence of load eccentricities has to be taken into account additionally and calculations for structural systems for which $L_{c,ef}$ is unknown are impossible, FE analyses can be applied universally for stability problems. Generally, the transfer of the design methods in FprEN 1995-1-1:2025 to other wood products than softwood without considering the negative effect of compressive plasticising should be questioned.

Finally, the flexural buckling design of columns is limited to analysis with beam theory, as FprEN 1995-1-1:2025 does not provide design methods or material properties for FE analysis with shell or solid elements. An extension of the design methods is currently being discussed in CEN/TC 250/SC 5/WG 11 "Numerical design" (of timber structures).

5 Conclusions

The in-plane buckling behaviour of timber columns is well understood, and extensively validated design models are available in FprEN 1995-1-1:2025. Various design methods are also available for the long-term behaviour of columns, although these are neither as differentiated nor as comprehensively validated. However, recent research highlighted the significant impact of the material-specific compressive plasticising on the load-bearing capacity of high-performance wood products such as beech GL and LVL, a factor not yet incorporated in FprEN 1995-1-1:2025. This omission is particularly critical for tall timber buildings.

The most significant development is arguably the paradigm shift from analytical to numerically based design, which has the potential to significantly improve the reliability of design through the use of two- and three-dimensional stress calculations and anisotropic elasto-plastic material models. This shift presents a highly promising outlook for the evolution of the Eurocodes.

References

1. Abeysekera, I., Feltham, I., Lawrence, A.: Buckling of slender timber beam-columns under combined loading, including creep. In: Proceedings of the 8th meeting of International Network on Timber Engineering Research (INTER), 54-2-1, online (2021)
2. Blaß, H.J.: Tragfähigkeit von Druckstäben aus Brettschichtholz unter Berücksichtigung streuender Materialeigenschaften. Ph.D. thesis, University of Karlsruhe (1987)
3. Brüninghoff, H., Klapp, H.: Stabilitätsnachweis im Holzbau - Biegedrillknicken mit Normalkraft. Bauen mit Holz. **10**, 27–33 (2005)
4. Buchanan, A.H., Johns, K.C., Madsen, B.: Column design methods for timber engineering. In: CIB-W18, 18-2-1, Beit Oren (1985)
5. Ehrhart, T.: European beech glued laminated timber. Ph.D. thesis, ETH Zurich (2019). https://doi.org/10.3929/ethz-b-000402805
6. Hartnack, R., Rautenstrauch, K.: Long-term load bearing of wooden columns influenced by climate -view on code. In: CIB-W18, 38-2-1, Karlsruhe (August 2005)
7. Hörsting, O.-P.: Zum Tragverhalten druck- und biegebeanspruchter Holzbauteile. Ph.D. thesis, Technische Universität Carolo-Wilhelmina zu Braunschweig (2008). https://doi.org/10.24355/dbbs.084-200807140200-0
8. Lindner, J., Giezelt, R.: Imperfektionsannahmen für Stützenschiefstellungen. Stahlbau. **53**(4), 97–101 (1984)
9. Narcy, A., Forêt, G., Pham, D.T., Lebée, A.: Comparison of CLT buckling strength criteria with experimental results. In: Proceedings of the 10th meeting of International Network on Timber Engineering Research (INTER), 56-2-1, Biel/Bienne (2023)
10. Petersen, C.: Statik und Stabilität der Baukonstruktionen: Elasto- und plasto-statische Berechnungsverfahren druckbeanspruchter Tragwerke; Nachweisformen gegen Knicken, Kippen, Beulen, 2nd edn. Friedr. Vieweg & Sohn, Braunschweig, Wiesbaden (1980)
11. Rautenstrauch, K., Becker, P.: Zur Berücksichtigung des Kriechens bei Druckstäben aus Holz. Bautechnik. **75**(11), 910–921 (1998). https://doi.org/10.1002/bate.199806030
12. Schänzlin, J.: Modeling the long-term behavior of structural timber for typical service class-II-conditions in South-West Germany. In: Habilitation. University of Stuttgart (2010). https://doi.org/10.18419/opus-348

13. Schänzlin, J., Bosch, G., Hamm, P.: kc-method – consideration of creep deformations. CEN/TC 250/SC 5/WG 3 N 354 (2022)
14. Taras, A.: Contribution to the Development of Consistent Stability Design Rules for Steel Members. Ph.D. thesis, Graz University of Technology, Graz (2010). https://doi.org/10.3217/1smsn-1bq83
15. Theiler, M.: Stabilität von axial auf Druck beanspruchten Bauteilen aus Vollholz und Brettschichtholz. Ph.D. thesis, ETH Zurich (2014). https://doi.org/10.3929/ethz-a-010273734
16. Töpler, J., Kuhlmann, U. In-plane buckling of beech LVL columns. In: Proceedings of the 9th meeting of International Network on Timber Engineering Research (INTER), 55-2-1, Bad Aibling (2022). https://doi.org/10.18419/opus-12610
17. Töpler, J.: Load-bearing capacity of imperfection-sensitive timber members under combined bending and compression (in preparation). Ph.D. thesis, University of Stuttgart (2025)
18. Toratti, T.: Creep of timber beams in a variable environment. Ph.D. thesis, Helsinki University of Technology, Helsinki (1992)
19. van der Put, T.A.C.M.: Discussion of failure criterion for combined bending and compression. In: CIB-W18, 24-6-1, Oxford (1991)
20. Zahn, J.J.: Design of wood members und combined load. J. Struct. Eng. **112**(9), 2109–2126 (1986). https://doi.org/10.1061/(ASCE)0733-9445(1986)112:9(2109)

Connections Design and Analysis: Numerical, Analytical and Code-Based Methods

Martina Sciomenta[1](✉) and Chiara Bedon[2]

[1] Department of Civil, Architectural and Building and Environmental Engineering, University of L'Aquila, L'Aquila, Italy
martina.sciomenta@univaq.it

[2] Department of Engineering and Architecture, University of Trieste, Trieste, Italy

Abstract. The design and performance of structural connections in timber engineering plays a critical role in the safety and functionality of timber structures, especially in applications involving tall buildings. In this chapter, the design and modelling strategies of structural connections in timber engineering are examined. The study highlights the significance of key parameters, including stiffness, strength, and ductility, and explores how these factors influence the behaviour and performance of various connection types under different loading conditions. Analytical and numerical methods for assessing the mechanical properties of connections are presented, including those outlined in widely used standards like Eurocode 5. The discussion highlights the limitations of these guidelines, particularly their inability to fully address complex behaviours such as non-linear deformation and long-term effects. The importance of experimental validation is also emphasized, as a critical step in refining models and improving design practices.

Keywords: Connections · Load-transfer mechanisms · Mechanical parameters · Guidelines · Modelling strategies

1 Introduction

In tall buildings, connection deformation can potentially cause serviceability failures through deflection of the structure [1, 2], and in slender structures could also lead to ultimate limit state failure through second-order P-delta effects [3]. Connection deformation also affects both serviceability and ultimate-limit state behaviour of mechanically jointed composite beams and slabs, including timber-concrete composites (TCC) [4] and dowel-laminated timber (DLT) [5]. Moreover, the stiffness of connections is a key parameter in modelling for vibration and acoustic serviceability [6, 7], to ensure, for example, occupant comfort or usability of sensitive instruments. Seismic design also requires modelling of connection stiffness and ductility. An overview of most recent studies and issues on connections is summarized in the following sections.

G. Fink et al. (Eds.): *Holistic Design of Taller Timber Buildings*, SPRTRCIENG, pp. 36–49, 2025.
https://doi.org/10.1007/978-3-032-02098-7_4

2 Fundamentals of Structural Connections

2.1 Types of Structural Connections

2.1.1 Mechanical Connections

Mechanical connections can be classified based on the type of fasteners in: (i) dowel-type connections and (ii) bearing connections. Fasteners, in turn, are divided into two groups depending on the force-transfer mechanism between the connected members. The main fastener group is represented by the dowel-type fastener (generally made of steel) including staples, nails, screws, bolts, dowels and threaded rods. Hardwood dowels [8] have been proposed as an alternative connector for DLT beams with the aim to reduce the use of adhesives in engineered wood products. The second group of fasteners is represented by the bearing-type or "surface-type" fasteners, such as toothed-plate connectors, punched metal plate connectors and split rings.

2.1.2 Bonded Connections

In this framework, just pure timber-timber joints are accounted as adhesive connections. Those connections are vital for the creation of laminated timber elements (i.e. glulam and LVL) and glued composed members [9] but are also used in structural joints between elements: end-joints for knee and reinforcements (i.e. scarf, lap or finger joints, knee and truss joints with gussets from plywood, LVL, etc.) [10].

2.1.3 Hybrid Connections

Hybrid connections are defined as joints combining two of the following jointing categories: carpentry joints, mechanical connections and bonded connections [11]. Bonded-in rods, often considered a subset of adhesive joints, represent a type of hybrid joint since they involve three materials (timber, rod-connector and adhesive bonding layer). Rods are usually made of steel, nevertheless, an alternative is represented by bonded-in fibre-reinforced polymer (FRP) rods, which use provide an improved resistance in corrosive environments, a lower weight and a reduced heat conduction into the joint in case of fire due to lower thermal conductivity. Another hybrid joint type entails the combination of adhesive bonding and steel plates. One option in this category is glued-in solid steel plates, and the epoxy-based adhesives seemed to be the most appropriate choice [12]. Another approach that conceptually eliminates the reliance on the adhesion between adhesive and steel is to use perforated steel plates [13]. To overcome adhesives 'short-coming related to limited gap-filling capabilities and strength-reducing effects in the bond-line during the initial hardening shrinkage, new hybrid connections using grouting technology with concrete-type adhesives (CTA) have been developed. Its application was proposed by [14] for hybrid steel-cross-laminated timber building systems consisting of threaded rods reinforced by a layer of epoxy grout encased into the CLT panel.

3 Key Parameters Influencing Connection Behaviour

3.1 Load-Transfer Mechanisms

3.1.1 Mechanical Connections

Connections with dowel-type fasteners transmit axial, shear, and even bending loads between structural members. They are categorised as shear connections since forces are mostly transferred between members via shear in the fasteners, but, to some extent, they also transmit forces through axial capacity of the fasteners [15]. Depending on the direction of the force relative to the fastener axis, dowel-type connectors can experience two types of loading: shear, when the force is applied perpendicular to the fastener axis, and withdrawal/pushing-in, when the force is applied along the fastener axis. Additionally, combined stresses may occur under certain conditions [9]. When subjected to lateral loads, dowels-type connectors may deform in bending, creating one or more plastic hinges within the fastener. In this situation, some parts of the load uptake also occur in tension. Depending on the surface and end anchorage of the fastener, the part carried in tension can be larger or smaller. Because of the loading, the dowel will press against the surrounding timber members, creating embedding pressure against the dowel.

Bearing-type connections transmit lateral loads only. The mechanism to transmit shear forces is through bearing on the connected materials. For double-sided connectors (e.g., split rings and double-sided toothed-plate connectors), forces are transferred through embedment stresses of the first member into the connector and then to the second member via the connector's shear resistance. Bolts manage any eccentricity to maintain joint stability. For single-sided connectors (e.g., shear plates and single-sided toothed-plate connectors), forces generate embedment stress between the connector and bolt. The bolt's shear resistance ensures force transfer, moving it to the second connector in timber-to-timber joints or directly to the steel member in steel-to-timber joints [9].

3.1.2 Glued Connections

The strength of glued joints is difficult to describe analytically. If the glue is ductile, (i.e. polyurethane) the shear strength describes its behaviour well. If it is brittle, the fracture energy, G_f, of the glue is the best descriptor. To characterize the brittleness, a brittleness ratio has been defined as f_v^2/G_f, [16]. One of the problems is that pure tension or shear is seldom present in a real joint, making it impossible to separate strength components even in testing. However, the resorcinol/phenol adhesives employed for the production of EWPs, are more brittle than wood itself, while polyurethane is less brittle than wood. Adhesive connections are generally considered to be rigid for practical engineering design purposes, but flexible adhesive joints have been proposed for structural joints in seismic regions [17, 18], and to increase the bending capacity of laminated timber [19, 20].

3.1.3 Hybrid Connections

The mechanical description of the load transfer in hybrid joints involving adhesives is additionally complicated by them is match of stiffness between the mechanical fasteners and the adhesive layer while these involving carpentry connections are affected by uncertainty related to friction and contact mechanisms. Taking as example the glued-in rods, which can be accounted as a mechanical joint or as combination of glued and mechanical connections. The force between the rod and the glue is generally seen as transferred either by compression or shear, however in some cases it is perceived as a combination of the two [21], depending on the combination of several influencing parameters (i.e. the ratio of the diameter of the hole to the diameter of the rod, the adhesive bond line thickness, the timber and rod surface features).

3.2 Stiffness, Strength, Ductility

The load carrying capacity of a dowel-type connection in shear is determined by three parameters: the embedding strength of the timber f_h, the dowel strength represented by its yield moment M_y and the anchorage capacity enabling tensile action in the dowel F_{ax}. The load carrying capacity of connections with dowel-type fasteners was first described by Johansen [22] which distinguished three different failure modes, depending on the relation between the embedding strength, the yield moment of the dowel and the thickness of the timber members. Nevertheless, this theory does not account for potential brittle failure modes. The other limitation of Johansen's formulation lies in the analysis method itself: the model can predict the ultimate failure load but cannot provide any indication of the deformability of the connection, nor, consequently, of its stiffness and ductility properties.

To properly account for the semi-rigidity of timber joints with mechanical fasteners it is necessary to evaluate its stiffness. It is possible by carrying out tests in accordance with some standardized procedures (i.e. EN 26891:1991 [23]), or alternatively by referring to different documents (i.e. the N.I.CO.LE document) [24] and Standards that provide simplified formulas for different types of connection, fastener diameter and timber density.

Connections are often intended to act as potential ductile elements contributing significantly to overall ductility and energy dissipation in case of overloading and allowing for safe load paths when construction tolerances are exceeded [25]. A ductile connection response is often associated with plastic fastener deformation, which allows for energy dissipation under reverse-cyclic loading and is crucial for seismic design [26], also known as "dynamic ductility".

Some authors have given empirical indications on the possible maximum and minimum values of ductility D [27]. Ductility is often defined as ratio the between the deformation corresponding to the maximum load (u_u) and the one at the elastic limit (u_y) and was correlated with the Johansen's failure modes; often the deformation at the point where the maximum load drops by 20% is considered. It can be experimentally estimated on mechanical fasteners by carrying out cyclic tests following EN 12512:2001/A1:2005 [28], by using the method proposed by Kobayashi and Yasumura [29] or the method presented in ASTM E2126–11 [30]. A novel method that resembles in some aspect the short procedure of EN 12512 [28] to determine the ductility class of dowel-type connections is the one proposed in the revised version of the standard EN 14592:2017 [31].

4 Design Standards and Guidelines

4.1 Codes

With changes and adaptations, the Johansen theory [22] is the basis for the so-called European Yield Model (EYM) which is used in the current generation of design codes EN 1995-1-1 [32]), National Design Specification for Wood Construction-NDS [33], Engineering Design in Wood-CSA O86 [34] and Standard for Design of Timber Structures-GB 50005 [35] for the estimation of the load-carrying capacity of connections in timber structures.

The EYM assumes simplified rigid-plastic material behaviour for the fastener when exposed to bending stresses and for the timber when exposed to embedment stresses [9] even though, various studies involving embedment tests on wood and tests on connections [36–38] revealed for both a highly non-linear behaviour. This can be attributed to the non-linear behaviour of steel dowels under bending and wood under embedment stress. EYM is also addressed as "local design criteria" since its applicable to the single connector and doesn't account for brittle failure modes. The applicability of this model is guaranteed by the application of the so called "global design criteria" which consists of spacing and distance rules between the single connectors, which ensure the global capacity of the cross section to withstand the forces transferred by the connectors [39].

Empirical formulations for the estimation of in-service stiffness of all dowel-type connections, which is identified as slip modulus, K_{ser} are provided in Sect. 8 of EC5. Slip modulus is a measure of joint stiffness, i.e. resistance to displacement, hence it has unit N/mm. The slip modulus of a joint at the ultimate limit state K_u is determined as 2/3 of the K_{ser}.

4.1.1 Limitations of Existing Guidelines

Connection stiffness is a critical aspect of structural performance, yet current design approaches exhibit significant limitations: thickness is not factored into stiffness calculations, leading to unrealistic assumptions that slim and thick connections share the same stiffness [40, 41]. For the design of multi-dowel connections, only general design rules but no design equations are given in the standard. In general, the load levels of the individual dowels for a specific load case (including the load-to-grain direction at each dowel) needs to be calculated in order to be able to compare these values to the corresponding single-connector strength. The compatibility of deformations is requested but only strongly simplified stiffness values for single-dowels not compatible with limit loads are provided [42].

Experimental studies (e.g., Sandhaas et al. [43]) have demonstrated that the EN 1995-1-1 equation for serviceability stiffness, K_{ser}, often fails to accurately predict dowel stiffness, especially for connections with multiple dowels or dowels with unconventional features. This issue becomes even more pronounced in large-scale structures like multi-story timber buildings, where deformation effects due to scale become significant. The non-linear deformation behaviour of dowel-type connections is another key challenge. Current parameters, such as K_{ser} and K_u (the connection stiffness at the ultimate limit states), do not account for the real non-linear, behaviour of connections, limiting the ability to design for robustness and ductility. Non-linear behaviour affects load distribution among fasteners, especially when connections experience combined rotational and translational forces. Furthermore, the stiffness's influence on force distribution within a structure cannot be realistically assessed, and modern techniques like interlayers (e.g., acoustic layers) are not accommodated in current formulas.

To optimize design, more comprehensive data on non-linear connection behaviour will be achieved and provided, including upper and lower stiffness limits alongside mean values. The existing rigid-ideally plastic limit design and empirical elastic stiffness formulas used in EC5 fail to address uncertainties in stiffness and ductility, will be overcome by carrying out further research and development in this area.

5 Modelling Strategies for Connections

5.1 Analytical Approaches

5.1.1 Closed-Form Equations for Shear Behaviour

Analytical closed-form models based on mechanical approaches contribute to understanding the factors influencing connection behaviour and are useful for designers.

The most widespread closed-form solution that has gained increasing acceptance in recent years is the European Yield Method (EYM) since compared to the other analytical formulations, it entails closed-form and rather simple equations. Uibel and Blaß [36] validated and expanded this formulation to enable the design of dowel-type connections in CLT. Specifically, they modified the timber embedment strength equations by incorporating corrective factors into Johansen's formulas. Other analytical solutions for the prediction of the load-displacement relationship for timber joints with dowel-type fasteners subjected to lateral loading have been presented by several research works. They considered a joint as a two-dimensional arrangement in a plane and represent the connector by a one-dimensional beam on a Winkler or discontinuous foundation. For a Winkler type foundation, the force per unit length beneath the connector is taken to be directly proportional to the displacement at all points along the length of the connector. This method was proposed also in more recent works [44–46] aiming at investigating the mechanical behaviour of different dowel-type connections.

To develop an exact theoretical design method for connections with dowel-type fasteners, many researchers have proposed adopting the theory of beam on foundation (BOF) for predicting the load-deformation behaviour of the fasteners since 1950, and the relevant research works can be mainly divided as linear and non-linear. In this case, the fasteners are assumed as elastic or elastoplastic beams supported by the linear or non-linear foundation equivalent from timber. For the linear model, the target is to solve the exact fastener deformation expressions by using the general solutions of the beam on elastic foundation (BOEF) model. Kuenzi [47] first introduced the BOEF model into dowel timber-to-timber connections with the consideration of the timber embedment properties and the connection geometric conditions, proving that the load-slip relationship of timber connections could be predicted with high accuracy at the elastic stage. However, challenges remain, with particular reference to the non-linear behaviour of materials which can't be easily accounted in closed-form equations. Additionally, theoretical solutions based on the traditional BOEF model are overly complex for direct application in practical engineering.

5.1.2 The Component Method

Although almost all timber connections should be treated as semi-rigid, EN 1995-1-1:2004 does not provide any methods for determination of the rotational stiffness of the connection. In this framework, the component method, which is one of the most used methods for the determination of the moment resistance and rotational stiffness of the steel and steel-concrete composite connection, can be successfully employed to investigate the characteristic properties of semi-rigid timber-steel connection. The originality of the component method is that it considers any joint as a set of "individual basic components". Each component is represented by an extensional spring, which is independent and can be connected to other springs or to rigid elements. Once the individual constitutive components are identified and characterized, the overall behaviour of the connection can be modelled through the so-called assembly procedures [48]. By using the component method, [49, 50] carried out detail investigation on ductile moment-resistant timber connections with the combination of glued-in rods and steel connecting studs. It was discovered that the method could well predict the joint response in terms of failure mode, ultimate resistance, stiffness, and rotation capacity.

5.2 Numerical Modelling Techniques

In parallel with these analytical approaches, developments in computational mechanics made it possible to develop numerical methods [51, 52], which accounted even for non-linear phenomena. These approaches have remained unused in practical design due to their complex implementation and their high running time, at the time of their invention, while today's computational resources strongly reduced corresponding limitations.

A large number of researchers have then focused on the non-linear BOF model by using the numerical methods. Dowel-type fasteners are numerically modelled as elastoplastic beams on a nonlinear foundation in engineered in wood-based products [53, 54]. For example, Hirai [52] proposed a non-linear controlling differential function solved by using the stepwise numerical linear approximation method, by adopting the elastoplastic behaviour of the dowel and the foundation.

5.2.1 Finite Element Modelling (FEM) for Connections

The accurate mechanical analysis of connections is highly complicated and challenging, due to the large number and type of structural details and influencing parameters that should be taken into account. Firstly, nonlinearity of the material behaviours of steel and timber (and possible other materials) has to be properly considered. Secondly, for dowel-type connections, there is contact between the dowel and the surface of the hole in the timber element, where normal and tangential stresses are transferred and friction between steel and wood has to be accurately taken into account. In this case, the possible presence of gaps and production tolerances must be also carefully considered. Thirdly, the load distribution is highly non-uniform within the timber parts loaded in tension, where the primarily applied compressive stresses are transformed via shear stresses into tensile stresses. The combination of these mutually affected processes occurs simultaneously and within a small spatial region [55].As a consequence, in most cases, the use of refined 3D numerical models is necessary. The intrinsic limitations of spring-based models compared to full 3D models, for collapse simulations, have been for example discussed in [56], while a vibration sensitivity study on timber-timber floors has been successfully carried out in [57] with spring-based 2D models. Many authors carried out various 3D models for connections, by taking into account specific computational features and strategies. Since the nonlinear material model is crucial to the development of accurate load-slip behaviour for timber connections, many authors proposed the use of a plasticity-based constitutive material formulation to model wood as elasto-plastic orthotropic according to the Hill yield criterion [58–60]. The most important aspect is the definition of a realistic damage and failure criterion [56]. Moreover, it is worth noting that the Hill yield criterion is symmetric in tension and compression and should be thus used with attention. To overcome this possible limitation, more advanced subroutines have been implemented to overcome this limitation [61].

Finally, to accurately account for the failure modes that significantly impact the estimation of the mechanical parameters of connections, various theories have been proposed specifically for anisotropic materials like wood. The most widely used are the maximum stress criterion [62], the Tsai-Wu stress criterion [63] and the Hoffman stress criterion [64]. Since the interaction between connector and timber is essential in finite element modelling, several authors (i.e. [65, 66]) introduced a fictitious 'soft layer' with cohesive damage interactions, at the interface between the dowel-type connector and the surrounding timber components. Compared to other modelling approaches, the Cohesive Zone Modelling (CZM) method has well-known intrinsic advantages, since it does not need: (i) pre-existing definition of cracks, (ii) prior assumptions for onset and growth of damage, (iii) complex moving mesh techniques, and (iv) a very dense mesh definition close to the cracks (to ensure local occurrence of infinite stress and strain peaks). Moreover, the CZM strategy can be also adapted to many other connection typologies for timber structures, such as notched connections [67] or bonded-in-rod connections [68].

5.2.2 Spring Elements and Equivalent Beam/Shell Elements

Connections are often effectively modelled as spring elements in numerous studies that use 2D finite element methods to validate analytical models describing the behaviour of various Lateral Load Resisting Systems (LLRS). For platform-type CLT buildings, [60, 69, 70] developed 2D models in which shear-walls consist of two-dimensional (i.e. area) rigid elements and the connections (e.g., hold-down, angle brackets and connections between adjacent CLT panels) were modelled as nonlinear elastic spring (link) elements. The adopted models differ each other mainly in the assumptions related to the implementation of the non-linear behavioural characteristics assigned to the mechanical anchors, including uni- vs bi-directional behaviour and tri-linear vs multi-linear behaviour. In analogy with this approach, [71] conducted a parametrical study to investigate the interaction between the floor-to-wall connections and lintel elements on the mechanical behaviour of CLT shear-walls with cut-out openings. The same approach was employed to analytically model the log-house timber walls under in-plane lateral loads by Sciomenta et al. [72]. In this case the typical corner joint is considered to be an elastic extensional spring. At the same time, static friction contributions are evaluated by means of classical Coulomb law, while the presence of possible gaps is properly considered in the lateral displacement evaluation. It is worth stressing out the crucial role of experimental test for the assessment of mechanical properties of connections. In particular, the calibration of the springs in the aforementioned numerical models was achieved via experimental test. One of the most cited and useful work is the one from Gavric et al. [73] which summarize the outcomes from an extensive experimental program on typical cross-laminated timber (CLT) screwed connections including in-plane monotonic and cyclic shear and withdrawal tests were performed on screwed wall-to-wall, floor-to-floor and wall-to-floor CLT connections.

5.3 Some Specific Challenges in Timber Connections

5.3.1 Long-Term Effects

Timber members exhibit creep: time-dependent deformations. This phenomenon is related to the duration of the action, the state of stress and hygrometric variations. According to EN 1995-1-1:2004, the additional deformation of joints caused by creep shall be calculated as the short-term deformation multiplied by the deformation factor (k_{def}). The short-term deformation shall be determined based on the slip modulus of the joint (K_{ser}) [74]. Using connectors that resist creep, such as grouted bolts with washers or specialized engineered wood connectors (i.e. dog screws or coach screws for timber-to-steel connections), can mitigate these issues. Temperature and moisture fluctuations can pose an additional challenge to connection performance, particularly in tension or compression members. Thermic and moisture variations can cause shrinking and swelling in timber leading to connection loosening or member splitting. In climates with significant temperature swings, connections that allow some movement (e.g., bolted joints or Z-clips) may perform better by accommodating these fluctuations without compromising the joint. The effect of temperature and humidity is particularly significative in hybrid connections involving glues. Kemmsies and Streicher [12] analysed the performances of different adhesive types and gluing techniques for bonded-in steel plates under temperature and humidity variations. One key finding was that epoxy-based adhesives seemed to be the most appropriate choice.

5.3.2 Group Effect and Connection Stiffness

EN 1995-1-1:2004 bases the design of multiple-fasteners connections on the behaviour of single fasteners. The connection slip modulus is determined by multiplying the slip modulus per shear plane of a single fastener, with the total number of fasteners and the number of shear planes. Nevertheless, it has been proved [75] that the assumption of a linear combination of the individual behaviours of each fastener is inaccurate, since the material and geometrical properties of the timber members and the number of fasteners in steel-to-wood connections influence their overall response as well as the connection failure mode. Jorissen and Fragiacomo [26] observed reduced shear capacity with decreasing spacing due to premature splitting of the connection. It was demonstrated by Hochreiner et al. [76] that the failure mechanism in the timber matrix in a connection could be related to the non-linear global moment-relative rotation behaviour of dowel groups. Hochreiner et al. [76] also demonstrated the potential for strain fields in the timber surrounding the dowels to overlap and interact with each other, thus alluding to the possibility of a group effect on the stiffness and failure modes of the connections [77].

References

1. Skidmore, Ownings, and Merril, LLP: *Gravity Framing Development of Concrete Jointed Timber Frame System* (2014)
2. Smith, R.: Deflection limits in tall buildings- Are they useful? In: *Structures Congress 2011 - Proceedings of the 2011 Structures Congress*, pp. 515–527 (2011). https://doi.org/10.1061/41171(401)45
3. Tesfamariam, T.: Performance-based design of tall coupled cross-laminated timber wall building. Earthq. Eng. Struct. Dyn. (2022)
4. Yeoh, D., Fragiacomo, M., Franceschi, M., Boon, K.: State of the art on timber-concrete composite structures: literature review. J. Struct. Eng. **137**, 1085–1095 (2011). https://doi.org/10.1061/(ASCE)ST.1943-541X.0000353
5. O'Ceallaigh, C., Harte, A.M., Mcgetrick, P.J.: Dowel laminated timber elements manufactured using compressed wood dowels. In: *CERI2022 - Civil Engineering Research in Ireland 2022*, pp. 222–227 (2022)
6. Edskär, I., Lidelöw, H.: Dynamic properties of cross-laminated timber and timber truss building systems. Eng. Struct. **186**, 525–535 (2019). https://doi.org/10.1016/j.engstruct.2019.01.136
7. Tulebekova, S., Malo, K.A., Rønnquist, A., Nåvik, P.: Modeling stiffness of connections and non-structural elements for dynamic response of taller glulam timber frame buildings. Eng. Struct. **261** (2022). https://doi.org/10.1016/j.engstruct.2022.114209
8. Fu, H., He, M., Alhaddad, W., Shen, Z.: Numerical simulation and design method of dowel-type timber joints connecting laminated veneer lumber. Structure. **62** (2024). https://doi.org/10.1016/j.istruc.2024.106172
9. Blaß, H.J., Sandhaas, C.: STEP - Timber Engineering - Principles for Design. KIT Scientific Publishing, Karlsruhe (2017)
10. Swedish Wood: *Design of Timber Structures, Volume 1: Structural Aspects of Timber Construction*, vol. 3, (2015)
11. T. Vallée, T. Tannert, and R. Meena. Dimensioning method for bolted, adhesively bonded, and hybrid joints involving fibre-reinforced-polymers'. *Proceedings of the 7th International Conference on FRP Composites in Civil Engineering,* CICE 2014, 2014
12. Kemmsies, M., Streicher, R.: Glued Timber-Steel Plate Joints, First RILE. RILEM Publications SARL, Paris (1999)
13. Schober, K.-U., Tannert, T.: Hybrid connections for timber structures. Eur. J. Wood Wood Prod. **74**(3), 369–377 (2016). https://doi.org/10.1007/s00107-016-1024-3
14. Shulman, S., Loss, C.: High-performance grout-reinforced shear connectors for hybrid steel-cross-laminated timber building systems: experimental study. J. Build. Eng. **67**, 106014 (2023). https://doi.org/10.1016/j.jobe.2023.106014
15. Wilkinson, G., Augarde, C.: A serviceability investigation of dowel-type timber connections featuring single softwood dowels. Eng. Struct. **260**, 114210 (2022). https://doi.org/10.1016/j.engstruct.2022.114210
16. Aicher, S.: Structural adhesive joints including glued-in bolts. Timber Eng., 333–363 (2003)

17. Sebera, V., Pečnik, J.G., Azinović, B., Milch, J., Huč, S.: Wood-adhesive bond loaded in mode II: experimental and numerical analysis using elasto-plastic and fracture mechanics models. Holzforschung. **75**(7), 655–667 (2021). https://doi.org/10.1515/hf-2020-0141
18. Śliwa-Wieczorek, K., Zając, B., Kozik, T.: Tests of polymeric adhesive joints in aspect of their application in prefabricated timber structures. Arch. Civ. Eng. **66**, 113–125 (2020). https://doi.org/10.24425/ace.2020.131778
19. Brunner, M., Lehmann, M., Kraft, S., Fankhauser, U., Richter, K., Conzett, J.: A flexible adhesive layer to strengthen glulam beams. J. Adhes. Sci. Technol. **24**, 1665–1701 (2010). https://doi.org/10.1163/016942410X507759
20. Pečnik, J.G., et al.: Mechanical performance of timber connections made of thick flexible polyurethane adhesives. Eng. Struct. **247** (2021). https://doi.org/10.1016/j.engstruct.2021.113125
21. Tlustochowicz, G., Serrano, E., Steiger, R.: State-of-the-art review on timber connections with glued-in steel rods. Mater. Struct. **44**(5), 997–1020 (2011). https://doi.org/10.1617/s11527-010-9682-9
22. Johansen, K.W.: Theory of timber connections. In: *IABSE: International Association of Bridge and Structural Engineering*, vol. 9, pp. 249–262 (1949)
23. CEN (European Committee for Standardization): EN 26891:1991 - Timber Structures - Joints made with mechanical fasteners - General principles for the determination of strength and deformation characteristics (ISO 6891:1983). Structural Engineer. (1991)
24. Ceccotti, A.: Progettare la sicurezza. La Norma Italiana per le COstruzioni in LEgno (NICOLE). In: Sicurezza e comfort nelle abitazioni con strutture di legno (2001)
25. Ottenhaus, L.M., Jockwer, R., van Drimmelen, D., Crews, K.: Designing timber connections for ductility – a review and discussion. Constr. Build. Mater. **304** (2021). https://doi.org/10.1016/j.conbuildmat.2021.124621
26. Jorissen, A., Fragiacomo, M.: General notes on ductility in timber structures. Eng. Struct. **33**(11), 2987–2997 (2011). https://doi.org/10.1016/j.engstruct.2011.07.024
27. Racher, P.: Mechanical timber joints-general. Timber Eng. Lect. C. **1**, 1–10 (1995)
28. European Committee for Standardization (CEN): EN 12512:2001 Timber structures - test methods - cyclic testing of joints made with mechanical fasteners. **3**(1) (2005)
29. Kobayashi, K., Yasumura, M.: Evaluation of plywood sheathed shear walls with screwed joints tested according to ISO 21581. In: *Proceedings of 44th CIB-W18 Meeting*, p. 44-15–8 (2011)
30. American Society for Testing and Materials (ASTM): ASTM E2126-11 test methods for cyclic (reversed) load test for shear resistance of vertical elements of the lateral force resisting systems for buildings. (2011)
31. E. C. for S.- CEN: EN 14592: Timber structures — Dowel-type fasteners — Requirements'. Brussels, Belgium (2017)
32. CEN: EN 1995-1-1:2004 - Eurocode 5: Design of timber structures - part 1-1: Common rules and rules for buildings. **1**(2005) (2011)
33. American Wood Council: National Design Specification (NDS) for Wood Construction (2024)
34. Canadian Standards Association: Engineering Design in Wood. In: *Csa-O86*, p. 264, Mississauga, Canada (2009) Available: http://shop.csa.ca/en/canada/structures/o86-09-consolidation/invt/27013742009
35. SAC: GB 50005: Code for Design of Timber Structures, Beijing, China (2017)
36. Blaß, H.J., Uibel, T.: Tragfähigkeit von stiftförmigen Verbindungsmitteln in Brettsperrholz (2007). https://doi.org/10.5445/KSP/1000006318
37. Bader, T., Schweigler, M., Serrano, E., Dorn, M., Enquist, B., Hochreiner, G.: Integrative experimental characterization and engineering modeling of single-dowel connections in LVL. Constr. Build. Mater. **107** (2016). https://doi.org/10.1016/j.conbuildmat.2016.01.009

38. Zarnani, P., Quenneville, P.: Wood block tear-out resistance and failure modes of timber rivet connections: a stiffness-based approach. J. Struct. Eng. **140**, 4013055 (2013). https://doi.org/10.1061/(ASCE)ST.1943-541X.0000840
39. M. U. Pedersen, 'Dowel Type Timber Connections', 2002
40. Dorn, M., de Borst, K., Eberhardsteiner, J.: Experiments on dowel-type timber connections. Eng. Struct. **47**, 67–80 (2013). https://doi.org/10.1016/j.engstruct.2012.09.010
41. C. Sandhaas, J. Munch-Andersen, and P. Dietsch, Design of Connections in Timber Structures, a State-of-the-Art Report. 2020
42. M. Dorn and T. Bader, Non-linear Connection Models in Timber Engineering. 2016
43. Sandhaas, C., van de Kuilen, J.-W.: Strength and stiffness of timber joints with very high strength steel dowels. Eng. Struct. **131** (2016). https://doi.org/10.1016/j.engstruct.2016.10.046
44. Gelfi, P., Giuriani, E., Marini, A.: Stud shear connection Design for Composite Concrete Slab and Wood Beams. J. Struct. Eng. **128** (2002). https://doi.org/10.1061/(ASCE)0733-9445(2002)128:12(1544)
45. Symons, D., Persaud, R., Stanislaus, H.: Slip modulus of inclined screws in timber-concrete floors. Proceed. Inst. Civ. Eng. Struct. Build. **163**(4), 245–255 (2010). https://doi.org/10.1680/stbu.2010.163.4.245
46. Di Nino, S., Gregori, A., Fragiacomo, M.: Experimental and numerical investigations on timber-concrete connections with inclined screws. Eng. Struct. **209**, 109993 (2019). https://doi.org/10.1016/j.engstruct.2019.109993
47. Kuenzi, E.W.: Theoretical Design of a Nailed or Bolted Joint Under Lateral Load, Madison,WI (1955)
48. Yang, H., Liu, W., Ren, X.: A component method for moment-resistant glulam beam–column connections with glued-in steel rods. Eng. Struct. **115**, 42–54 (2016). https://doi.org/10.1016/j.engstruct.2016.02.024
49. Tomasi, R., Crosatti, A., Piazza, M.: Theoretical and experimental analysis of timber-to-timber joints connected with inclined screws. Constr. Build. Mater. **24**(9), 1560–1571 (2010). https://doi.org/10.1016/j.conbuildmat.2010.03.007
50. Andreolli, M., Piazza, M., Tomasi, R., Zandonini, R.: Ductile moment-resistant steel-timber connections. Proc. Inst. Civ. Eng. Struct. Build. **164**(2), 65–78 (2011)
51. Foschi, R.O.: Load-slip characteristics of nails. Wood Sci. **7**(1), 69–77 (1974)
52. Hirai, T.: Nonlinear load-slip relationship of bolted wood-joints with steel side-members. II. Application of the generalized theory of a beam on an elastic foundation. J. Jpn. Wood Res. Soc. **29**(12), 839–844 (1983)
53. Sawata, K., Yasumura, M.: Estimation of yield and ultimate strengths of bolted timber joints by nonlinear analysis and yield theory. J. Wood Sci. **49**(5), 383–391 (2003). https://doi.org/10.1007/s10086-002-0497-3
54. Hochreiner, G., Bader, T.K., De Borst, K., Eberhardsteiner, J.: Stiftförmige Verbindungsmittel im EC5 und baustatische Modellbildung mittels kommerzieller Statiksoftware. Bauingenieur. **88**, 275–289 (2013)
55. Dorn, M., Hofstetter, K.: 3D-Modeling of Dowel-Type Timber Connections. Universität Stuttgart (2010)
56. Bedon, C., Sciomenta, M., Fragiacomo, M.: Correlation approach for the push-out and full-size bending short-term performances of timber-to-timber slabs with self-tapping screws. Eng. Struct. **238** (2021). https://doi.org/10.1016/j.engstruct.2021.112232
57. Mazelli, A., Bedon, C.: Sensitivity of human-induced vibrations and comfort analysis to partially rigid STS joints in timber-to-timber composite floors. Measurement. **228**, 114396 (2024). https://doi.org/10.1016/j.measurement.2024.114396
58. Kharouf, N., McClure, G., Smith, I.: Elasto-plastic modeling of wood bolted connections. Comput. Struct. **81**(8), 747–754 (2003). https://doi.org/10.1016/S0045-7949(02)00482-0

59. Bedon, C., Fragiacomo, M., Amadio, C., Sadoch, C.: Experimental study and numerical investigation of Blockhaus shear walls subjected to in-plane seismic loads. J. Struct. Eng. **141** (2015). https://doi.org/10.1061/(ASCE)ST.1943-541X.0001065
60. Sciomenta, M., Fanti, R., Doudak, G., Polastri, A., Casagrande, D.: Predicting the non-linear behaviour of cross laminated timber shearwalls with cut-out openings. Structure. **68**, 107138 (2024). https://doi.org/10.1016/j.istruc.2024.107138
61. Sandhaas, C., Van De Kuilen, J.W., Blass, H.J.: Constitutive model for wood based on continuum damage mechanics. In: World Conference on Timber Engineering 2012, WCTE 2012, vol. 1, pp. 159–167 (2012)
62. Patton-Mallory, M., Cramer, S.M., Smith, F.W., Pellicane, P.J.: Nonlinear material models for analysis of bolted wood connections. J. Struct. Eng. **123**(8), 1063–1070 (1997). https://doi.org/10.1061/(ASCE)0733-9445(1997)123:8(1063)
63. Carradine, D.M., Dolan, J.D.: Development of the "displaced volume model" to predict failure for multiple-bolt timber joints. In: *CIB-W18 Timber Structures*, p. 37-7-1 (2004)
64. Mascia, N., Simoni, R.: Analysis of failure criteria applied to wood. Eng. Fail. Anal. **35**, 703–712 (2013). https://doi.org/10.1016/j.engfailanal.2013.07.001
65. Avez, C., Descamps, T., Serrano, E., Léoskool, L.: Finite element modelling of inclined screwed timber to timber connections with a large gap between the elements. Eur. J. Wood Wood Prod. **74**(3), 467–471 (2016). https://doi.org/10.1007/s00107-015-1002-1
66. Chiara, B., Massimo, F.: Numerical analysis of timber-to-timber joints and composite beams with inclined self-tapping screws. Compos. Struct. **207**, 13–28 (2019)
67. Bedon, C., Fragiacomo, M.: Three-dimensional modelling of notched connections for timber-concrete composite beams. Struct. Eng. Int. **27**(2), 184–196 (2017). https://doi.org/10.2749/101686617X14881932435295
68. Bedon, C., Rajcic, V., Barbalic, J., Perkovic, N.: CZM-based FE numerical study on pull-out performance of adhesive bonded-in-rod (BiR) joints for timber structures. Structure. **46**, 471–491 (2022). https://doi.org/10.1016/j.istruc.2022.10.084
69. Mestar, M., Doudak, G., Polastri, A., Casagrande, D.: Investigating the kinematic modes of CLT shear-walls with openings. Eng. Struct. **228**, 111475 (2021). https://doi.org/10.1016/j.engstruct.2020.111475
70. Casagrande, D., Doudak, G., Vettori, M., Fanti, R.: Proposal for an equivalent frame model for the analysis of multi-storey monolithic CLT shearwalls. Eng. Struct. **245**, 112894 (2021). https://doi.org/10.1016/j.engstruct.2021.112894
71. Ruggeri, E.M., D'Arenzo, G., Cavoli, D.L., Cottonaro, R.D., Fossetti, M.: Experimental investigation on the lateral performance of CLT shear walls connected to perpendicular walls. Procedia Struct. Integr. **44**, 464–471 (2022). https://doi.org/10.1016/j.prostr.2023.01.061
72. Sciomenta, M., Bedon, C., Fragiacomo, M., Luongo, A.: Shear performance assessment of timber log-house walls under in-plane lateral loads via numerical and analytical modelling. Buildings. **8**(8) (2018). https://doi.org/10.3390/buildings8080099
73. Gavrić, I., Fragiacomo, M., Ceccotti, A.: Cyclic behavior of typical screwed connections for cross-laminated (CLT) structures. Eur. J. Wood Wood Prod. **73** (2015). https://doi.org/10.1007/s00107-014-0877-6
74. van de Kuilen, J.-W., Dias, A.: Creep factors for timber-timber and timber-concrete joints. Studi. Res. **34** (2015)
75. Jockwer, R., Fink, G., Köhler, J.: Assessment of the failure behaviour and reliability of timber connections with multiple dowel-type fasteners. Eng. Struct. **172** (2018). https://doi.org/10.1016/j.engstruct.2018.05.081
76. Hochreiner, G., Bader, T., Schweigler, M., Eberhardsteiner, J.: Structural behaviour and design of dowel groups experimental and numerical identification of stress states and failure mechanisms of the surrounding timber matrix. Eng. Struct. **131** (2016). https://doi.org/10.1016/j.engstruct.2016.10.043

77. Gikonyo, J., Schweigler, M., Bader, T.: Beam-on-foundation modelling of dowel-type single fastener connections in cross laminated timber. Eng. Struct. **303** (2024). https://doi.org/10.1016/j.engstruct.2024.117519

High-Performance Connections

Giacomo Iovane[1(✉)], Saeid Javidi[2], Beatrice Faggiano[1], Michele Mirra[3], Valentino Nicolussi[4], and Giuseppe D'Arenzo[5]

[1] University of Naples Federico II, Naples, Italy
giacomo.iovane@unina.it
[2] Birmingham City University, Birmingham, UK
[3] Delft University of Technology, Delft, The Netherlands
[4] University of Bologna, Bologna, Italy
[5] University of Enna Kore, Cittadella Universitaria, Enna, Italy

Abstract. The structural performance of Taller Timber Buildings (TTBs) heavily depends on the efficiency and reliability of connections, which play a crucial role in transferring loads, ensuring stability, and enhancing energy dissipation under dynamic actions such as seismic and wind loads. Traditional timber connections, while effective for low-rise structures, face significant limitations in TTB applications due to increased mechanical demands, potential brittle failure mechanisms and reduced energy dissipation capacity. To overcome these challenges, High-Performance Connections (HPCs) have been developed to enhance strength, stiffness, ductility and cyclic resistance.

This study reviews HPC solutions across different structural systems, including timber-framed, panelized and hybrid structures. Various advanced connection techniques, such as self-centering post-tensioned joints, friction-based dissipative devices and hybrid steel-timber interfaces are analyzed to assess their mechanical properties and effectiveness in improving structural resilience. Additionally, the study discusses the key challenges associated with the adoption of HPCs, including standardization, seismic performance, long-term durability and constructability.

The study contributes to advancing the design and implementation of HPCs, promoting the widespread adoption of timber as a viable material for sustainable high-rise construction.

Keywords: Connections · Stiffness · Strength · Ductility

1 Introduction

In Taller Timber Buildings (TTB), connections are essential for transferring forces and accommodating displacements between structural elements. They influence stiffness and structural behaviour and, when specifically designed for seismic regions, connections can provide ductility and energy dissipation, since timber elements are typically designed to respond elastically. During severe earthquakes, cyclic loading imposed on connection components can result in stiffness and strength degradation, as well as irreversible damage. Consequently, the selection of appropriate connections is considered crucial for the overall seismic response of TTBs.

G. Fink et al. (Eds.): *Holistic Design of Taller Timber Buildings*, SPRTRCIENG, pp. 50–63, 2025.
https://doi.org/10.1007/978-3-032-02098-7_5

The types of connections in TTBs are strictly related to the Lateral Load Resisting System (LLRS) adopted to withstand the seismic actions, such as timber framed structures (made with one-dimensional elements, like post, beams and braces [1]), panelised structures (made with Cross-Laminated Timber (CLT) or lightweight timber frames) and hybrid structures (made of timber-to-steel or timber-to-concrete structural systems).

In seismic regions, dowel-type timber connections are typically adopted to ensure energy dissipation through the yielding of steel metal fasteners, e.g. nails, bolts and screws. In panelised structures, mechanical anchors, such as hold-down and angle brackets, are used to connect timber shear walls to the floor and to the foundation to restrain rocking and sliding movements. While certainly effective in low-rise timber buildings, these connections face limitations in TTBs, where structural demands in strength, stiffness and energy dissipation are greater [2]. Moreover, brittle failures may be induced by traditional connections, negatively affecting energy dissipation under seismic loading. These limitations have led to the development of High-Performance Connection (HPC) systems, specifically designed to enhance stiffness, cyclic resistance, ductility and fatigue durability, thus significantly improving the seismic resilience of TTBs.

This chapter reviews HPCs applied in different lateral load-resisting systems, such as framed structures, panelized structures and hybrid structures.

Different techniques and devices aimed at enhancing joint mechanical characteristics, such as stiffness, strength and ductility, have been introduced and discussed in comparison to traditional joints. In particular, the peculiarities and advancements in achieving targeted performance objectives to overcome specific limitations associated with traditional timber connections are highlighted.

High-Performance Connections (HPCs) are integral components of structural lateral load-resisting systems. Besides providing enhanced joint characteristics, HPCs may significantly contribute to the overall ductility and energy dissipation capacity of timber structures subjected to lateral loads. Consequently, to fully understand their role in structural resilience, the behaviour of HPCs should be evaluated not only individually but also within the context of the entire lateral load-resisting system.

2 Connection for Timber Framed Structures

Timber-framed structures are characterized by one-dimensional (1D) structural elements, such as beams, columns and braces. It is possible to distinguish 2 categories of joints connecting primary members (excluding the column-to-foundation joints): (1) *beam-to-column joints*; (2) *joints between brace and beam/column.*

In timber framed structures, particularly in taller buildings, the number of connections is relatively high and the stiffness of the joint can affect global deformations, as well as the strength of the joint influences the collapse modes of the structures. Therefore, the joint behavior should be carefully understood and characterized. Currently, the standards for timber constructions do not include a mechanical classification method for timber joints, which is essential for the correct design of the joints and for the prediction of the actual global behavior of the structures. To fill this gap, in literature, a method for the classification of beam-to-column joints in timber structures in terms of rotational stiffness, bending strength and ductility has been proposed in Iovane et al.

[3], based on the moment-rotation relationship of the joints. According to this method, the joints can be classified considering the internal actions they transfer. When timber connections are not designed to transfer bending moments, they have to be regarded as pinned joints. Conversely, when they are designed to transfer bending moment further to axial loads and shear loads, the rotational stiffness of the joints should be taken into account. Therefore, in terms of mechanical properties, they can be classified with reference to stiffness as rigid, semi-rigid or pinned joints, with reference to strength as full or partial strength, intending that full strength joints are able to restore the continuity of the member, by transferring internal forces equivalent to the strengths, while partial strength joints are able to transfer the design internal forces [3]. It can be highlighted that in rigid frames, traditional moment-resisting connections are difficult to be achieved in practice for spatial frame configurations [4]. Moreover, traditional timber connections exhibit some significant limitations, including stiffness and strength degradation under cyclic loads and brittle failure mechanisms, as identified in many research on the seismic performance [5]. Therefore, some authors have studied and proposed special design details and devices aimed at improving the joint mechanical features, such as stiffness, strength and dissipative capacity, thus enhancing the overall structural performance.

With regards to the first category of the joints, such as the beam-to-column joints, four types of joints available in literature can be identified, depending on the type of connection components used (e.g. Fig. 1) [3].

Type 1. Connection with internal or external vertical steel plates (rectangular- or T-shaped) and connectors (generally bolts, screws or pins [6–11]);
Type 2. Connection with circular or rectangular arrangement of fasteners (generally bolts or screws [12–15]);
Type 3. Connection with top and bottom plates or brackets and fasteners (generally bolts or screws [4, 16]);
Type 4. Post-tensioned connection with internal pre-stressed cable and dissipative devices [17, 18].

Among Type 1 joints, studies concern the analysis of the mechanical behavior both of the connections with internal steel plate and on the connections with double external steel plates. The joints with internal steel plate can be characterized by a single plate [6, 19] or a T-shaped plate that connects the column to the beam [7, 8]. Generally, bolts orthogonally to the plate are used. Some studies have shown that the use of reinforcement screws arranged orthogonally to the grain of the beam and column contribute to increase the overall strength and stiffness of the joint [6, 19]. In particular, the authors have studied the influence of the distribution, number and diameter of screws, as well as the spacing and distances, on the mechanical behavior of the joints, generally obtaining higher stiffness, through minimized gaps and reduced splitting, and strength, as the screws absorb tension perpendicular to the grain, reducing the risk of splitting, while ductility remains almost unchanged.

Among Type 2 joints, most of the studies concern connections with a circular arrangement of connectors [12–14], while few studies are carried out on connections with a rectangular arrangement of connectors [15]. Generally, both joint types are characterized by a single column and a coupled beam equipped, in most cases with screws or, as alternative with dowels or bolts. The authors have demonstrated that the use of glued connectors and shear reinforcement plates at the column-beam interface can increase the rotational stiffness of the joint, avoiding the brittle failures of timber, thereby improving overall mechanical behavior compared to typical realizations.

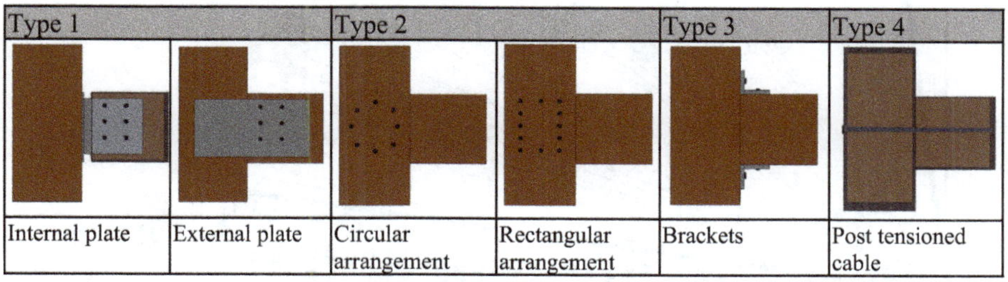

Fig. 1. Types of beam-to-column joints [3].

With regards to Type 3 joints, Kasal et al. studied aluminum brackets [4] and timber plate [16], connected to the beam and column by screws, as an alternative to traditional steel elements. Both types of joints have a significant influence on the behavior in terms of stiffness and resistance, but with a reduction in dissipative capacity.

Contrary, Type 4 joints were conceived to provide the structure with re-centering capabilities following seismic events, by adopting post-tensioned cables, and shifting the ductility demand away from metallic connection elements (plates, fasteners, etc.), instead delegating it to dedicated devices specifically developed for this purpose, such as fluid-viscous or friction devices. Additionally, the lateral stiffness of the frame increases with the number of post-tensioned tendons [17, 18].

With regards to the second category of joints, such as joints between brace and beam/column, recent advancements have focused on developing high-performance connection types to specifically improve strength and ductility. Yousef-beik et al. [20] adapted the Resilient Slip Friction Joint (RSFJ) introduced by Zarnani and Quenneville [21]. This is a damage-free connection featuring serrated plates compressed by a bolt and disc spring system, which increases ductility via friction dissipation and restores the connection to the original position after deformation.

By applying the classification method of beam-to-column joints in terms of stiffness proposed by Iovane et al. [3] on a study sample of 46 joints, the results in terms of dimensionless Moment-Rotation curves (e.g. Fig. 2) show that Type 1 joints is the least rigid, classified as pinned joint, followed by Type 4 and Type 2 which have a progressively higher rotational stiffness, classified as semi-rigid joint. No joint among those analyzed is classified as rigid.

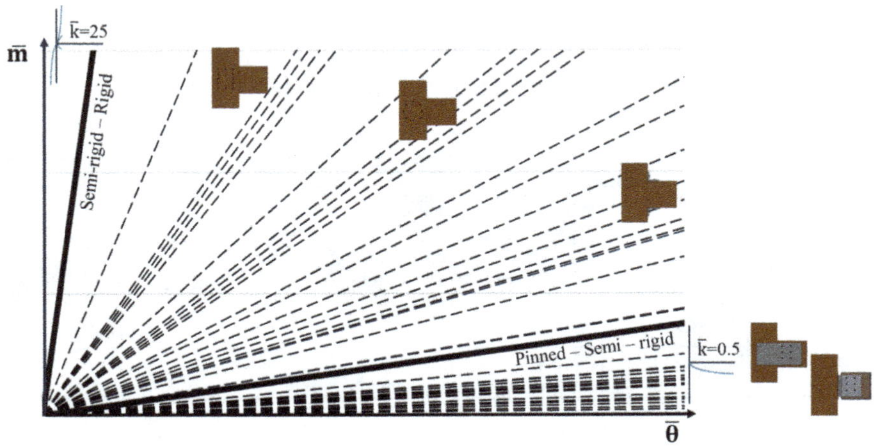

Fig. 2. Classification by stiffness of study joints [3].

3 Connections for Panelized Structures

Panelized structural systems, primarily composed of engineered wood products, such as cross-laminated timber (CLT), glued-laminated timber (Glulam) and laminated veneer lumber (LVL), play a fundamental role in ensuring the lateral stability of taller timber buildings. The connections employed within these systems are critical for achieving structural integrity, stiffness and energy dissipation, particularly under dynamic loading conditions, such as seismic or wind-induced forces. The effectiveness of these connections significantly influences the overall safety, resilience and durability of the structure.

Traditional connection systems in panelized structures, including hold-downs, angle brackets and screw-based fasteners, are widely used in low-rise timber construction. However, the application in taller timber buildings presents significant limitations, including restricted load capacity, brittle failure mechanisms and inadequate energy dissipation [22]. In seismic-prone regions, traditional hold-downs and angle brackets may exhibit strength degradation, pull-out failures and timber splitting due to limited ductility and reliance on steel fasteners. Additionally, the repeated cyclic loading characteristic of seismic events can cause extensive damage, compromising both safety and reparability. These challenges have driven extensive research efforts to develop more robust, high-performance connections that enhance mechanical performance while maintaining constructability and durability.

In the case of panelized structures, it is possible to distinguish 3 categories of connections: (1) *Wall-to-Floor and Wall-to-Foundation connections*; (2) *Panel-to-Panel connections between parallel panels*; (3) *Panel-to-Panel connections between perpendicular panels* (e.g. Fig. 3).

Connections between walls and floors or foundations (Type 1) are essential for transferring overturning moment and shear forces, ensuring the lateral stability of panelized structures. Traditional hold-downs and angle brackets, although commonly used, often exhibit limitations in taller timber buildings due to insufficient load-bearing capacity and brittle failure mechanisms.

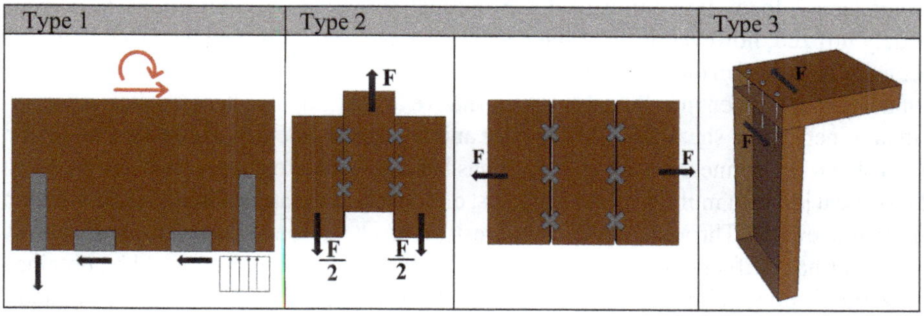

Fig. 3. Types of connections in panelized structures.

To overcome these challenges, various innovations have been developed and investigated over the past decade. One approach has involved the introduction of high-capacity hold-downs and reinforced brackets incorporating mixed-angle self-tapping screws, which have shown enhanced stiffness, load carrying capacity and ductility [23, 24]. Additionally, modifications to traditional hold-downs have been proposed, including the integration of hourglass-shaped steel plates designed to yield before the embedment of timber. This solution improves energy dissipation by promoting a more controlled failure mechanism [25]. Bi-directional connectors have been developed to enhance the rocking resistance of CLT shear walls by utilizing multiple connections rather than single hold-downs [26, 27]. These connectors resist both tensile and shear forces through the use of inclined fully-threaded screws, which increase tensile capacity. Furthermore, the use of thick metal washers and increased bracket thickness prevents brittle failure mechanisms commonly observed in traditional hold-downs and angle brackets [28]. Another innovative approach involves slip-friction connectors, also referred to as slotted-bolted connections [29, 30]. These systems consist of steel plates with slotted bolt holes that allow controlled sliding and friction-based energy dissipation, thereby improving the overall seismic response of panelized timber buildings. Additionally, the High-Force-to-Volume (HF2V) damping device has been developed as a substitute for conventional hold-downs, offering enhanced energy dissipation and self-centering capability to minimize post-seismic residual deformations [31]. Further advancements in seismic-resistant connections consist in resilient slip friction joints (RSFJs [32]) and uplift friction dampers (UFDs [33]), which provide controlled slip and self-centering capabilities, significantly reducing seismic damage accumulation. A new system recently proposed is the tube connector [34], which consists of a hollow steel tube inserted into pre-drilled holes in CLT panels and secured with a threaded rod or adhesive bonding. These connectors enhance pull-out resistance and structural integrity, improving seismic resilience and stress distribution while maintaining a concealed aesthetic. The ability to provide higher load-bearing capacity makes them an efficient alternative to conventional fasteners in modern taller timber construction.

Panel-to-panel connections between parallel panels (Type 2) are fundamental for ensuring continuity and effective load transfer within panelized structures [35]. The influence of these connections on the diaphragm actions and the overall structural response under lateral loads is significant. Conventional screw-based connections have been widely utilized; however, limited shear capacity have been observed when a diaphragm behavior is desired in multi-panel walls and floors. To address these limitations, shear-key connectors have been introduced as an alternative to traditional fasteners, allowing a significant increase in shear transfer capacity and an improvement in diaphragm behavior. Such shear-key connector with slots [36] has been designed to enhance shear transfer in the vertical joints of multi-panel shear walls, effectively mitigating damage accumulation in seismic events. Through experimental tests, it has been found that a single shear-key connector has stiffness and strength values comparable to those that would be obtained for vertical spline joints with approximately thirteen couples of 6 mm screws. In addition to shear-key solutions, prefabricated multi-directional point connectors have been introduced to improve the assembly precision and mechanical performance of parallel panel connections. The X-Rad connection system, for example, consists of steel plates with pre-installed inclined screws, which provide enhanced strength and stiffness by linking adjacent cross-laminated timber (CLT) panels in multiple directions [37]. These connections allow for efficient on-site installation while ensuring high load-bearing capacity. Another prefabricated solution, the SHERPA-CLT Connector, includes metal plate connectors installed at angle, T-joints and longitudinal panel junctions [38]. The pre-fabrication of these connectors enables rapid assembly without the need for extensive scaffolding, reducing construction time and labor demands. Further advancements have focused on improving failure mechanism and energy dissipation. The X-Bracket, a steel bracket connection system, has been developed to increase the ductility and energy dissipation capacity of CLT panels [39]. This system distributes shear and tensile forces more efficiently, mitigating the risk of brittle failure and reducing damage accumulation in timber elements. Similarly, hybrid connection solutions integrating steel components with bonded adhesives have been introduced. The Holz-Stahl-Komposit (HSK) system, for example, involves the insertion of steel plates into timber panels, which are bonded using a chemical adhesive [40]. To ensure a controlled yielding mechanism, duct tape is applied in designated areas, preventing full bonding and creating a "weak zone" that functions as a yielding fuse. This configuration improves the ductility of the connection while maintaining high strength. For applications requiring enhanced seismic resilience, post-tensioned hybrid shear walls have been implemented. These systems incorporate U-shaped flexural plates (UFPs) in the vertical joints of wall panels, enabling energy dissipation and re-centering capabilities [41]. Additionally, friction-based energy dissipative systems, such as slotted steel plates with bolted connections (e.g., Slip-Friction Connectors), have been developed to improve hysteretic performance and reduce strength degradation under cyclic loading [32, 33]. These solutions enhance the overall structural performance of multi-panel timber systems, particularly in seismic-prone regions.

Connections between perpendicular panels (Type 3) are essential for ensuring the structural integrity and load transfer efficiency at the intersections between walls and floors in panelized timber structures. These connections must provide sufficient stiffness, strength and deformation capacity, while minimizing stress concentrations that could compromise structural performance. Traditional solutions, including steel brackets and screw-based fasteners, have been widely utilized in connections between perpendicular walls and floors and walls. Among these, angle brackets and self-tapping screws have been commonly employed. These connections contribute to generate structural interactions, which modify the structural behavior, providing increased lateral stiffness and load-bearing capacity [42, 43]. However, in floor-to-wall connections, the withdrawal stiffness of the connections may contribute to reduce the deformation capacity and the energy dissipation, which can be detrimental in seismic applications. To address this issue, angle brackets with slotted holes have been introduced, which allow for controlled shear transfer while permitting a free rocking mechanism of the wall, thereby improving ductility and mitigating damage accumulation under cyclic loading conditions [44].

4 Connections for Composite and Hybrid Structures

Timber-steel hybrid structures are characterized by the intelligent and more efficient use of the two materials to improve the global and local behavior of all-timber or all-steel constructions. Despite the ever-increasing interest worldwide in steel-timber hybrid structures, research is still very recent and rather limited nowadays.

From the analysis of the state of the art, three types of connections can be distinguished (e.g. Fig. 4 [45]): (1) *Steel beam-to-column joints to CLT floors*; (2) *Timber beam-to-steel column joints*; (3) *Timber beam-to-timber column and/or timber or steel brace joints with steel devices.*

Type 1 refers to the continuity joints of CLT panel slab around the steel column. The common configurations have the following features: (1.1) steel cover plates [46–49]; (1.2) timber cover plates or half-lap connections [46, 49, 50]; (1.3) embedded steel continuity bars [46–50]; (1.4) glued butt joints [46]. The continuity of CLT slabs offers generally a significant contribution to the bare steel joint, in terms of bending strength, rotational stiffness and ductility. In particular, in terms of bending strength, in moment resisting joints the contribution can double the load capacity. Specifically, overall, the 1.1 and 1.3 configurations are the most efficient, followed by the 1.2 and then 1.4 configurations.

Type 2 joint common configurations have the following features: (2.1) top and seat steel angle brackets bolted to the timber beam through long bolts and to the steel column through short special bolts [51, 52]; (2.2) T-stub pre-welded to the column and slotted vertically into the timber beam by means of bolts placed perpendicularly to the slotted steel plate [51]; (2.3) steel box-shaped bracket connected to the timber beam through screwed-in threaded rods and to the steel column through steel bolts [53]. The steel box devices provide in overall the best performance in term of initial rotational stiffness, static ductility and failure mode without timber damage. The T-stub offers the highest initial rotational stiffness, but a fragile failure in the timber beam, even though with large plastic deformation. The steel angle brackets show the smallest initial rotational stiffness, with ductile failure mode, without timber damage.

In Type 3 joints dissipative capacity is delegated to ad hoc devices located at the ends of timber members or at the connection between timber beam and columns. The common configurations have the following features: (3.1) steel link connected to the timber beam by means of glued threaded bars and to the steel/timber column via bolts [54–56]; (3.2) steel reduced beam section (RBS) element, placed at the beam end, connected to the timber beam and to the upper and lower timber columns by means of inclined self-tapping screws and bolted to a steel box [57]; (3.3) intermediate steel panel box in the column connected to the upper and lower timber columns through glued steel bars and to the timber beam via pre-stressed steel cable [58]; (3.4) intermediate steel brace connection among timber beam, column and steel brace by means of glued bars [59].

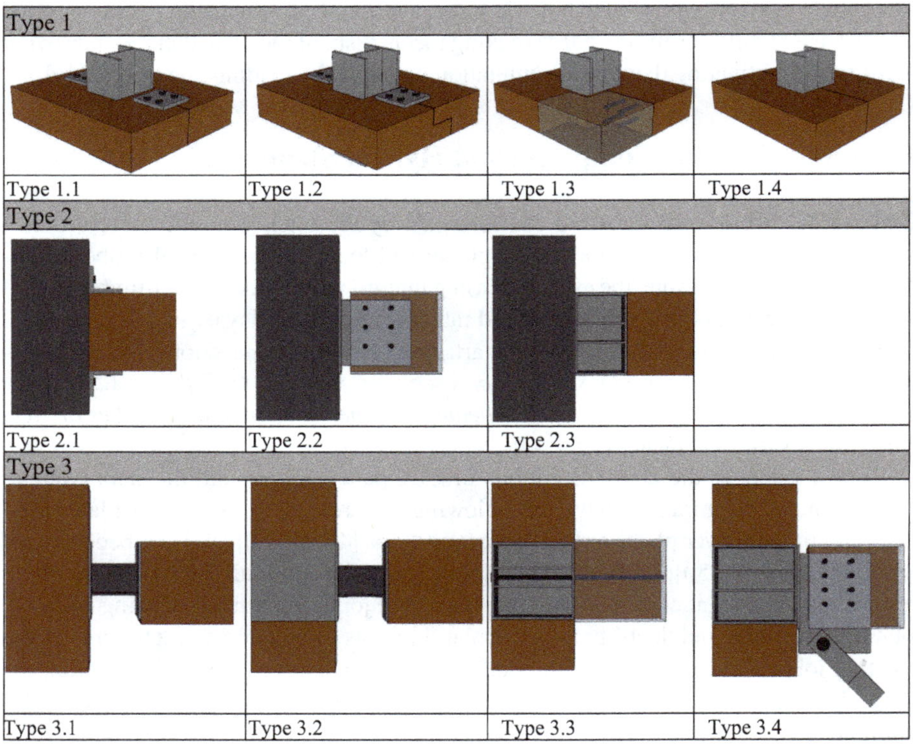

Fig. 4. Types of joints in hybrid structures.

All the joint configurations studied are characterized by a more stable hysteretic curve as respect to the traditional joints, with a larger dissipative capacity, stiffness and strength and absence of fragile failures for timber members. In particular, these joints are designed so that the energy dissipation is provided entirely by the steel links, while timber elements and screws remain elastic under large rotation demands.

5 Challenges and Outlook

The development and implementation of high-performance connections (HPCs) for Taller Timber Buildings (TTBs) present several challenges that need to be addressed to enhance their effectiveness, reliability and applicability. While significant advancements have been made in the design of HPCs for different structural systems - such as timber-framed, panelized and hybrid structures - several issues remain open for future research and development.

One of the primary challenges is the mechanical characterization and standardization of HPCs. Unlike steel or concrete structures, where joint classification methods are well established, timber connections lack a universally accepted mechanical classification framework. The absence of a standardized methodology for defining stiffness, strength and ductility parameters leads to inconsistencies in design approaches and hinders the direct comparison of different connection types. Future research should focus on the development of robust classification criteria to facilitate the design and performance assessment of HPCs in TTBs.

Another critical issue is the seismic performance and resilience of connections. While several studies have proposed dissipative solutions, such as resilient slip-friction joints and post-tensioned tendons, further experimental and numerical investigations are necessary to evaluate their long-term reliability under cyclic loading. The influence of repeated seismic events, potential degradation of mechanical properties and residual deformations should be extensively examined to ensure that HPCs contribute to the overall resilience and reparability of timber structures. Moreover, research on self-centering connections capable of minimizing post-event maintenance is crucial for improving the sustainability of TTBs in earthquake-prone regions.

In terms of material compatibility and durability, the integration of timber with steel or composite elements requires careful consideration of differential deformations, moisture sensitivity and long-term creep effects. The performance of hybrid timber connections under varying environmental conditions, including humidity fluctuations and temperature variations, needs further investigation. Additionally, the introduction of novel materials, such as high-strength adhesives and engineered timber composites, offers promising possibilities for improving connection performance, but their long-term behavior and fire resistance require thorough assessment.

In conclusion, while high-performance connections hold great potential to improve the structural integrity and seismic resilience of TTBs, ongoing research is required to address standardization, durability, constructability and long-term monitoring. Advancements in these areas will contribute to the broader adoption of timber as a viable material for sustainable high-rise construction, further promoting innovation in the field of structural engineering.

References

1. Ascione, F., Esposito, F., Iovane, G., Faiella, D., Faggiano, B., Mele, E.: Sustainable and efficient structural systems for tall buildings: exploring timber and steel–timber hybrids through a case study. Buildings. **14**(2), 524 (2024). https://doi.org/10.3390/buildings14020524
2. Nicolussi, V., Pozza, L., Polastri, A., Casagrande, D.: Structural limitations in the use of earthquake-resistant proprietary connections in multi-storey Clt buildings: a multi-parametric numerical study. In: *18th World Conference on Earthquake Engineering* (2024)
3. Iovane, G., Noviello, C., Mazzolani, F.M., Landolfo, R., Faggiano, B.: A proposal for the mechanical classification of beam-to-column joints for timber structures. In: *World Conference on Timber Engineering (WCTE 2023)*, 19-22 June 2023, Oslo, Norway, vol. 5, pp. 2995–2300 (2023a). https://doi.org/10.52202/069179-0390

4. Kasal, B., Heiduschke, A., Pospisil, S., Urushadze, S., Ceccotti, A.: Shake table tests of a three-story spatial timber frame with moment connections. In: *World Conference on Timber Engineering (WCTE2010)*, vol. 2, pp. 917–922 (2010)
5. Popovski, M., Karacabeyli, E., Prion, H.: Dynamic response of braced timber frames. In: Eighth Canadian Conference on Earthquake Engineering (1999)
6. He, M.J., Liu, H.F.: Comparison of glulam post-to-beam connections reinforced by two different dowel-type fasteners. In: *Construction and Building Materials*, pp. 99–108. Elsevier Ltd. (2015)
7. Salem, S.: Experimental testing of steel-wood-steel glulam frame bolted connections. In: *World Conference on Timber Engineering (WCTE2016)*, 22-25 August, Vienna, Austria (2016) ISBN: 978-3-903039-00-1
8. Salem, S., Petrycki, A.: Experimental testing of wood-steel-wood moment resisting bolted connections. In: *World Conference on Timber Engineering (WCTE2016)*, 22–25 August, Vienna, Austria (2016) ISBN: 978-3-903039-00-1
9. Zhang, E., Chen, G., Wang, C., Zhu, W., Yang, W.: Experimental investigation on lateral performance of southern pine nailed connections under monotonic and cyclic loading. Structure. **69**(2024), 107334 (2024). https://doi.org/10.1016/j.istruc.2024.107334
10. Rebouças, A., Mehdipour, Z., Branco, J., Lourenço, P.B.: Ductile moment-resisting timber connections: a review. Buildings. **12**(240), 2022 (2022). https://doi.org/10.3390/buildings12020240
11. Li, Z., Wang, Y., Liu, X., Ou, J., Gan, Z.: Mechanical performance of glulam beam-to-column joints with geometrical modulus: experimental investigation and analytical approach. J. Build. Eng. **84**(2024), 108437 (2024). https://doi.org/10.1016/j.jobe.2024.108437
12. Beltran, G.G.: Local energy dissipators for skeleton structures. In: *World Conference on Timber Engineering (WCTE2016)*, 22–25 August, Vienna, Austria (2016) ISBN: 978-3-903039-00-1
13. Polastri, A.: Caratterizzazione del comportamento di giunti semi-rigidi per strutture lignee in zona sismica. PhD Thesis, Supervisor: Prof. Maurizio Piazza, Prof. Ian Smith. University of Trento, University of Brescia, University of Padova, University of Trieste, University of Udine, University of Venezia (2010)
14. Johanides, M., Kubíncová, L., Kubíncová, D., Lokaj, A., Sucharda, O., Mynarcík, P.: Analysis of rotational stiffness of the timber frame connection. Sustainability. **13**, 156 (2021). https://doi.org/10.3390/su13010156
15. Wang, X.T., Zhu, E.C., Niu, S., Wang, H.J.: Analysis and test of stiffness of bolted connections in timber structures. Constr. Build. Mater. **303**, 124495 (2021). https://doi.org/10.1016/j.conbuildmat.2021.124495
16. Kasal, B., Asce, M., Guindos, P., Polocoser, T., Asce, S.M., Heiduschke, A., Urushadze, S., Pospisil, S.: Heavy laminates timber frames with rigid three-dimensional beam-to-column connections. J. Perform. Constr. Facil. **28**(6), A4014014 (2014). https://doi.org/10.1061/(ASCE)CF.1943-5509.0000594
17. Rahmzadeh, A., Iqbal, A.: Numerical modelling and analysis of post-tensioned timber beam-column joint. Eng. Struct. **245**(9), 112762 (2021). https://doi.org/10.1016/j.engstruct.2021.112762
18. Ding, Y., Zhou, Z., Huang, L., Si, Y.: Seismic performance of self-centering glulam frame with friction damper. Eng. Struct. **245**, 112857 (2021)
19. Wang, M., Song, X., Gu, X., Zhang, Y., Luo, L.: Rotational behavior of bolted beam-to-column connections with locally cross-laminated glulam. J. Struct. Eng. ASCE. **141**, 04014121 (2015) ISSN 0733-9445/04014121(7)/$25.00
20. Yousef-beik, S.M.M., Bagheri, H., Veismoradi, S., Zarnani, P., Hashemi, A., Quenneville, P.: Seismic performance improvement of conventional timber brace using re-centring friction connection. In: *Structures*, vol. 26, pp. 958–968. Elsevier (2020)

21. Zarnani, P., Quenneville, P.: New design approach for controlling brittle failure modes of small-dowel-type connections in cross-laminated timber (CLT). Constr. Build. Mater. **100**, 172–182 (2015)
22. Gavric, I., Fragiacomo, M., Ceccotti, A.: Cyclic behaviour of typical metal connectors for cross-laminated (CLT) structures. Mater. Struct. Constr. **2015**(48), 1841–1857 (2014). https://doi.org/10.1617/s11527-014-0278-7
23. Krauss, K., Li, M., Lam, F.: Influence of mixed-angle screw installations in CLT on the cyclic performance of commercial hold-down connections. Constr. Build. Mater. **364**, 129918 (2022). https://doi.org/10.1016/j.conbuildmat.2022.129918
24. Wright, T., Li, M., Moroder, D., Lim, H., Carradine, D.: Cyclic behaviour of CLT shear wall hold-down connections using mixed angle self-tapping screws. Eng. Struct. **286**, 116123–116123 (2023). https://doi.org/10.1016/j.engstruct.2023.116123
25. Latour, M., Rizzano, G.: Cyclic behavior and modeling of a dissipative connector for cross-laminated timber panel buildings. J. Earthquake Eng. **19**, 137–171 (2015)
26. D'Arenzo, G., Casagrande, D., Polastri, A., Fossetti, M., Fragiacomo, M., Seim, W.: CLT shear walls anchored with shear-tension angle brackets: experimental tests and finite-element modeling. J. Struct. Eng. ASCE. **147**(7) (2021). https://doi.org/10.1061/(asce)st.1943-541x.0003008
27. Azinović, B., Pazlar, T., Kržan, M.: The influence of flexible sound insulation layers on the seismic performance of cross laminated timber walls. J. Build. Eng. **43**, 103183 (2021). https://doi.org/10.1016/j.jobe.2021.103183
28. Casagrande, D., Polastri, A., Sartori, T., Loss, C., Chiodega, M.: Experimental campaign for the mechanical characterization of connection systems in the seismic design of timber buildings. In: World Conference on Timber Engineering (WCTE), 22–25 August, Vienna, Austria (2016) ISBN: 978-3-903039-00-1
29. Fitzgerald, D., Miller, T.H., Sinha, A., Nairn, J.A.: Cross-laminated timber rocking walls with slip-friction connections. Eng. Struct. **220**, 110973 (2020)
30. Fitzgerald, D., Sinha, A., Miller, T.H., Nairn, J.A.: Axial slip-friction connections for cross-laminated timber. Eng. Struct. **228**, 111478 (2021)
31. Wrzesniak, D., Rodgers, G.W., Fragiacomo, M., Chase, J.G.: Experimental testing of damage-resistant rocking glulam walls with lead extrusion dampers. Constr. Build. Mater. **102**, 1145–1153 (2016). https://doi.org/10.1016/j.conbuildmat.2015.09.011
32. Hashemi, A., Zarnani, P., Masoudnia, R., Quenneville, P.: Seismic resistant rocking coupled walls with innovative resilient slip friction (RSF) joints. J. Constr. Steel Res. **129**, 215–226 (2017)
33. Tatar, A., Dowden, D.M.: Analytical and numerical investigation of a low-damage uplift friction damper for self-centering cross-laminated timber rocking walls. Eng. Struct. **254**, 113836 (2022)
34. Schneider, J., Tannert, T., Tesfamariam, S., Stiemer, S.: Experimental assessment of a novel steel tube connector in cross-laminated timber. Eng. Struct. **177**, 283–290 (2018)
35. Gavric, I., Fragiacomo, M., Ceccotti, A.: Cyclic behavior of typical screwed connections for cross-laminated (CLT) structures. Eur. J. Wood Wood Prod. **73**, 179–191 (2015). https://doi.org/10.1007/s00107-014-0877-6
36. Polastri, A., Casagrande, D.: Mechanical behaviour of multi-panel cross laminated timber shear-walls with stiff connectors. Constr. Build. Mater. **332**, 127275 (2022). https://doi.org/10.1016/j.conbuildmat.2022.127275
37. Polastri, A., Giongo, I., Angeli, A., Brandner, R.: Mechanical characterization of a pre-fabricated connection system for cross-laminated timber structures in seismic regions. Eng. Struct. **2018**(167), 705–715 (2018)

38. Shao, F., Wang, Y., Lian, W., Benjeddou, O.: Experimental and numerical investigation on withdrawal connectors usage for lateral resistance of timber shear walls structure. J. Build. Eng. **44**, 103266 (2021)
39. Trutalli, D., Marchi, L., Scotta, R., Pozza, L.: Capacity design of traditional and innovative ductile connections for earthquake-resistant CLT structures. Bull. Earthq. Eng. **17**(4), 2115–2136 (2018). https://doi.org/10.1007/s10518-018-00536-6
40. Zhang, X., Popovski, M., Tannert, T.: High-capacity hold-down for mass-timber buildings. Constr. Build. Mater. **2018**(164), 688–703 (2018)
41. Iqbal, A., Pampanin, S., Palermo, A., Buchanan, A.H.: Performance and design of LVL walls coupled with UFP dissipaters. J. Earthq. Eng. **19**, 383–409 (2015). https://doi.org/10.1080/13632469.2014.987406
42. D'Arenzo, G., Ruggeri, E.M., Fossetti, M.: Lateral performance of cross-laminated timber shear walls connected to perpendicular walls: experimental tests and analytical modeling. J. Struct. Eng. **150**(8) (2024). https://doi.org/10.1061/jsendh.steng-13151
43. Ruggeri, E.M., D'Arenzo, G., Fossetti, M.: Investigating the effect of the structural interactions due to floors and lintels on the lateral response of multi-storey cross-laminated timber shear walls. Eng. Struct. **290**, 116327–116327 (2023). https://doi.org/10.1016/j.engstruct.2023.116327
44. Masroor, M., Tannert, T.: Validation and application of analytical design approach for multi-storey platform-type CLT shear walls. Eng. Struct. **327**, 119655 (2025). https://doi.org/10.1016/j.engstruct.2025.119655
45. ECCS TC11 WG 6, State of Art on Steel-Timber-(Concrete) Structures: *Interim Report of ECCS TC11/WG6: Steel-Timber Hybrid Structures*, N°145, 1st edn, (2024) ISBN: 978-92-9147-199-7
46. Keipour, N., Valipour, H.R., Bradford, M.A.: Experimental study of steel-timber composite (STC) beam to steel column joints having a flush end-plate. Eng. Struct. **174**, 906–918 (2018). https://doi.org/10.1016/j.engstruct.2018.08.009
47. Ataei, A., Valipour, H.R., Bradford, M.A., Chiniforush, A.A.: Experimental study of steel-timber composite beam-to-column joints with extended end plates. Constr. Build. Mater. **226**, 636–650 (2019). https://doi.org/10.1016/j.conbuildmat.2019.07.154
48. Nouri, F., Valipour, H.R., Bradford, M.A.: Structural behaviour of steel-timber composite (STC) beam-to-column connections with double angle web cleats subjected to hogging bending moment. Eng. Struct. **192**(1), 1–17 (2019a). https://doi.org/10.1016/j.engstruct.2019.04.092
49. Keipour, N., Valipour, H.R., Bradford, M.A.: Structural behaviour of steel-timber versus steel-concrete composite joints with flush end plate. Constr. Build. Mater. **262**, 120885 (2020). https://doi.org/10.1016/j.conbuildmat.2020.120885
50. Nouri, F., Valipour, H.R.: Semi-rigid partial-strength steel-timber composite (STC) connections with mechanically anchored steel rods. J. Constr. Steel Res. **158**, 560–575 (2019b). https://doi.org/10.1016/j.jcsr.2019.04.017
51. Karagiannis, V., Málaga-Chuquitaype, C., Elghazouli, A.Y.: Behaviour of hybrid timber beam-to-tubular steel column moment connections. Eng. Struct. **131**, 243–263 (2017)
52. Sirumbal-Zapata, L.F., Málaga-Chuquitaype, C., Elghazouli, A.Y.: Experimental assessment and damage modelling of hybrid timber beam-to-steel column connections under cyclic loads. Eng. Struct. **200**, 109682 (2019)
53. Liu, L., Yang, H.: Parameter identification of ductile timber beam-to-steel column connections: procedure and assessment. Eng. Struct. **259**, 114179 (2022). https://doi.org/10.1016/j.engstruct.2022.114179
54. Faggiano, B., Iovane, G.: First considerations on the design approach and criteria for seismic resistant and bracing timber frames. In: *World Conference on Timber Engineering*

(WCTE2016), 22--25 August, Vienna, Austria, pp. 4549–4556 (2016) ISBN: 978-3-903039-00-1. Full paper ID1094.
55. Iovane, G., Oliva, V., Faggiano, B.: Design and analysis of dissipative seismic resistant heavy timber frame structures equipped with steel links. Procedia Struct. Integr. **44**, 1864–1869 (2023). https://doi.org/10.1016/j.prostr.2023.01.239
56. Iovane, G., Rodrigues, L., Branco, J., Faggiano, B.: Monotonic tests on beam-to-column joint with steel link for timber seismic resistant structures. In: *World Conference on Timber Engineering (WCTE 2023)*, 19–22 June 2023a, Oslo, Norway (2023). https://doi.org/10.52202/069179-0288
57. Gohlich, R., Erochko, J., Woods, J.E.: Experimental testing and numerical modelling of a heavy timber moment-resisting frame with ductile steel links. Earthquake Eng. Struct. Dyn. **47**(6), 1460–1477 (2018)
58. Chen, F., Li, Z., Li, M., Shu, Z.: Experimental testing of post-tensioned steel-timber hybrid frames equipped with energy-dissipating braces. Eng. Struct. **296**, 116960 (2023)
59. Gilbert, C.F., Erochko, J.: Development and testing of hybrid timber-steel braced frames. Eng. Struct. **198**, 109495 (2019)

Floor Design for Taller Timber Buildings

Joonas Jaaranen[1(✉)], Sebastian Krug[2], Vlatka Rajcic[3], Martina Sciomenta[4], and Chiara Bedon[5]

[1] Department of Civil Engineering, School of Engineering, Aalto University, Espoo, Finland
joonas.jaaranen@aalto.fi

[2] Institute for Timber Design, University of Applied Sciences Biberach, Biberach, Germany

[3] Department of Structures, University of Zagreb, Zagreb, Croatia

[4] Department of Civil, Architectural and Building and Environmental Engineering, University of L'Aquila, L'Aquila, Italy

[5] Department of Engineering and Architecture, University of Trieste, Trieste, Italy

Abstract. Floors have two fundamental structural functions in buildings: (i) transferring the vertical loads (that they are subjected to) to the vertical load bearing structures, and (ii) acting as diaphragms in the overall lateral load resisting system. In this Chapter the focus is on the former function, concentrating especially on their out-of-plane behaviour. To this aim, there are in fact various floor systems that can be used in taller timber buildings (TTBs). Among other aspects, the choice of the suitable floor system for a particular timber building depends widely on several influencing parameters, such as the span, the assigned loads, the requirements of the floor for the intended use, and the floor layout. The floors are subjected to various loads, and their design is very often governed by serviceability aspects, such as static deflections and dynamic responses (e.g., vibrations). A robust assessment of the serviceability performance requires the evaluation response of the floor system to the imposed loads, and the application of corresponding design criteria (in the region the building is situated in). In this section, aspects such as different floor systems, relevant loads, mechanical behaviour and serviceability criteria in different regions shall be discussed.

Keywords: Timber floor systems · Taller timber buildings (TTBs) · Hybrid floor systems · Serviceability performance · Vibration and deflection criteria

1 Floor Systems in TTBs

Floors usually consist of multiple components, such as the load-bearing structural systems and the non-load-bearing layers (e.g., flooring on top and ceiling underneath the floor structure). In the following, commonly encountered options for both are shortly presented as the basis for further discussion.

1.1 Load-Bearing Systems

The load-bearing system transfers vertical loads to the supporting structures and is mainly responsible for the structural performance of the floor. The structural systems can be

G. Fink et al. (Eds.): *Holistic Design of Taller Timber Buildings*, SPRTRCIENG, pp. 64–75, 2025.
https://doi.org/10.1007/978-3-032-02098-7_6

differentiated by (i) their materials into timber-only and hybrid systems and (ii) their geometrical arrangement into joist- and slab-based systems, as illustrated in Fig. 1.

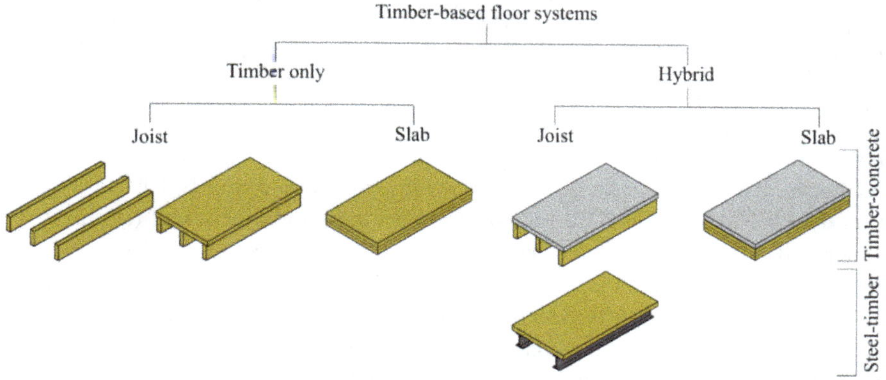

Fig. 1. Schematic illustrations of different timber-based floor systems categorized by materials and geometries.

The traditional structure is a joisted floor, in which the structure consists of parallel joists spanning from one support to another. These joists can be made of sawn timber, laminated veneer lumber (LVL), glued-laminated timber (GLT) or prefabricated I-joists. However, pure joisted floors are less common in TTBs, due to limited performances. Especially sawn timber joists are normally used for short spans only, around 3–5 m [1], due to the small available cross-sections.

Modern timber constructions often use wood-based composite elements that consist of timber joists (webs) connected to timber-based top and/or bottom plates (flanges). Typical examples are ribbed-panels (joists and top plate) and hollow-box systems (joists, top and bottom plates). The elements are usually prefabricated utilizing structural glueing between the basic components, although mechanical connections (such as inclined screws) can be also used, if they provide adequate composite action between the components. The main benefit of composite systems is that – due to the high structural efficiency – they allow to reach significantly longer spans, compared to conventional joisted floors with relatively low material volume. Further discussion can be found for example in [2, 3].

An alternative that has recently become very popular is the use of mass timber panels for the load-bearing structure. The most common panel type is the cross-laminated timber (CLT), which is made of an odd number of glued layers of timber boards, rotated 90° with respect to adjacent ones [4]. Due to their intrinsic structure, CLT panels exhibit also plate-like biaxial bending behaviour. Recently, there has been interest providing connections between CLT panels that can ensure effective biaxial behaviour for whole floors [5–7] and significantly improve their efficiency. Other suitable mass timber products are nail-laminated timber (NLT) elements, in which timber boards are laminated horizontally by nailing [8], and dowel-laminated timber (DLT), see for example [9]. The latter is an adhesive-free alternative to CLT laminated vertically using wooden dowels.

In hybrid systems, the structure combines timber with other materials to achieve the preferable mechanical behaviour. Among the most prevalent hybrid systems in timber buildings, are timber-concrete composite (TCC) floors. In the floor applications, TCC structures typically consist of timber components, a concrete slab placed on top, and the shear connection between them (Fig. 1). This configuration seeks to combine the advantages of timber and reinforced concrete construction. Depending on design requirements and construction methods, various types of shear connectors can be employed, such as notched connectors, screws, or metal plates. At the European level, the Technical Specification CEN/TS 19103:2022 [10] was published as a guideline for designing TCC floors. It is expected to be integrated into Eurocode 5 as EN 1995-3 by 2028. Currently, CEN/TS 19103:2022 regulates shear connectors, including (i) dowel-type fasteners, (ii) steel reinforcement bars embedded in timber perpendicular to the shear plane, and (iii) notched connections (see Fig. 2). However, alternative connection systems may also be employed, provided that their characteristic load-bearing capacity and stiffness are known. Given the increasing adoption of CLT-concrete composite floors in hybrid timber buildings, it is important to account for the influence of cross-layers in CLT panels on the stiffness of notched connections. The stiffness reduction caused by these cross-layers should be considered in the design process, as suggested in DIN CEN/TS 19103/NA:2024 [11]. These additional provisions are currently being discussed as part of the process of converting CEN/TS 19103:2022 into EN 1995-3 and are expected to be incorporated. Moreover, it should be noted that the design of TCC floors is often governed by serviceability requirements related to deformation. One strategy to enhance deformation serviceability is the design of continuous TCC floor systems [12]. Precast TCC systems represent an important advancement in the dissemination of TCC technology. These elements are often supplied with pre-determined structural capacities, simplifying the design process and fostering adoption. Several companies have demonstrated how the use of precast elements supports rapid construction, which is particularly valued in TTB where investors often prioritize short construction times [13]. In conclusion, TCC plays an important role in the transition from conventional concrete construction to more sustainable building methods and materials. By combining the strengths of timber and concrete, TCC systems contribute to reducing the environmental footprint of modern construction while meeting high performance standards.

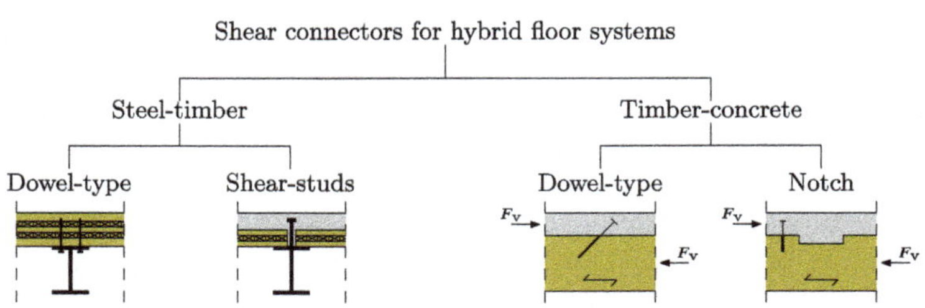

Fig. 2. Schematic illustrations of common shear connectors used in hybrid floor systems, including steel–timber and timber–concrete configurations.

The hybrid steel-timber systems allow for the combination of high strength and ductility of steel with high stiffness and light weight of timber. However, currently hybrid steel-timber structures are built usually with hot-rolled steel elements and timber diaphragms. In hybrid steel-timber systems, connection plays the key role in their seismic performance [14]. Well-known hybrid solutions are the steel-timber flooring and wall systems. The combination of steel, particularly cold-formed steel (CFS) members, and timber plate materials enable lightweight construction with structural, cost, and environmental benefits for low- and high-rise construction [15]. CFS sections can outperform most alternative construction materials due to their cost, strength-to weight ratio, and ease of erection. However, the sections are slender, thin-walled, and susceptible to different types of buckling, such as lateral torsional, local, distortional, crippling, and bearing failure [16]. These failures are mostly due to the thin plate elements of CFS section beams, which can be avoided or delayed by using timber plate acting as stiffeners [17] or by prestressing of the CFS members [18]. These structural elements are of low mass and beneficial in terms of earthquake resistance, while resistance to uplift loading due to high wind loads can be ensured by massive panels, and high-capacity connectors, which connect the panels to the vertical structural components at regular intervals. However, very limited information exists on the influence of timber plate and demountable shear connections on other forms of buckling in CFS floor systems using prestressed CFS sections [19, 20]. Hybrid CFS-timber structural designs are based on traditional standards and regulation [21]. As this construction adopts unconventional connections, thus, the safety factors and design tool specified in the traditional design standards, EN1993-1-1:2005 [22] and EN1995-1-1:2004 [23] may lead to uncertainty in the robustness of the design, building stability and structural performance under loading conditions. Therefore, more reliable design tools are required which can be developed through experimental testing and numerical analysis. Connecting the timber plates and even boards to the CFS beams can result with the systems with significant load bearing capacity. The performance of these systems is strongly related to the resistance, slip modulus and spacing of the shear connectors (see Fig. 2). Existing research has demonstrated significant improvements to the floor flexural performance if the composite action between CFS and timber boards is optimised and exploited [20]. Such composite action relies on the efficacy of connection (slip modulus and shear stiffness), to enable the two components to work together as a composite system, to resist flexural loads [17]. The flexural performance of the CFS-timber flooring systems in terms of stiffness and load-carrying capacity can be significantly improved by introducing prestressed CFS sections, novel shear connectors with timber embedment resistance and improved connector bending performance [24].

1.2 Non-Load-Bearing Layers

Non-load-bearing layers in floors of taller timber buildings fulfil essential functions such as acoustic and thermal insulation, fire protection, and moisture control. Raised floors are particularly relevant in TTBs, as these structures are often used as office buildings, allowing technical installations to be efficiently integrated beneath the floor. Additionally, the mass of screed or similar materials enhances the vibration behaviour of floor systems, improving comfort for occupants. In purely timber floors, mass-increasing fill materials such as gravel or sand are frequently used to improve acoustic performance and dynamic behaviour. Other techniques to improve the impact sound insulation on timber floors were proposed in [25] that included the increase in mass or damping and the addition of a ceiling that could improve the sound insulation at low frequencies.

Concerning fire, the addition of gypsum board layers is widespread, in accordance with the encapsulation technique. When floors are constructed with open void spaces—such as those using engineered joists or counter battens beneath traditional solid joists—the risk of fire spreading through the floor void increases significantly. While the cavities of timber frame assemblies are completely filled with insulation materials in addition to the protective effect of cladding, the charring of a timber member is strongly dependent on the type of insulation used [26]. Any penetrations through the plasterboard, including downlighters, soil vent pipes, or ventilation duct heads, introduce vulnerabilities in the ceiling. These must be properly fire-stopped using fire collars, fire hoods, or other fire-rated products to maintain the integrity of the fire barrier.

A key aspect concerns the integration of Mechanical, Electrical, and Plumbing (MEP) services into mass timber high-rise buildings without compromising structural performance or aesthetics. The principal challenges related to the correct conception of MEP system concern the early design collaboration to minimize interferences (i.e. with soundproofing material), the use of integrated vertical and horizontal MEP zones to keep systems organized and the use of non-combustible enclosures for some systems and/or intumescent coating.

2 Loads on Floors

In taller timber buildings, long-span floors are subject to various actions that can significantly affect their performance. According to EN 1991-1-1 [27], loads are classified as "permanent" and "imposed". For floors, the first category of loads includes the combined self-weight of structural and non-structural elements acting permanently on the floors, while the second category includes variable actions (i.e., occupancy, furniture, partitions and machinery).

Both, the permanent and imposed loads can induce significant short- and long-term (creep) out-of-plane deformations on long-span floors. Those are also susceptible to "indirect actions" caused by fluctuations in environmental conditions (particularly changes in moisture and temperature), which can lead to differential vertical movements and affect the long-term performance. The fluctuations of moisture conditions have different sources as seasonal variations or indoor climate, which is influenced by human occupancy and their activities. Concerning the sources of thermal variations, this aspect is linked to the seasonal changes and sunlight exposure as to the type of ventilation, heating and cooling systems (i.e. underfloor heating) and the presence of adequate of the building's envelope.

Further imposed loads (e.g. wind, earthquake, or machinery actions) generate horizontal global deformations that can lead to floor drift and vibrations, potentially impacting serviceability and occupant comfort. Together, these indirect actions highlight the importance of considering time-dependent behaviours in the design of long-span timber floors, to ensure durability and keep comfort and functionality over the building's lifespan.

It is also worth to mentioning the walking-induced loads, which cause floor vibrations. Vibration levels at floor surfaces could be perceptible by humans depending on motion frequency contents and the types of activities being performed by observers or other people occupying a floor [28]. If the vibrations exceed certain limits, it can lead to discomfort and dissatisfaction among the occupants.

3 Short- and Long-Term Response to the Loads

In terms of serviceability, the most significant load-induced deformation is vertical deflection, which can be quasi-static (and occurs slowly, without the involvement of inertial effects) or dynamic (which occurs fast and involves inertial effects). The latter is often experienced by the users as vibrations. Furthermore, quasi-static deflection consists of instantaneous (elastic) and delayed (creep) responses to the loads. The detailed mechanical response of a floor depends significantly on the type of load and on the structural system.

3.1 Timber-Only Floors Systems

The conventional timber floors are designed as one-way systems, i.e., the loads are transferred by uniaxial bending and shear along the strong axis (span direction). However, even the simple joisted floors possess some lateral stiffness, and they are more accurately described as rib-stiffened plates [29], displaying biaxial bending and lateral load distribution between the joists. The amount of load distribution depends on the ratio between the lateral and longitudinal bending stiffness of the floor structure. In conventional floor, where the lateral stiffness is provided only by the subfloor and other non-structural layers, the lateral stiffness is low, and the lateral load distribution is limited. In composite systems, where the flanges are built of biaxial timber panels, load distribution between the joists can be clearly higher. In slab type floors, especially when biaxial panels, such as CLT, are used, the transverse stiffness is significant, leading to stronger biaxial behaviour.

Hygrothermal variations, i.e., variations of moisture content and temperature, cause time-dependent dimensional changes in timber. Most significant serviceability-related effect for timber floors from these dimensional changes are the bending deformations. Whether the variations cause significant deflection or not depends on the type of the structural system and the gradient of the variations over the cross-section. If the variations are close to uniform, they alone do not cause significant deflection of the floors. In the case of significant asymmetric moisture gradient and/or differences in hygrothermal properties of individual layers in timber-based composite systems, however, differential strains can induce significant bending deformations.

In timber structures, creep causes additional, time-dependent deflection for floors. It is a phenomenon, where deformation increases after instantaneous (elastic) deformation under sustained loads. The rate of creep is highest close to initial elastic deformation and decreases with time. Timber creep can be divided into two parts, viscoelastic and mechano-sorptive creep. In the viscoelastic part the rate depends on time and in the mechano-sorptive part on the variation of moisture content in wood. Therefore, under conditions where the moisture content of timber varies significantly, the creep rate can be significantly larger compared to constant moisture conditions. For a comprehensive review on timber creep, see [30].

3.2 Additional Effects on Hybrid Floor Systems

In hybrid systems, components with different creep and hygrothermal characteristics are joined together. As a result, these systems develop time-dependent deformations that are usually negligible in timber-only floors. In the case of TCC floors, timber and concrete are both subject to dimensional changes due to variations of temperature and moisture content. However, they respond to changes with different rates due to their different transport and expansion properties. Additionally, concrete shrinks significantly during and after curing. Due to these effects, differential inelastic strain develops between the components (even in a constant indoor environment due to the concrete shrinkage), which causes additional stresses and deformations in the structure [31]. Especially, the concrete shrinkage is relevant to all TCC floors as it causes permanent downward deflection. The inelastic strains from other sources change according to the varying environmental conditions and can cause upward or downward deflection depending on the situation. A proposal to account for the effects of the environmental variations and creep in TCCs in the context of Eurocodes are presented in CEN/TS 19103:2022 [10].

4 Analysis Approaches for Deflections and Vibrations

Analytical methods can be used to estimate load-induced deformations and vibrations in timber floors. In practice, their use is limited to rectangular floor layouts with uniform structure over the floor area. For one-way floors subjected to uniform loads, deflections can be obtained using Euler-Bernoulli beam theory with effective width and bending stiffness. In CLT floors, low rolling shear stiffness of the transverse layers causes significant shear deformation in bending. Their effects can be accounted for in analytical calculation by utilizing γ-method, shear analogy method or Timoshenko beam theory [4].

However, as mentioned, even floors designed as one-way system, have lateral stiffness to some extent. To avoid significant over-design, responses to point loads or dynamic responses are usually estimated based on plate theory using equivalent bending and twisting stiffnesses. Especially the dynamic response can be sensitive to the support conditions, their arrangements and stiffnesses, and should be accounted for. Simplified methods to account for the support flexibility are proposed, for example, in the revised version of EN 1995-1-1 [32] that will be published in the near future. The intrinsic limitations of simplified methods for the vibration serviceability assessment have been discussed in many research contributions. In [33], for example, the effects of rough modelling assumptions both in terms of floor components and connections, and in terms of human-induced loads have been properly emphasized, pointing out the importance of a robust methodology of general applicability to timber floors.

The long-term deformations due to creep are normally calculated by utilizing the effective modulus method, where the modulus of elasticity (and possible other stiffness properties) is decreased based on the relevant creep factors, such as in standard EN 1995-1-1 [23]. In composite systems, the differential inelastic strains between the components can cause significant deflection. Their effects, with addition of creep, can be accounted for in the analytical calculations for TCCs according to methods presented in CEN/TS 19103:2022 [10].

Specialized experimental, numerical (static deflections and dynamic vibration analyses) and analytical methods are adopted to carry out deflection and vibration analysis of long-span timber floors for taller buildings. The main aim is to check their compliance with the performance requirements [34] in terms of: (i) ultimate limit state or strength capacity, undergoing the gravity and lateral loads, and (ii) serviceability limit state, controlling the deflection and vibration. Numerical simulations on floors are carried out by using both design-oriented software and general-purpose programs.

Design-oriented software packages are commonly used by practitioners for conventional structural analysis of timber floors due to (i) their compliance with design standards such as EN 1995-1-1 [23] and other regional codes, (ii) the ability to define the floor system accurately, either as a beam-slab configuration (e.g., joisted floors) or as a panel system (e.g., cross-laminated timber), and (iii) the capability to perform both static and dynamic analyses, also considering nonlinear and fire-related effects. For more advanced applications, such as seismic, blast, or fire scenarios, general-purpose finite element programs are typically preferred, as they provide a wide range of material models, element types, and solver options (e.g., implicit or explicit solvers). To promote consistency in modelling strategies, several initiatives have been launched, including the Modelling Guide for Timber Structures [35] and a working group WG11 under CEN/TC 250/SC 5, which is currently preparing dedicated finite element modelling guidelines as part of the next generation of Eurocode 5.

Great attention is paid to hybrid systems which are widely used and investigated. For numerical investigations, finite element models are generally calibrated against experimental results and are utilized for parametric studies for flooring systems. One of the key points is represented by the connections' stiffness since it strongly influences the coupling degree among slab and beams and so the overall stiffness of the composite system [36]. propose a calculation procedure to achieve the accurate values of maximum force and stiffness of a given timber-to-timber slab joined with Self-Tapping Screws (STSs) and loading condition. The procedure accounts for the effective shear force and stiffness of screws in a given position on the slab and provides it in a simplified way for STSs stiffness and force formulations (correlation coefficients).

5 Design Criteria for Controlling Deflection and Vibrations According to Different Design Standards

The design of floors in TTBs must account for various serviceability criteria, with particular focus on controlling deflections and vibrations. While international standards vary in detail, the fundamental principles are largely consistent. Most standards prescribe deflection limits as a function of the span length and loading conditions and include criteria for vibration performance, often based on natural frequency, point load deflection, and dynamic response (e.g. acceleration or velocity). The commonly applied standards are briefly presented below, and for further insights, please see [37], among others.

The European standard EN 1995-1-1 [23] defines deformation limits for timber floors based on the type of floor and boundary conditions. These limits address both live loads and combined dead plus live loads, ensuring that vertical deflections remain within acceptable ranges to maintain structural performance and user comfort. The new draft version of EN 1995-1-1 [32] introduces three specific criteria for vibration control: (i) minimum fundamental natural frequency, (ii) maximum deflection under a point load, and (iii) acceleration and velocity limits. These align with the existing approach by [38] but expand upon it through the introduction of floor performance levels. These levels explicitly define vibration limit values, allowing floor systems to be designed more precisely to meet user requirements. For further details on the proposed performance levels and vibration criteria, see [39].

Swiss standards for timber structures, as outlined in SIA 265:2012 [40], reference serviceability limits defined in SIA 260:2013 [41], addressing static deflection under concentrated loads as well as velocity response and acceleration criteria. In the United States, the ANSI/AWC NDS [42] and International Building Code (IBC) [43] specify deflection limits for floor systems based on live and combined loads, but do not explicitly regulate vibration performance, leaving this aspect to other guidelines or project-specific considerations. Similarly, the Canadian standard CSA O86:19 [44] includes deflection criteria for live and total loads while also incorporating vibration assessments based on natural frequency and static deflection under point loads, with further guidance provided in the Canadian CLT Handbook [45] for optimising CLT floor performance. Australian standards, such as AS 1720.1-2010 [46] and AS 1170.1-2002 [47], define deflection limits depending on floor usage but do not explicitly address vibration control in timber floors. In contrast, the Chinese standard GB 50005:2017 [48] specifies deflection limits for various floor types and loading conditions, while the "Technical Standards for Vibration Comfort of Building Floors" JGJ/T 441-2019 [49] provide additional criteria for vibration performance, including minimum frequency requirements.

References

1. Zhang, B., Rasmussen, B., Jorissen, A., Harte, A.: Comparison of vibrational comfort assessment criteria for design of timber floors among the European countries. Eng. Struct. **52**, 592–607 (2013)
2. Weckendorf, J., Toratti, T., Smith, I., Tannert, T.: Vibration serviceability performance of timber floors. Eur. J. Wood Wood Prod. **74**, 353–367 (2016)
3. Bazli, M., Heitzmann, M., Ashrafi, H.: Long-span timber flooring systems: a systematic review from structural performance and design considerations to constructability and sustainability aspects. J. Build. Eng. **48**, 103981 (2022)
4. Brandner, R., Flatscher, G., Ringhofer, A., Schickhofer, G., Thiel, A.: Cross laminated timber (CLT): overview and development. Eur. J. Wood Wood Prod. **74**, 331–351 (2016)
5. Zöllig, S., Frangi, A., Franke, S., Muster, M.: Timber structures 3.0 - new technology for multi-axial, slim, high performance timber structures. In: WCTE 2016, World Conference on Timber Engineering, Vienna, Austria (2016)
6. Zhang, S., Chui, Y.H.: Characterizing flexural behaviour of panel-to-panel connections in cross-laminated timber floor systems. Structure. **28**, 2047–2055 (2020)
7. Asselstine, J., Lam, F., Zhang, C.: New edge connection technology for cross laminated timber (CLT) floor slabs promoting two-way action. Eng. Struct. **233**, 111777 (2021)

8. Derikvand, M., Jiao, H., Kotlarewski, N., Lee, M., Chan, A., Nolan, G.: Bending performance of nail-laminated timber constructed of fast-grown plantation eucalypt. Eur. J. Wood Wood Prod. **77**, 421–437 (2017)
9. Sotayo, A., et al.: Review of state of the art of dowel laminated timber members and densified wood materials as sustainable engineered wood products for construction and building applications. Dev. Built Environ. **1**, 100004 (2020). https://doi.org/10.1016/j.dibe.2019.100004
10. CEN/TS 19103: Technical Specification. Eurocode 5: Design of Timber Structures – Structural Design of Timber-Concrete Composite Structures – Common Rules and Rules for Buildings. DIN German Institute for Standardisation, DIN (2022)
11. DIN CEN/TS 19103 / NA: National Annex - Nationally Determined Parameters - Eurocode 5: Design and Construction of Timber Structures - Calculation of Timber-Concrete Composite Elements - General Rules and Rules for Buildings. DIN (2024)
12. Krug, S., Jaaranen, J., Fink, G., Schänzlin, J.: Determination of the moment–rotation relation of continuous timber–concrete composite floors based on the component method. Structure. **69**, 107365 (2024)
13. Steenbakkers, P., Verheggen, B., Vola, M., Vermeulen, D.J.: Reaching new heights in timber-hybrid design: designing the. Int. J. High-Rise Build. **13**(2), 195–204 (2024)
14. Gohlich, R.J.: Development of an Innovative Hybrid Timber Steel Moment Resisting Frame for Seismic Resistant Heavy Timber. Carleton University, Ottawa, Canada (2016)
15. Chen, B., Roy, K., Fang, Z., Uzzaman, A., Chi, Y., Lim, J.B.: Web crippling capacity of fastened cold-formed steel channels with edge-stiffened web holes, un-stiffened web holes and plain webs under two-flange loading. Thin-Walled Struct. **163**, 107666 (2021)
16. Trahair, N.S., Bradford, M.A., Nethercot, D., Gardner, L.: The Behaviour and Design of Steel Structures to EC3. CRC Press, London (2017)
17. Karki, D., Al-Hunaity, S., Far, H.: Composite connections between CFS beams and plywood panels for flooring systems: testing and analysis. Structure. **40**, 771–785 (2022)
18. Hadjipantelis, N., Gardner, L., Wadee, A.M.: Prestressed cold-formed steel beams: concept and mechanical behaviour. Eng. Struct. **172**, 1057–1072 (2018)
19. Romero, A., Odenbreit, C.: Experimental investigation on novel shear connections for demountable steel-timber composite (STC) beams and flooring systems. Eng. Struct. **304**, 117620 (2024)
20. Bomba, D.V., Dance, G., Chira, A., Walker, M.G., Nagy, Z.: Structural response of hybrid timber - cold formed steel floors. ce/papers. **6**(3–4), 1853–1858 (2023)
21. Navaratnan, S., Small, D.W., Gatheeshgar, P., Poologanathan, K., Thamboo, J.T., Mensin, P.: Development of cross laminated timber-cold-formed steel composite beam for floor system to sustainable modular building construction. Structure. **31**, 681–690 (2021)
22. EN 1993-1-1: Eurocode 3: Design of Steel Structures - Part 1-1: General Rules and Rules for Buildings. CEN (2005)
23. EN 1995-1-1: Eurocode 5: Design of Timber Structures – Part 1-1: General – Common Rules and Rules for Buildings. CEN (2014)
24. Sahin, B., Bravo-Haro, M.A., Elghazouli, A.Y.: Assessment of cyclic degradation effects in composite steel-concrete members. J. Constr. Steel Res. **192**, 107231 (2022)
25. Emms, G., Chung, H., Mcgunnicle, K., Dodd, G.: Improving the impact insulation of light timber floors. In: 1st Australasian Acoustical Societies' Conference 2006, ACOUSTICS 2006: Noise of Progress, Christchurch, New Zealand (2006)
26. Tiso, M., Just, A.: Design criteria for insulation materials applied in timber frame assemblies. J. Struct. Fire Eng. **9**(3), 252–263 (2018)
27. EN 1991-1-1:2002: Eurocode 1: Actions on Structures - Part 1–1: General Actions - Densities, Self-Weight, Imposed Loads for Buildings. CEN (2002)

28. ISO 2631–2:1989(E): Evaluation of Human Exposure to Whole-Body Vibration – Part 2: Continuous and Shock-Induced Vibration in Buildings (1 to 80 Hz). Internatioonal Organization for Standardization (1989)
29. Hu, L.J., Chui, Y.H., Onysko, D.M.: Vibration serviceability of timber floors in residential construction. Prog. Struct. Eng. Mater. **3**(3), 228–237 (2001)
30. Morlier, P. (ed.): Creep in Timber Structures. CRC Press, London (1994)
31. Fraciacomo, M., Ceccotti, A.: Long-term behavior of timber–concrete composite beams. I: Finite element modeling and validation. J. Struct. Eng. **132**(1), 113–122 (2006)
32. FprEN 1995-1-1:2025: *Eurocode 5 — Design of timber structures — Part 1–1: General rules and, CEN/TC 250* (2024)
33. Mazelli, A., Bedon, C.: Sensitivity of human-induced vibrations and comfort analysis to partially rigid STS joints in timber-to-timber composite floors. Measurement. **228**, 114396 (2024)
34. Crews, K., Buchanan, A., Quenneville, P.: Development of high performance structural timber systems for non-residential buildings in New Zealand and Australia. Procedia Eng. **14**, 1582–1589 (2011)
35. Chen, Z., Tung, D., Karacabeyli, E.: Modelling Guide for Timber Structures, Special Publication SP-544, 1st edn. Quebec, FPInnovations (2022)
36. Bedon, C., Sciomenta, M., Fragiacomo, M.: Correlation approach for the push-out and full-size bending short-term performances of timber-to-timber slabs with self-tapping screws. Eng. Struct. **238**, 112232 (2021)
37. Zhang, S., Zhou, J., Chui, Y.H.: Deflection criteria for controlling timber floor vibrations: a 200-year evolution. Eng. Struct. **326**, 119370 (2025)
38. Winter, S., Hamm, P., Richter, A.: Vibration and damping behaviour of timber and timber-concrete composite floors (original in German) (2010)
39. Ruf, J., Hamm, P., Schneider, M.: Comparison of the calculated and measured vibration behaviour of long-span timber floors. In: *Proceedings of the World Conference on Timber Engineering (WCTE 2025)*, Brisbane (2025)
40. SIA 265:2012: Design of Timber Structures. SIA Swiss Society of Engineers and Architects (2012)
41. SIA 260:2013: Basis of Design and Actions on Structures. SIA Swiss Society of Engineers and Architects (2013)
42. ANSI/AWC: National Design Specification (NDS) for Wood Construction. American Wood Council's (AWC) (2018)
43. 2021 International Building Code (IBC) (2021) Available: https://codes.iccsafe.org/content/IBC2021P2
44. CSA O86:19: Engineering Design in Wood. Canadian Standards Association (2019)
45. Karacabeyli, E., Gagnon, S.: Cross Laminated Timber (CLT) Handbook. FPInnovations, Vancouver (2019)
46. AS 1720.1-2010: Timber structures, Part 1: Design Methods, \ (2010)
47. AS 1170.1-2002: Structural Design Actions – Part 1: Permanent, Imposed and Other Actions. Standards Australia (2002)
48. GB 50005:2017: Standard for Design of Timber Structures. China Construction Industry Press, Beijing (2017)
49. JGJ/T 441-2019: Technical Standard for Human Comfort of the Floor Vibration. China Construction Industry Press, Beijing (2019)

Vibration-Based Serviceability and Acoustic Assessment of Timber Floors

Angelo Aloisio[1,4(✉)], Dag Pasquale Pasca[2], Thomas Hillberger[3], Roberto Tomasi[4], Thomas P. S. Reynolds[8], Chiara Bedon[5], Sven Vallely[6], Rok Prislan[7], Kemal Edip[9], Mohammadreza Salehi[4], Carlos Martín de la Concha Renedo[10], and Iván M. Díaz[10]

[1] Department of Civil, Construction-Architectural and Environmental Engineering, Università degli Studi dell'Aquila, L'Aquila, Italy
angelo.aloisio@univaq.it

[2] Norsk Treteknisk Institutt (Norwegian Institute of Wood Technology), Oslo, Norway

[3] Institute for Construction and Material Sciences – Unit of Timber Engineering, University of Innsbruck, Innsbruck, Austria

[4] Faculty of Science and Technology, Norwegian University of Life Sciences (NMBU), Ås, Norway

[5] Department of Engineering and Architecture, Università degli Studi di Trieste, Trieste, Italy

[6] Swiss Federal Laboratories for Materials Science and Technology (Empa), Dübendorf, Switzerland

[7] Faculty of Civil and Geodetic Engineering, University of Ljubljana, and InnoRenew CoE/University of Primorska, Koper–Izola, Slovenia

[8] School of Engineering, University of Edinburgh, Edinburgh, UK

[9] Ss. Cyril and Methodius University in Skopje, Institute of Earthquake Engineering and Engineering Seismology-IZIIS, Skopje, North Macedonia

[10] Department of Continuum Mechanics and Theory of Structures, ETSI Caminos, Canales y Puertos, Universidad Politécnica de Madrid, Madrid, Spain

- a_{peak}: Peak acceleration response
- a_{rms}: Root mean square acceleration
- a_{rmq}: Root mean quad acceleration
- c: Damping coefficient
- $k_{e,1}$: Frequency factor in case of a double span floor on rigid supports. In case of a single span $k_{e,1} = 1.0$
- $k_{e,2}$: Frequency factor to consider the effect of the transverse floor stiffness. In case of a one-way spanning floor $k_{e,2} = 1.0$
- $(EI)_L$: Longitudinal bending stiffness of the floor per unit of length
- $(EI)_T$: Transverse bending stiffness of the floor span per unit of length
- μ: Resonant buildup factor, which may be taken as $\mu = 0.4$
- ζ: Modal damping ratio
- b: Floor width
- b_{ef}: Effective width
- c_f: Fourier coefficient of the dynamic load factor, which can be assumed equal to $c_f = 0.0714$
- F: Force applied in the point with greatest deflection

G. Fink et al. (Eds.): *Holistic Design of Taller Timber Buildings*, SPRTRCIENG, pp. 76–93, 2025.
https://doi.org/10.1007/978-3-032-02098-7_7

- F_p: Person weight walking on the floor equal to $F_p = 700$ N according to EC5
- F_{dyn}: Vertical dynamic force caused by a walking person, equal to $F_{dyn} = 50$ N according to EC5
- f_1: Fundamental frequency in Hz
- f_{lim}: Limit frequency between the resonant or a transient response in Hz
- f_w: Walking frequency in Hz
- h: Cross-section height
- $I_{mod,mean}$: Mean modal impulse, in N·s
- k_{black}: Reduction factor equal to $k_{black} = 0.7$ according to EC5
- k_{res}: Factor to account for higher modes of vibrations, according to EC5
- l: Floor span (the longest span in case of double span floor)
- m: Floor mass per unit of area
- M^*: Floor modal mass
- $v_{1,peak}$: Peak velocity response for the fundamental mode
- w: Load deflection
- K_p: Peak factor
- b_b: Width of the building
- h_b: Height of the building
- $q_{z,ref}$: Reference wind load
- c_{fw}: Force coefficient
- K: Dimensionless coefficient
- $\Phi_{1,x}(z)$: First mode shape
- M_1: Modal mass of the first mode
- R: Resonance response factor
- SEL: Sound exposure level
- VDV: Vibration dose value
- w_{1kN}: Load deflection under a concentrated 1kN load
- B: Floor width
- l_2: Shorter span of a double-span floor
- $(EI)_i$: Bending stiffness of the i-th member of a composite floor per unit of length
- $(EA)_i$: Axial stiffness of the i-th member of a composite floor per unit of length
- a_i: i-th member center of mass distance from the composite section center of mass
- γ_i: i-th member partial composite action factor
- S: Percentage of decrease in structural rigidity
- t: Elapsed time in years from timber bridge construction

1 Introduction

Timber floors are particularly susceptible to noise and vibration issues when compared to other construction methods, such as concrete floors. This is due to the relatively high stiffness-to-mass-density-ratio of timber floors, which enables so-called ‘lightweight’ floor constructions, in contrast to ‘heavyweight’ floor constructions consisting, for example, of concrete slabs and steel girders. Quantified, lightweight floors, as defined in ISO 10140-5:2021, are considered as floors whose area density is less than 150 kg/m^2 [1]. A heavyweight reference floor is specified in Annex C of ISO 10140-5:2021 [1]. The heavyweight reference floor is specified to be of reinforced concrete and have a minimum thickness of 100 mm, but preferably a thickness of 140 mm, meaning a lower bound mass density of 250 kg/m^2 and a preferred reference mass density of 350 kg/m^2. While timber floor constructions have the benefit of being relatively stiff and light, which is a positive from a static structural perspective, these lightweight structures are prone to higher vibration amplitude and noise transmission due to the reduced mass of the structure.

The vibration and noise issues of timber floors may be divided into two parts based on the frequency range of consideration with some overlap between the two. Vibration-based serviceability of floors considers generally frequencies from 1 Hz to 80 Hz [2,3], while acoustics generally considers a frequency range from 50 Hz to nominally 5000 Hz [4–14]. However, research has suggested that acoustic frequencies as low as 20 Hz should be a consideration for lightweight constructions [15–17], indicating that a frequency range of 20 Hz to 5000 Hz should ideally be considered. Accordingly, this subsection is divided into two parts:

1. Vibration-based serviceability of floors
2. Acoustics of floors

2 Human-Induced Vibrations Issues

Timber floors are particularly susceptible to human-induced vibrations, which can give rise to various issues affecting their performance and occupant comfort. One significant issue is the potential discomfort experienced by occupants due to excessive vibrations. When timber floors are excited by walking or other human activities, floor vibration can be perceived by occupants, causing discomfort and even affecting the usability of the space. Excessive vibrations can lead to discomfort while walking, using furniture, or performing tasks that require stability [18]. It is crucial to address this issue to ensure the satisfactory serviceability of timber floors [19–24].

Human-induced vibrations in timber floors can also impact the functionality and performance of sensitive equipment and installations. In spaces where delicate instruments, equipment, or machinery are present, excessive vibrations can cause operational issues, measurement inaccuracies, or even damage to the equipment. This issue is particularly relevant in environments such as laboratories, healthcare facilities, and industrial settings where precise measurements and stable conditions are necessary [25].

Another concern is the potential fatigue and degradation of timber elements caused by prolonged exposure to excessive vibrations. Over time, repeated dynamic loading can lead to fatigue failure, and reduced strength, and durability of the timber floor system. This issue can compromise the long-term structural integrity of the floor and may require maintenance or strengthening interventions.

Additionally, the perception of vibrations in timber floors can vary among individuals, and some occupants may be more sensitive to vibrations than others. This issue highlights the importance of considering occupant comfort and wellbeing in the design and assessment of timber floors, as individual sensitivity can influence the acceptability of vibrations in different contexts [26].

Addressing these issues requires a comprehensive understanding of the dynamic behaviour of timber floors and the factors influencing human-induced vibrations. Experimental studies, such as impact tests and field measurements, play a crucial role in assessing the vibrational performance of timber floors and identifying potential mitigation strategies.

By considering these specific issues related to human-induced vibrations in timber floors, researchers and designers can develop effective design approaches, including appropriate damping measures, structural optimization techniques, and user guidelines, to ensure the satisfactory serviceability, occupant comfort, and long-term performance of timber floor systems [27–31].

3 Acoustics of Floors

The EU Environmental Noise Directive of 2002 specifically addresses the prevention of environmental noise pollution. The Directive focuses on noise to which humans are exposed, particularly in built-up areas. In 2008, the Directive was extended to include vibration as a form of pollution [32]. The Directive considers the direct or indirect influence of vibration, heat, or noise as pollution. However, despite the Directive, noise and vibration are often treated separately in the context of sustainable indoor comfort. International and national standards typically address these factors individually. In the literature and relevant standards, the perception of vibration in buildings has been extensively analyzed and studied over the past decades [33,34]. However, many authors suggest that the main discomfort experienced in buildings is related to a combined effect of noise and vibration. Noise and vibration co-occur in buildings, and even if the acoustic or vibration thresholds meet legal or standard limits, occupants can still report annoyance [35]. Nering et al. [36] proposed the evaluation of exposure to simultaneous events based on research by Howarth and Griffin [37]. They presented a graphical representation of the annoyance level as a function of sound exposure level (SEL) and vibration dose value (VDV). Physically, vibrations can be linked to the radiated sound power of a planar structure when expressed in terms of velocities. However, the sound power levels (W) emitted from a structure are linked to not just the structure's vibration levels, but also the so-called radiation efficiency of the structure. An expression for a planar element radiating into a fluid half-space is [38],

$$W = \frac{\sigma \rho c S \overline{|v|^2}}{2}. \tag{1}$$

In the case of a floor radiating sound into the room above or below, S would be the surface area of the floor, ρ and c would respectively be the density and speed of sound in air, $\overline{|v|^2}$ would be the time- and spatially-averaged mean square velocity of the radiating surface, and σ would be the radiation efficiency of the floor. A noteworthy observation of Eq. 1, is that for a given floor, $W \propto \sigma\overline{|v|^2}$. This implies that reducing vibration levels may not necessarily equate to reduced radiated sound power levels if the radiation efficiency of the structure is increased in the process. The radiation efficiency of a vibrating structure is related to its response frequency and the vibratory velocity distribution on its surface. Typically, an element radiates with low efficiency below its critical frequency, $f < f_c$, where the critical frequency may be calculated according to Eq. (2) [38]

$$f_c = \frac{c^2}{2\pi}\sqrt{\frac{m''}{B}}. \tag{2}$$

m'' is the mass density and B is the bending stiffness of the element per unit width. For orthotropic plate elements, such as timber, two critical frequencies are calculated for each planar direction, i.e. $f_{c,x}$ and $f_{c,y}$. At the critical frequency, $f = f_c$ the radiation efficiency is at a maximum, and approaches a value of unity for $f > f_c$. A consequence of this, is that, while increasing the stiffness of a floor may shift problematic resonances from the human sensitive range of vibrations, this increase in stiffness can consequently increase the radiation efficiency of the floor, decreasing its acoustic performance. Care must also be given when treating undesirable noise and vibration levels by the addition of mass, that the vibration velocity distribution across the floor is not altered in such a way as to negate the effect of the additional mass on the radiated sound power levels of the structure.

Acoustic problems are prevalent in lightweight constructions [39–41]. Typical annoying sounds in buildings include people talking, television noise, footsteps. Transmission between rooms can occur through airborne or structure-borne paths. The paths themselves may be either direct path or flanking paths. An illustration of some potential flanking paths is illustrated in Fig. 1 for one room situated directly above another. For the case of a structure-borne sound source, there are three primary transmission paths, while for the case of an airborne sound source, there are 7 primary transmission paths. However, this schematic is for illustrative purposes, with the number of possible transmission paths depending on the structure and configuration. For example, there is a similar case for rooms adjacent to each other, when considering Fig. 1 as a top-down perspective. Accordingly, sound can reach the listener's room through various transmission paths, such as direct radiation from the separating wall, transmission of vibrations to adjacent walls, or transmission of vibrations from side walls of the source room to the partition or side walls of the listener room [42]. Flanking sounds refer to all sounds propagating through partition walls or floors, and solving flanking transmission problems is crucial for effective sound insulation.

Olsson [43] recently investigated the impact sound transmission of lightweight timber floors. The study focused on transmission and insulation without reverberation using fluid elements connected to reflection-free boundaries. The results

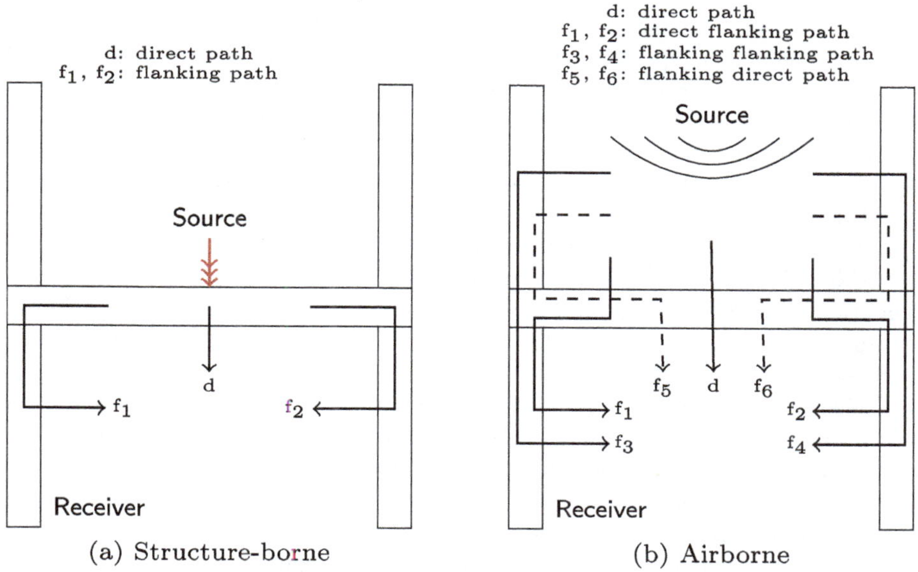

Fig. 1. Sound transmission paths between two rooms (excluding indirect transmission) separated by a floor element. (a) Structure-borne sound transmission (b) Airborne sound transmission.

showed that a floor model with a hard screed surface exhibited higher impact force compared to a softer floor, although this effect was less pronounced at the lowest frequencies.

Reducing flanking transmission involves limiting the vibrations transmitted to the walls and floors in the source room. The goal is to minimize sound radiation from the walls and floors of the receiving room and reduce vibrations transmitted from the source room to the receiving room [44]. Achieving this requires a complete separation of the structural and non-structural parts of the adjoining apartments.

Separation is typically achieved using a layer more compliant than the timber of the structural elements that are to be connected. In the vertical direction, a compliant interlayer is used between overlapping walls, floors, and bearing walls, floors, and bearing walls. However, these layers are often subjected to static loads, so stiff layers should be used in the load-bearing structure. One drawback is that stiff layers can increase coupling and reduce the effectiveness of sound insulation [45,46]. Soundproof steel angle brackets can also be employed to prevent acoustic bridges, where rigid parts are separated by interlayers. Elastomers, such as closed cellular polyurethane (CCP) and mixed cellular polyurethane (MCP), are frequently used to reduce low-frequency noise [47,48].

A quantity known as the Vibration Reduction Index (K_{ij}) is a standardised quantity used in acoustics to characterise junction attenuation and determine the contribution of flanking sound transmission to the overall sound transmission into a space. The vibration reduction index is defined in the ISO 12354 series [13,14]. Schoenwald et al. experimentally demonstrated the efficacy polyurethane interlayers with a design frequency of $f_0 = 20\,\mathrm{Hz}$ for improving the vibration reduction index and flanking sound transmission performance of a mass timber structure with CLT walls and glulam floors. A best-case scenario showed an improvement of the vibration reduction index up to 17 dB when considering structure-borne sound transmission across a T-junction between vertically adjacent rooms. The best-case scenario compared a reference configuration (a): no elastic interlayers and steel angle brackets; and a best-case scenario configuration (b): no angle brackets and an elastic interlayer between the mass timber elements. The introduction of a third configuration (c), which contained an elastic interlayer between the mass timber elements and decoupled angle brackets with elastic interlayers showed an improvement of up to 12 dB. The vibration reduction improvements of configurations (b) and (c) resulted in 13 dB and 10 dB improvements of the airborne sound insulation of the floor system (R'_W), respectively. The addition of decoupled angle brackets, required for structural supports, reduced the airborne sound insulation performance by 3 dB over the best-case scenario. This result indicates that further optimisation of the noise and vibration performance of the decoupled angle brackets for the investigated connection configuration is limited. However, this does not exclude the possibility of new decoupling techniques and technologies to further reduce flanking sound transmission [49].

There are alternative measures to reducing flanking sound transmission, however, these measures come with their own drawbacks. [50]

While separation is an effective solution for sound insulation, it can increase the overall deformability of the building, potentially compromising its stability. The influence of flexible sound insulation layers on the seismic performance of Cross-Laminated Timber (CLT) walls has been studied by Azinović et al. [48]. The study showed that the bedding insulation layer under the wall negligibly affected the load-bearing capacity under lower vertical loads. However, the stiffness of the wall decreased to less than 40% of the un-insulated wall due to additional lateral deformations caused by the insulation. Experiments also indicated that a higher vertical load substantially increased the lateral load bearing capacity and stiffness of the shear wall due to the associated increase in friction. The cyclic response of insulated steel angle brackets used for Cross-Laminated Timber (CLT) connections was assessed by Kržan et al. [51]. The tests revealed that insulation under the angle bracket had a marginal influence on the load-bearing capacity but significantly affected the stiffness characteristics, resulting in a reduction of effective stiffness by 22% and 45% in pure shear and tensile loading, respectively. Furthermore, the insulated specimens exhibited lower relative energy dissipation and equivalent viscous damping coefficient compared to the uninsulated samples, although this difference decreased with increasing displacements and repeated cycles.

The mainstream research on sound insulation of timber buildings focuses on two aspects:

- Modelling approaches. See the recent papers by Fox et al. [52], Paolini et al. [53], De Santis et al. [54], ~~and~~ Wang et al. [55], and Vallely et al. [56]. Fox et al. [52] developed a composite model structure for predicting low-frequency vibration in light timber floors, considering the coupling via air cavity. Paolini et al. [53] proposed a method for avoiding hexahedral meshing for the thin elastomer layer. De Santis et al. proposed an analytical model for the stiffness prediction of screw connection with deformable interlayers [54]. Wang et al. [55] implemented state-of-the-art approaches for predicting impact forces, structural vibration and radiated sound power of timber joist floors. Vallely et al. [56] proposed a homogenisation method for Cross-Laminated Timber (CLT) elements. The method is based on first-order shear deformation theory and allows for determination of broadband frequency-independent bending, extensional, shear, and bending-extensional material stiffness matrices. From the material stiffness matrices, effective orthotropic engineering constants can be derived, however, with the assumption that the behaviour of the bending and extensional deformation of the elements remained uncoupled. This limitation is generally not an issue, as Cross-Laminated Timber (CLT) elements are typically composed of symmetric stacking sequences and flexural vibrations dominate the acoustic response of the structure [57]. In cases where the coupled bending-extensional behaviour of CLT elements would be considered non-negligible, the full material stiffness matrices should be implemented, as considered in Vallely et al. [58]. The homogenisation method has been validated against experimental measurements for both modal and forced-response models up to 5500 Hz [56,58].
- Experimental investigations, especially on the role of the floor coverings, see Lietzen [59], who studied the effect of floor coverings on impact sound insulation.

Regarding the experimental investigations on floor coverings, Huang et al. [60] investigated the performance of three kinds of elastic cushion materials for timber floors: Portuguese cork, foam, and polypropylene plastic foam board. They found that foam boards exhibit the highest performance. More details about the studies on acoustic issues will be detailed when addressing specific aspects of different floor typologies in the following paragraphs.

4 Serviceability Criteria

Serviceability requirements for timber floors specifically focus on the performance of lightweight floor systems. Standards and guidelines in this regard consider parameters such as velocity and acceleration to assess floor response. The root mean square values (v_{rms} and a_{rms}) are commonly used as they provide an average measure of the response, accounting for the excitation duration. Another metric, known as root-mean-quad (a_{rmq}), is employed for cumulative measurements, particularly in the analysis of vibration dose values (VDVs) according to standards like BS 6472 and ISO 2631 [3,61].

These serviceability requirements are defined by threshold values expressed using the response factor R. The response factor R serves as a multiplier applied to the base curve value, indicating the level of vibration perceptible to an average human. Different multiples of R, such as 4, 8, and 48, establish various performance levels for the floor system [2].

In the assessment of timber floors, the literature proposes different approaches, considering these parameters individually or in combination. Basaglia [62] distinguishes two main approaches for serviceability assessment based on the floor's intended use:

- Residential Timber Floors: these floors are typically characterized by smaller spans and lighter loads, resulting in higher frequencies. Basaglia suggests employing a pass-fail criterion that considers various parameters such as static deflection, peak velocity, root-mean-quad acceleration, or a combination of these factors.
- Office Timber Floors: these floors generally have larger spans and heavier loads, leading to lower frequencies compared to residential floors. For low-frequency office floors, more detailed procedures have been proposed.

Chang et al. [63] introduced an approach that combines the Response Factor and Vibration Dose Value (VDV) methods to assess the performance of these floors. Additionally, rules developed by Hamm et al. [64,65] and Abeysekera et al. [66], which are currently incorporated in the draft of Eurocode 5 (EC5), provide further guidance. The synoptic table in Table 1, Fig. 2, and Table 2 summarize these criteria.

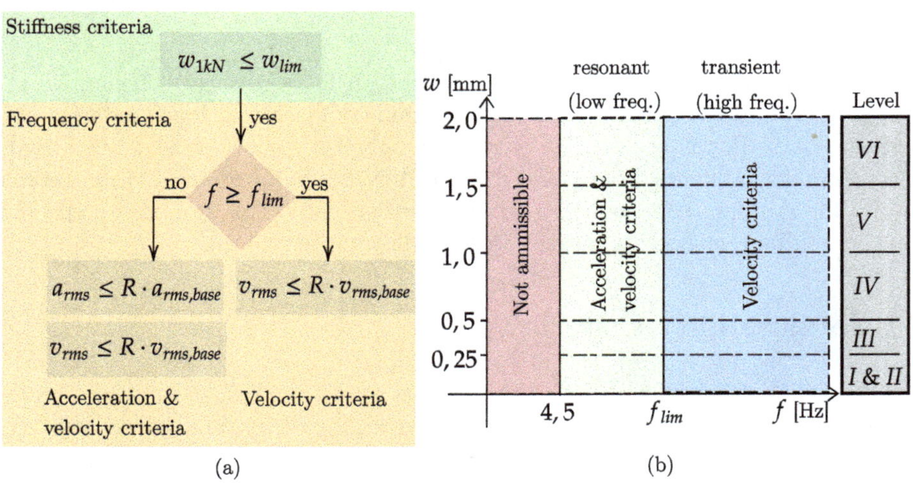

Fig. 2. (a) Pass/fail design approach for limiting vibration in timber floor according to Eurocode 5 draft; (b) Floor classification based on the serviceability criteria.

The verification is based on two sequential criteria, stiffness-based and frequency-based. The first step, see Fig. 2a, is a stiffness-based verification: the designer must verify the floor deflection under a concentrated 1 kN load (w_{1kN}), is below a given threshold. If this criterion is satisfied, the designer must also prove that the first natural frequency of the floor is beyond a certain threshold f_{lim}, representing the limit between the resonant and the transient response. The typical human walking pace has a dominant frequency (f_w) ranging between 1.5 to 2.5 Hz. Still, to account for the contribution of the higher harmonics, f_{lim} is typically set as four times the walking frequency [67].

Suppose the inequality in frequency is not satisfied. In that case, the designer must verify that the root mean square acceleration and velocity are below the thresholds defined in terms of the response factor R (see Table 2). The verification can be limited to a velocity check if the inequality is satisfied. The first two verification steps, stiffness and frequency-based, are illustrated in the diagram in Fig. 2(b) where the y and x-axes show the floor deflection and the first natural frequency and three regions are identified. If the first natural frequency of the floor is below 4.5 Hz, the floor behaviour is not admissible (red). For low-frequency floors, with the first natural frequency between 4.5 Hz and f_{lim} (green region), the designer must satisfy both acceleration and velocity criteria. Only the velocity criterion must be verified in the case of high-frequency floors (light blue region). In the absence of more accurate predictions, EC5 provides simplified expressions of the main parameters (f_1, w_{1kN}, a_{rms} and $v_{1,peak}$) needed to verify the vibration performance of a timber floor.

The synoptic Table 1 summarises all the expressions provided by the current EC5 draft to verify the vibration performance of a timber floor. Equation (3) and Eq. (5) present the simplified formulations for estimating the first natural frequency and vertical deflection. Table 3 provides the values for the $k_{e,1}$ factor used to calculate the fundamental frequency in case of a double-span floor on rigid supports. The remaining equations estimate the simplified acceleration and velocity responses in the absence of more accurate predictions.

Empirical design criteria for lightweight floors, shown in Table 1 based on static displacement and the fundamental natural frequency, were not validated against the dynamic response of CLT floors [68]. So far, no specific serviceability criteria have been proposed with a particular reference to CLT floors under multiple occupancy classifications. While traditional lightweight timber floors are prevalent in one-way systems, CLT floors exhibit a plate-like behaviour, being supported on all four sides. The scientific literature highlights two relevant aspects.

Some scholars affirm that using the current vibration criteria for the CLT floor design leads to conservative estimates; see Zhang et al. [69] and Hu et al. [70]. In this sense, Hu et al. [70] proposed a serviceability design criterion based on experimental tests on CLT strips behaving like simply-supported beams. According to Hu et al. [70], CLT strips can be considered the worst scenario since they neglect the effect of four-side support, see the recommendations of the CLT Handbook sponsored by the Canadian forest industry [71].

On the other hand, the human-induced dynamic response of CLT panels is associated with a significant contribution of higher modes, neglected by the EC5 formulation [72–75]. If all sides of the panel are supported, the number of participating modes can increase significantly up to 100 Hz [76]. This phenomenon is magnified by semi-rigid support conditions, intra-slab joints and a plan aspect ratio close to one. Recently, Milojevic et al. [77] numerically assessed the effect of connections, proving their stronger influence on high-frequency floors rather than low-frequency floors. Also, the effect of multiple people activities appears crucial for CLT floors. Wang [78] showed that the vibration amplitude of CLT floors under multi-person loadings was almost double that under single-person.

Table 1. Synoptic table of the mathematical formulation enclosed in the Eurocode 5 draft for assessing the serviceability of timber floors.

Serviceability criteria according to the new Eurocode 5 draft

Frequency

$$f_1 = k_{e,1} k_{e,2} \frac{\pi}{2l^2} \sqrt{\frac{(EI)_L}{m}} \tag{3}$$

with

$$k_{e,2} = \sqrt{\left(1 + \frac{\left(\frac{l}{b}\right)^4 (EI)_T}{(EI)_L}\right)} \tag{4}$$

Deflection

$$w_{1kN} = \frac{Fl^3}{48\,(EI)_L\, b_{ef}} \tag{5}$$

with:

$$b_{ef} = \min\left\{0,95 \left(\frac{(EI)_T}{(EI)_L}\right)^{0,25}; b\right\} \tag{6}$$

Acceleration

$$a_{rms} = \frac{k_{res} \mu F_{dyn}}{\sqrt{2} \cdot 2\zeta M^*} \tag{7}$$

with

$$k_{res} = \max\left\{0,192 \left(\frac{b}{l}\right) \left(\frac{(EI)_T}{(EI)_L}\right)^{0,25}; 1\right\} \tag{8}$$

and

$$F_{dyn} = c_f F_p \tag{9}$$

Velocity

$$v_{1,peak} = k_{black} \frac{I_{mod,mean}}{(M^* + 70kg)} \tag{10}$$

with

$$I_{mod,mean} = \frac{42 f_w^{1,43}}{f_1^{1,3}} \tag{11}$$

and

$$M^* = \frac{m \cdot l \cdot b}{4} \tag{12}$$

Table 2. Floor vibration criteria according to the floor performance level.

	Floor performance level					
Criteria	I	II	III	IV	V	VI
Response factor R	4	8	12	24	36	48
Upper deflection limit $w_{1kN} \leq w_{lim}$ [mm]	0.25	0.5	1	1.5	2	
Stiffness criteria for all floors	$w_{1kN} \leq w_{lim}$ [mm]					
Frequency criteria for all floors	$f_1 \geq 4.5$ Hz					
Acceleration criteria for resonant vibration design situations ($f_1 < f_{1,lim}$)	$a_{rms} \leq 0.005\text{R m/s}^2$					
Velocity criteria (for all floors)	$v_{rms} \leq 0.0001\text{R m/s}$					

Table 3. Factor $k_{e,1}$ to calculate the fundamental frequency in case of a double span floor on rigid supports. l is the longer span, l_2 is the shorter span of a double span floor in m.

l_2/l	1	0.9	0.8	0.7	0.6	0.5	0.4	0.3	0.2
$k_{e,1}$	1	1.09	1.16	1.21	1.25	1.28	1.32	1.36	1.41

Thus, multi-person activities are more likely to cause the occupants discomfort, although the serviceability criteria in Table 2 are satisfied. Kozar et al. [79] explored the vibrations caused by human action on five-layer Cross-Laminated Timber (CLT) panels with a height of 14 cm by considering different combinations of thicknesses and spans. The authors discussed the serviceability requirements, highlighting that if the minimum required natural frequency of the CLT panel is 8 Hz, the spans could go up to 6 m.

5 Conclusions

Timber floors present unique challenges in the built environment due to their distinctive physical properties. With a high stiffness-to-mass-density ratio, these lightweight structures offer considerable structural advantages but simultaneously face significant performance issues related to vibration and sound transmission when compared to heavyweight alternatives like concrete floors.

Lightweight timber floors, whose area density is typically below 150 kg m^{-2} [1], have become a key contributor to low-carbon construction. Their reduced mass, however, makes them more susceptible than heavyweight alternatives to both human-induced vibrations in the 1–80 Hz band [2,3] and airborne or impact sound in the 20–5000 Hz band [4,6]. A central theme of this chapter has been the relationship between these phenomena: increasing stiffness to push fundamental modes above the dominant walking range may simultaneously raise radiation efficiency, so that radiated sound power does not fall even as vibration amplitudes diminish, as implied by Eq. (1). Conversely, adding mass can lower both vibration and noise but may be impractical at floor–wall junctions or may compromise the carbon footprint advantage of timber.

The sequential stiffness–frequency–vibration checks in the draft Eurocode 5 (EC5) offer a pragmatic design route, yet experimental and numerical studies on plate-like Cross-Laminated Timber (CLT) panels indicate that these rules can be conservative because they neglect higher-order modes that contribute significantly up to at least 100 Hz [73,76]. Although some researchers report that current criteria over-predict occupant discomfort for CLT floors [69,70], others show that multi-person activities can nearly double vibration amplitudes relative to single-person excitation and thus trigger complaints even when EC5 limits are met [78]. At the same time, field surveys confirm that occupants often perceive combined noise–vibration annoyance even when each parameter separately satisfies its guideline value [36], underscoring the need for unified metrics that merge sound-exposure level and vibration dose value.

Modelling capabilities continue to advance: efficient CLT homogenisation [56], hybrid finite-element/statistical-energy-analysis schemes for joist floors [55] and contact-free air-cavity coupling techniques [52] now permit broadband predictions to 5 kHz within practical time frames, provided that connection stiffness and damping are characterised accurately. Mitigation solutions such as resilient interlayers, elastomeric angle brackets and floating screeds can trim flanking transmission by 6–12 dB, but they simultaneously reduce in-plane stiffness by up to 45 % and may affect seismic performance [48,51]. Tuned floor coverings, for example foam-based boards, have demonstrated impact-sound improvements of up to 15 dB with negligible influence on global floor dynamics [60].

Looking ahead, research should converge on four priorities: first, the formulation of a harmonised limit state that spans 20–5000 Hz and reflects combined noise–vibration perception; second, the development of stochastic multi-person and rhythmic load models suitable for assembly and educational buildings; third, standardised high-frequency characterisation of semi-rigid connections and inter-panel gaps; and fourth, lifecycle-aware mitigation strategies that balance acoustic comfort with embodied-carbon targets. For practitioners, an integrated workflow that pairs rapid early-stage analytical models with detailed vibro-acoustic simulations and prototype testing remains essential; response-factor limits can guide preliminary sizing, but plate-like floors should always be checked with modal superposition or time-history methods that capture higher modes. Combining modest mass additions with resilient detailing at floor–wall interfaces typically provides the best trade-off between vibration control, impact-sound insulation and structural efficiency, provided that long-term creep of isolators is considered and that floor performance is verified post-occupancy with wireless accelerometers and sound-level meters.

As sustainable building practices continue to drive increased adoption of timber construction, these serviceability and acoustic considerations will remain crucial for ensuring acceptable levels of occupant comfort, particularly as regulations and standards evolve to better address the complex interaction between vibration and acoustic performance in lightweight floor systems.

Acknowledgments. The CA20139 - Holistic design of taller timber buildings (HELEN) is gratefully acknowledged for networking support.

References

1. ISO 10140-5:2021, "Acoustics – Laboratory measurement of sound insulation of building elements – Part 5: Requirements for test facilities and equipment," standard, International Organization for Standardization, Geneva, CH (2021)
2. ISO 10137:2007, "Bases for design of structures — serviceability of buildings and walkways against vibrations," standard, ISO, Geneve, CH (2007)
3. BS 6472:2008, "Guide to evaluation of human exposure to vibration in buildings," standard, BS - BRITISH STANDARD, London, UK (2008)
4. ISO 717-1:2020, "Acoustics – Rating of sound insulation in buildings and of building elements – Part 1: Airborne sound insulation," standard, International Organization for Standardization, Geneva, CH (2020)
5. ISO 717-2:2020, "Acoustics – Rating of sound insulation in buildings and of building elements – Part 2: Impact sound insulation," standard, International Organization for Standardization, Geneva, CH (2020)
6. ISO 10140-2:2021, "Acoustics – Laboratory measurement of sound insulation of building elements – Part 2: Measurement of airborne sound insulation," standard, International Organization for Standardization, Geneva, CH (2021)
7. ISO 10140-3:2021, "Acoustics – Laboratory measurement of sound insulation of building elements – Part 3: Measurement of impact sound insulation," standard, International Organization for Standardization, Geneva, CH (2021)
8. ISO 15186-1:2000, "Acoustics – Measurement of sound insulation in buildings and of building elements using sound intensity – Part 1: Laboratory measurements," standard, International Organization for Standardization, Geneva, CH (2000)
9. ISO 15186-2:2003, "Acoustics – Measurement of sound insulation in buildings and of building elements using sound intensity – Part 2: Field measurements," standard, International Organization for Standardization, Geneva, CH (2003)
10. ISO 15186-3:2002, "Acoustics – Measurement of sound insulation in buildings and of building elements using sound intensity – Part 2: Laboratory measurements at low frequencies," standard, International Organization for Standardization, Geneva, CH (2002)
11. ISO 16283-1:2014/Amd 1:2017, "Acoustics – Field measurement of sound insulation in buildings and of building elements – Part 1: Airborne sound insulation," standard, International Organization for Standardization, Geneva, CH (2017)
12. ISO 16283-2:2020, "Acoustics – Field measurement of sound insulation in buildings and of building elements – Part 2: Impact sound insulation," standard, International Organization for Standardization, Geneva, CH (2020)
13. ISO 12354-1:2017, "Building Acoustics – Estimation of acoustic performance of buildings from the performance of elements – Part 1: Airborne sound insulation between rooms," standard, International Organization for Standardization, Geneva, CH (2017)
14. ISO 12354-2:2017, "Building Acoustics – Estimation of acoustic performance of buildings from the performance of elements – Part 2: Impact sound insulation between rooms," standard, International Organization for Standardization, Geneva, CH (2017)

15. Ljunggren, F., Simmons, C., Hagberg, K.: Correlation between sound insulation and occupants' perception-proposal of alternative single number rating of impact sound. Appl. Acoust. **85**, 57–68 (2014)
16. Ljunggren, F., Simmons, C., Öqvist, R.: Correlation between sound insulation and occupants' perception-proposal of alternative single number rating of impact sound, part ii. Appl. Acoust. **123**, 143–151 (2017)
17. Ljunggren, F., Simmons, C.: Correlation between sound insulation and occupants' perception-proposal of alternative single number rating of impact sound, part iii. Appl. Acoust. **197**, 108955 (2022)
18. Opazo-Vega, A., MuÃoz-Valdebenito, F., Oyarzo-Vera, C.: Damping assessment of lightweight timber floors under human walking excitations. Appl. Sci. (Switzerland) **9**(18), 3759 (2019)
19. Homb, A., Kolstad, S.: Evaluation of floor vibration properties using measurements and calculations. Eng. Struct. **175**, 168–176 (2018)
20. Zhang, Z., Wang, S., Zhou, H.: Effects of stud partitions on the vertical vibration performances of metal plate connected timber. J. Forest. Eng. **7**(2), 167–173 (2022)
21. Zhao, X., Zhang, B., Kilpatrick, T., Sanderson, I.: Numerical analysis on global serviceability behaviours of tall clt buildings to the Eurocodes and UK national annexes. Buildings **11**(3), 124 (2021)
22. Ebadi, M., Doudak, G., Smith, I.: Evaluation of floor vibration caused by human walking in a large glulam beam and deck floor. Eng. Struct. **196**, 109349 (2019)
23. Ebadi, M., Doudak, G., Smith, I.: Vibration responses of glulam beam-and-deck floors. Eng. Struct. **156**, 235–242 (2018)
24. Robertson, M., Holloway, D., Taoum, A.: Vibration of suspended solid-timber slabs without intermediate support: assessment for human comfort. Aust. J. Struct. Eng. **19**(4), 266–278 (2018)
25. Yokoyama, Y., Koyama, Y., Nishitani, S., Fukuda, S.: Effect of deflection on evaluation of walking vibration of long-span timber floor from the viewpoint of habitability. J. Environ. Eng. (Japan) **87**(797), 379–390 (2022)
26. Nesheim, S., Malo, K., Labonnote, N.: Competitiveness of timber floor elements: an assessment of structural properties, production, costs, and carbon emissions. For. Prod. J. **71**(2), 111–123 (2021)
27. Shahnewaz, M., Dickof, C., Zhou, J., Tannert, T.: Vibration and flexural performance of cross-laminated timber - glulam composite floors. Compos. Struct. **292**, 115682 (2022)
28. Xin, Z., et al.: Determining elastic constants of full-size cross laminated timber panel supported on four nodes using a vibration method. Constr. Build. Mater. **323**, 126513 (2022)
29. Arnold, M., Dietsch, P., Maderebner, R., Winter, S.: Diagonal laminated timber–experimental, analytical, and numerical studies on the torsional stiffness. Constr. Build. Mater. **322**, 126455 (2022)
30. Huang, H., Gao, Y., Chang, W.-S.: Human-induced vibration of cross-laminated timber (clt) floor under different boundary conditions. Eng. Struct. **204**, 110016 (2020)
31. Sun, J., Niederwestberg, J., Cheng, F., Chui, Y.: Frequencies prediction of laminated timber plates using ann approach. J. Renew. Mater. **8**(3), 319–328 (2020)
32. European Committee for Standardization. Guide to the application of the european directive relating to the assessment and management of environmental noise. Technical report (2008)
33. British Standards Institution. Guide to Measurement and Evaluation of Human Exposure to Whole-Body Mechanical Vibration and Repeated Shock (1987)

34. International Organization for Standardization. Mechanical vibration and shock–evaluation of human exposure to whole-body vibration (1997)
35. Nering, W., Jabben, J.: Annoyance caused by environmental noise and vibrations: a systematic review and meta-analysis. Environ. Impact Assess. Rev. **83**, 106425 (2020)
36. Nering, K., Kowalska-Koczwara, A.: Determination of vibroacoustic parameters of polyurethane mats for residential building purposes. Polymers **14**(2), 314 (2022)
37. Howarth, H.V., Griffin, M.J.: Annoyance caused by vibration and noise from construction sites. J. Sound Vib. **145**(3), 411–422 (1991)
38. Vigran, T.E.: Building Acoustics. CRC Press, Boca Raton (2014)
39. Fossen, K., Björn, E.: Acoustics in timber buildings-a design guide. J. Build. Apprais. **4**(4), 323–335 (2008)
40. Olsson, A.: Acoustic performance of timber-framed multi-storey residential buildings: a literature review. Build. Environ. **149**, 216–226 (2019)
41. Caniato, M., Machimbarrena, M., Schelfhout, S., Torfs, D.: Design for acoustic performance of lightweight buildings: a state-of-the-art review. Build. Environ. **205**, 108395 (2022)
42. International Organization for Standardization. Building acoustics–estimation of acoustic performance of buildings from the performance of elements–part 12: Laboratory measurement of the impact sound insulation of floors (2017)
43. Olsson, J., Linderholt, A.: Low-frequency impact sound of timber floors: a finite element-based study of conceptual designs. Build. Acoust. **28**(1), 17–34 (2021)
44. Ljunggren, F., Ågren, A.: Elastic layers to reduce sound transmission in lightweight buildings. Build. Acoust. **20**(1), 25–42 (2013)
45. Negreira, J., Irusta, R., Astigarraga, H., Alday, F., Garcia, A.: Characterisation of elastic insulating layers for the reduction of impact sound insulation of lightweight timber floors. Appl. Acoust. **86**, 112–119 (2014)
46. Reichelt, F., Strecker, P., Winter, S.: Characteristics of the acoustic performance of elastic wood wool cement boards. Appl. Acoust. **104**, 105–114 (2016)
47. Medved', J., Ingeleare, B., De Geetere, L.: Impact sound insulation concept for lightweight timber floor. Adv. Mater. Res. **855**, 245–251 (2014)
48. Azinović, B., Dolšek, M., Planinc, I., Klančar, G.: Influence of flexible insulation layers on seismic performance of cross-laminated timber walls. Eng. Struct. **243**, 112876 (2021)
49. Gibson, B., Nguyen, T., Sinaie, S., Heath, D., Ngo, T.: The low frequency structure-borne sound problem in multi-storey timber buildings and potential of acoustic metamaterials: A review. Build. Environ. **224**, 109531 (2022)
50. Homb, A., Guigou-Carter, C., Rabold, A.: Impact sound insulation of cross-laminated timber/massive wood floor constructions: Collection of laboratory measurements and result evaluation. Build. Acoust. **24**(1), 35–52 (2017)
51. Kržan, M., Govekar, E., Turk, G.: Cyclic response of insulated steel angle brackets for clt connections. Eng. Struct. **242**, 112507 (2021)
52. Fox, C., Nakamura, M. Vibration prediction in light timber floors with air cavity using a composite model structure. Build. Environ. **157**, 831–839 (2019)
53. Paolini, R., Perillo, G., Masoero, M., Bignozzi, V.: Prediction of airborne sound transmission through multi-layered thin elastic components by using a computational mesh-free method. Build. Environ. **153**, 23–36 (2019)
54. De Santis, S., Boscato, G., Cigana, D., Fragiacomo, M.: Timber screwed connections with elastic interlayer: analytical model for stiffness prediction. Eng. Struct. **244**, 112869 (2021)

55. Wang, D., et al.: Predicting impact forces, structural vibration and radiated sound power of timber joist floors. Build. Environ. **186**, 107307 (2020)
56. Vallely, S., Schoenwald, S.: An efficient analytical method to obtain the homogenised frequency-independent elastic material properties of cross-laminated timber elements. J. Sound Vib. **546**, 117424 (2023)
57. Cremer, L., Heckl, M.: Structure-borne sound: structural vibrations and sound radiation at audio frequencies. Springer, Heidelberg (2005)
58. Vallely, S., Schoenwald, S.: Higher-order modal parameter estimation and verification of cross-laminated timber plates for structural-acoustic analyses. Acta Acustica **8**, 52 (2024)
59. Lietzen, M.: Effect of floor coverings on impact sound insulation. Build. Environ. **198**, 122221 (2022)
60. Huang, X., He, X., Fan, H., Lv, Y., Zhou, T., Ren, P.: Performance of elastic cushion materials for impact sound insulation of timber floors. Build. Environ. **202**, 108329 (2021)
61. ISO 2631-1:1997. Mechanical vibration and shock—evaluation of human exposure to whole-body vibration—part 1: General requirements. Standard, ISO, Geneve, CH (1997)
62. Basaglia, B.M.: Dynamic Behaviour of Long-Span Timber Ribbed-Deck Floors. PhD thesis (2019)
63. Chang, W., Goldsmith, T., Harris, R.: new design method for timber floors—peak acceleration approach. In: INTER - International Network on Timber Engineering Research, vol. 51-20-1 (2018)
64. Hamm, P., Richter, A., Winter, S.: Floor vibrations – new results. In: WCTE - World Conference of Timber Engineering, vol. Riva del Garda (2010)
65. Hamm, P., Marcroft, J., Toratti, T.: Vibrations of floors – comparison of measured data and suggested design. In: INTER - International Network on Timber Engineering Research, vol. 53-20-1 (2020)
66. Abeysekera, I., Hamm, P., Toratti, T., Lawrence, A.: Development of a floor vibration design method for eurocode 5. In: INTER - International Network on Timber Engineering Research, vol. 51-20-2 (2018)
67. EN12512. Timber structures. test methods. cyclic testing of joints made with mechanical fasteners (2022)
68. Ussher, E., Arjomandi, K., Weckendorf, J., Smith, I.: Predicting effects of design variables on modal responses of clt floors. Structures **11**, 40–48 (2017)
69. Zhang, S., Zhou, J., Niederwestberg, J., Chui, Y.: Effect of end support restraints on vibration performance of cross laminated timber floors: an analytical approach. Eng. Struct. **189**, 186–194 (2019)
70. Hu, L.: Serviceability of next generation wood buildings: laboratory study of vibration performance of cross-laminated-timber (clt) floors. Project 301006159: report 2012/13 (2013)
71. Gagnon, S., Pirvu, C.: Clt handbook, special pub. sp-528e (2011)
72. Weckendorf, J., Smith, I.: Dynamic characteristics of shallow floors with cross-laminated-timber spines, vol. 1, pp. 176–185 (2012)
73. Maldonado, S., Chui, Y.: Effect of end support conditions on the vibrational performance of cross-laminated-timber slabs. In: Proceedings of 13th World Conference on Timber Engineering (2014)
74. Ussher, E., Aloisio, A., Pasca, D.P., Hansen, S.L., Tomasi, R.: Effect of partition walls on the vibration serviceability of cross-laminated timber floors. J. Build. Eng. **95**, 110001 (2024)

75. Aloisio, A., Pasca, D.P., Owolabi, D., Loss, C.: Vibration serviceability of hybrid clt-steel composite floors based on experimental and numerical investigations using random walk models. Eng. Struct. **304**, 117600 (2024)
76. Weckendorf, J., Toratti, T., Smith, I., Tannert, T.: Vibration serviceability performance of timber floors. Eur. J. Wood Wood Prod. **74**(3), 353–367 (2016)
77. Milojević, M., Racic, V., Marjanović, M., Nefovska-Danilović, M.: Influence of inter-panel connections on vibration response of clt floors due to pedestrian-induced loading. Eng. Struct. **277**, 115432 (2023)
78. Wang, C., Chang, W.-S., Yan, W., Huang, H.: Predicting the human-induced vibration of cross laminated timber floor under multi-person loadings. Structures **29**, 65–78 (2021)
79. Kozaric, L., Purcar, M., Zivkovic, S.: Vibrations of cross-laminated timber floors [vibracije križno lameliranih drvenih medukatnih panela]. Drvna Industrija **70**(3), 307–312 (2019)

Vibration-Based Structural Health Monitoring of Timber Structures

Angelo Aloisio[1,5(✉)], Dag Pasquale Pasca[2], Pier Francesco Giordano[3], Marco Civera[4], Marco Martino Rosso[4], Maria Pina Limongelli[3], Thomas P. S. Reynolds[8], Chiara Bedon[6], and Blaž Kurent[7]

[1] Department of Civil, Construction-Architectural and Environmental Engineering, Università degli Studi dell'Aquila, L'Aquila, Italy
[2] Norsk Treteknisk Institute (Norwegian Institute of Wood Technology), Oslo, Norway
[3] Department of Architecture, Built Environment and Construction Engineering, Politecnico di Milano, Milan, Italy
[4] Department of Structural, Geotechnical and Building Engineering, Politecnico di Torino, Turin, Italy
[5] Faculty of Science and Technology, Norwegian University of Life Sciences, Ås, Norway
angelo.aloisio@univaq.it
[6] Department of Engineering and Architecture, Università degli Studi di Trieste, Trieste, Italy
[7] Faculty of Civil and Geodetic Engineering, University of Ljubljana, Ljubljana, Slovenia
[8] School of Engineering, University of Edinburgh, Edinburgh, UK

1 Introduction

Buildings and other civil structures can be "interrogated" through their vibration signatures: because natural frequencies, mode shapes and damping depend on mass, stiffness and energy-dissipation mechanisms, measuring ambient or forced responses reveals those properties without destructive tests. Two complementary approaches are in routine use: ambient/operational modal analysis, which exploits traffic, wind, human activity or micro-tremors as unmeasured inputs, and experimental modal analysis, which relies on shakers or vibrodynes to provide known forces [1–3]. Data are gathered with temporary or permanent networks of accelerometers and loggers; short campaigns suffice for finite element model calibration, whereas long-term installations capture the influence of temperature, humidity or occupancy on modal behaviour. The same information underpins vibration-based damage detection and post-event safety checks. Yet research on timber, especially multi-storey and hybrid systems, lags behind other materials, even though wood's moisture- and temperature-sensitive properties, and its susceptibility to biological degradation, make continuous structural health monitoring (SHM) highly desirable for verifying the flexibility and strength assumed at design.

G. Fink et al. (Eds.): *Holistic Design of Taller Timber Buildings*, SPRTRCIENG, pp. 94–103, 2025.
https://doi.org/10.1007/978-3-032-02098-7_8

2 Applications of Vibration-Based Monitoring to Timber Structures

Due to the relatively immature state of the research in this specific sector, there is a lack of standardised guidelines and best practices for implementing vibration-based SHM systems in timber structures. This strongly hampers practitioners to adopt these technologies consistently. Apart from these general considerations, each timber-based solution presents its own specific challenges and requires dedicated precautions and arrangements.

For CLT structures, since the most critical failure mode (rolling shear failure [4]) involves delamination, adhesive failure, and/or cohesive failure between two adjacent CLT layers [5], researchers have investigated the integration of different kinds of SHM sensors (measuring both the static or dynamic behaviour of the structure and the environmental conditions) directly within the layers of CLT panels. However, embedding sensors within timber members, even relatively small ones such as strain gauges or fiber optic sensors, can be labor-intensive and may require specialized techniques to ensure proper installation and long-term durability. Furthermore, embedding sensors within glued layers can be challenging, as the adhesive layer may interfere with the sensor's performance – for instance, leading to stress concentrations that can cause premature failure of the sensor [6]. Degradation of the adhesive or the wood-adhesive interface (due to ageing and/or environmental conditions, such as moisture, temperature variations, and ultraviolet radiation) can lead to false alarms or missed detection [6]. Nevertheless, embedded sensors have been proven time and time again to be very effective, especially for moisture content monitoring at various depths and locations [7,8]. In this specific regard, several researchers (e.g. [9] and [10]) emphasize the importance of carefully positioning moisture sensors both at critical locations (like joints, edges, and other areas prone to moisture exposure) and at different depths inside the material. Monitoring moisture at different depths provides insights into the moisture distribution through the CLT cross-section; in fact, timber members' response to changes in the surrounding climate is slower for larger members and, consequently, moisture gradients can develop and cause moisture-induced stresses [11]. The same considerations should be extended to vibration-measuring sensors, such as the classic piezoelectric or MEMS-based accelerometers, also to account for the high variability of mechanical properties in wood, to strategically place the sensors at critical locations.

Hybrid timber-based structures (including CLT hybrid structures), which combine timber with other materials such as steel or concrete, present additional challenges for SHM [12]. SHM sensors and techniques need to address the unique challenges posed by the interaction between different materials and their respective responses to loading and environmental conditions. That is achieved by monitoring the performance of the interfaces between timber and other materials to ensure the local structural integrity of the hybrid system. Due to this and the previous point, it is clear that (i) timber structures require, more than other construction with other building materials, sensors at many locations; (ii) sensors for static and dynamic monitoring should be always paired with environmental sensors, to adjust the measurements for the local environmental and operational factors (EOFs). Again, this aspect is more relevant in timber than in other materials. Therefore, the integration of multiple sensors into 'sensing nodes' is beneficial; furthermore, to solve the practical issues related to cabling, wireless sensor networks should be preferred, to allow optimal placement without technical constraints. Wireless communication protocols, such as LoRa and other low-power wide-area network technologies, are becoming more and more adopted for such SHM applications thanks to their long-range, low-power data transmission capabilities.

Hence, it is crucial to establish how far research on analysing the dynamic behaviour of timber buildings has gone to highlight critical issues and knowledge gaps. The literature on this topic is reviewed, highlighting similarities and differences to other structures and identifying future research needs and directions. First, an overview of the modal identification of timber buildings is provided. After that, the state of the research on damage identification and model calibration is investigated.

The literature on ambient vibration tests (AVTs) and forced vibration tests (FVTs) on civil structures is vast. According to the authors' knowledge, the first AVT on a timber building was performed by Ellis and Bougard in 2001 [13]. The tests were conducted on a full-size, six-storey timber framed structure constructed inside BRE's Cardington laboratory. They performed both FVTs and AVTs at different stages of construction, which made it possible to evaluate the contribution to the global stiffness of the timber frame alone, the staircase, and the finishing and cladding (bricks). The results of their research indicate that the building's non-structural components play a large role in the contribution to the lateral stiffness of the building at service levels. More recently, some other researchers have attempted to extract the modal properties of mid-rise timber buildings (Reynolds et al. [14,15], Feldmann et al. [16]) using operational modal analysis (OMA) methods. The research conducted by Reynolds and colleagues constitutes probably the largest database of AVT performed on timber structures in Europe to date. They tested different timber structural archetypes: post and beam, timber-framed, pure CLT and hybrid timber-concrete structures. It is also worth mentioning the tests performed in Germany and Austria on eight timber observation towers (with a height of up to 45 m), a 100 m tall wind turbine and three multi-storey residential timber buildings (with a height of up to 26 m). The findings of all these testing campaigns have allowed for assessing the simplified relationship between height and natural frequency for multi-storey buildings given in Eurocode 1.

In North America, where there is a deeply-rooted tradition of wooden frame housing, efforts have been made to understand the dynamic behaviour of smaller low-rise residential buildings, see Mugabo et al. [17], and all the infield investigations on light-frame wood buildings by Hafeez et al. [18–20]. Kim et al. [21] combined vibration and force measurements. They used load cells between the column and foundation stone to measure axial column force, and ambient vibration tests were performed to measure natural vibration frequency and mode shape.

The results of these campaigns shed some light on the dynamic behaviour of tall timber buildings, providing viable information concerning stiffness and damping of the tested structures to designers and stakeholders [22]. Nevertheless, extensive dynamic tests on mid-rise and high-rise timber structures represent a missing part, although some research has been conducted [23–26]. This will aid in learning important lessons and enhance the engineering community's confidence towards using this material.

Damage, such as cracks and reduction of cross sections in structural members, can modify the dynamic response of structures [27]. Thus, vibration data are processed to extract damage-sensitive features (DSFs) and ultimately obtain damage indices which alert about variations of DSFs between a reference and the current state of the structure [28]. According to the available resources (e.g., the number and location of sensors) different levels of damage identification can be attained [29], namely: (i) damage detection (alert about the existence of damage); (ii) damage localization (find the position of damage); (iii) damage quantification (assess the gravity of damage); (iv) damage prognosis (forecast the evolution of damage). The dynamic properties of a structure do not vary only due to the occurrence of damage. EOFs, such as ambient temperature, humidity, wind conditions and structural usage, can modify the properties of healthy structures. This is a major concern in damage identification.

Variations of EOFs might hide the effect of structural anomalies and hamper damage identification. Besides, variations in the dynamic behaviour due to EOFs might be erroneously attributed to damage. Therefore, the influence of EOFs on different materials and structural typologies must be carefully understood and considered for damage identification purposes.

In addition to temperature and relative air humidity, one of the main concerns in the case of timber buildings is the moisture content (MC), which influences several properties of timber, such as strength, density, and elastic modulus. Larsson et al. [30] were the first to carry out long-term monitoring of a hybrid timber building and investigated the relationship between environmental factors and the dynamic response of a hybrid timber-concrete building. The authors have tracked the modal parameters, i.e., natural frequencies, modal shapes, and damping ratios, for three years, together with hygrothermal parameters, i.e., temperature, relative humidity, absolute humidity, and moisture content. The results of the long-term monitoring show that modal frequencies change with the temperature, showing maximum and minimum values in early autumn and early spring, respectively. In contrast, damping ratios did not present seasonal variations. Furthermore, it is observed that the modal frequencies decrease in the first year after construction due to the drying out of timber elements. In [31], the results of a 3-year monitoring campaign on a Pres-Lam building are presented. In this case, the results show that temperature and relative humidity, as well as post-tensioning losses, do not affect the structure's dynamic behaviour. Recently, Aloisio et al. [32] investigated the effect of MC on the dynamic properties of an eight-storey CLT building and related the moisture content variation to the shear modulus G through model updating, based on a model developed in [26].

Nevertheless, the specific effects of varying moisture content (MC) on the dynamic properties of timber-made structures and infrastructures remain largely underinvestigated. MC monitoring, per se, is quite commonly performed, at least since the late 2000s. This is also referred to as hygrothermal monitoring (see e.g. [33]). These embedded sensing capabilities represent the natural progression from portable pin-type moisture meters, which can be used to manually measure moisture content at the surface or at different depths of the CLT elements during construction, but require human technicians and can only provide a snapshot of moisture levels at specific points in time. Especially for timber infrastructures, the technology was successfully tested on several wood bridge case studies in Switzerland [34,35] and several similar field tests in Germany, with applications to 21 large-span timber structures [36] and other bridges and structures [37,38]. In one case, the MC monitoring system reportedly remained operational for five years without the need for maintenance. Similar MC applications were tested on 17 modern and historical wooden bridge constructions in central Europe [39], in the USA during the Development of the 'Smart Timber Bridge' program [40], in Norway [41], Finland [42], and Sweden (both in road bridges and pedestrian ones)[43]. However, out of all these examples, only the last one [43] was paired with accelerometric readings. Other noteworthy and successful field applications of combined MC-dynamic monitoring are represented by the George W. Peavy Forest Science Complex (or "Peavy Hall") a large timber building monitored during its construction, where researchers assessed the building's long-term performance including monitoring moisture levels and structural vibrations in their study [44]; the House of Natural Resources (HoNR) in Zürich, Switzerland [45]; and the University of British Columbia (UBC) Tallhouse [46]. Therefore, the moisture-related EOFs on modal parameters and derived DSFs will require further experimental investigations.

It is also worth mentioning that MC and dynamic monitoring can (and should) be paired, as internal moisture is notoriously a potential leading cause for damage, for different reasons: sustained exposure to moisture content levels above 20%, due to moisture/water intrusion, localized moisture accumulation, water entered during construction or assembly and trapped inside, and/or other phenomena, can lead to damage by several species of wood-decaying fungi and mould growth; while in-service drying can cause the development of cracks [47]. In certain cases, these occurrences may happen in the inner layers of CLT roof and walls, without signs noticeable through visual inspections on the external surfaces. Therefore, without proactive MC monitoring and management, the negative effects could be either direct, due to the insurgence of moisture-related problems, or indirect, due to the damage-unrelated perturbations in the identified modal parameters.

One of the most promising techniques in the field of structural health monitoring (SHM) for timber structures is vibration-based analysis. This method involves the use of sensors to detect and analyze vibrations within a structure, which can reveal critical information about its integrity and any potential damage. Vibration-based SHM is particularly beneficial for timber structures as it provides a non-invasive means to continuously monitor their condition, thus helping to prevent catastrophic failures and prolong their lifespan.

The research by Suzuki et al. [48] explores the use of machine learning techniques to enhance the accuracy of SHM systems for timber structures. By employing piezoelectric sensors to capture vibration waveforms and neural networks to classify damage locations, the study achieved an impressive 83.8% accuracy in identifying damage even in previously unlearned timber pieces. This demonstrates the potential of advanced machine learning methods to generalize across different timber samples, making SHM systems more robust and reliable. Similarly, Chunyu et al. [49] have developed a specialized device for monitoring vibrations in ancient timber structures. This innovation addresses the practical challenges of securing sensors in delicate, historical buildings. The device's design ensures stable and accurate vibration monitoring, which is essential for detecting structural issues early and implementing timely maintenance. Zhibin et al. [50] introduced a health monitoring device designed to improve the accuracy and ease of installation for vibration monitoring in ancient wooden buildings. Their approach enhances the practicality of long-term SHM by simplifying the alignment of monitoring components, thereby ensuring consistent and reliable data collection. This method is particularly useful for historical structures where precision and minimal invasiveness are critical. Another significant contribution is from Oiwa et al. [51], who proposed an AI-based system for continuous monitoring of timber health using piezoelectric sensors. By applying machine learning techniques such as k-nearest neighbor and support vector machine, their system demonstrated strong classification performance in identifying structural damage. This integration of AI and SHM underscores the potential for highly automated, accurate, and efficient monitoring systems that can operate with minimal human intervention.

3 Conclusions

Vibration-based structural health monitoring has proved to be a powerful, non-destructive tool for characterising, calibrating and safeguarding civil structures, yet its systematic application to timber remains in its infancy. The studies reviewed in this chapter show that modal testing (operational or forced) can reveal stiffness, damping and mass changes caused by moisture cycling, duration-of-load effects and biological decay that are unique to wood, while also supporting damage localisation and finite-element model updating. Progress is nevertheless constrained by the pronounced anisotropy and spatial heterogeneity of timber, by the strong, often coupled influence of temperature and especially moisture content on identified modal parameters, and by the absence of standardised protocols for sensor placement, data correction and feature normalisation under environmental and operational factors. Recent field campaigns on mid-rise buildings and bridges confirm that multi-year datasets, coupled with embedded moisture probes and wireless accelerometer networks, are essential for disentangling true structural change from reversible hygro-thermal trends [30,32]. At the same time, machine-learning classifiers trained on high-frequency vibration signatures, sometimes augmented with local piezoelectric actuation, have reached damage-identification accuracies above 80% on unlearned specimens [48,51], signalling a viable path toward automated, low-maintenance monitoring of both modern CLT and heritage timber. To translate these advances into routine practice, future work must establish harmonised guidelines that link sensor networks with moisture and temperature measurements, quantify acceptable environmental correction ranges for modal features, and define reliability targets for algorithmic decision support. Only then can vibration-based SHM fulfil its dual promise of ensuring the safety of ever-taller timber buildings and preserving the embodied carbon advantage that makes them attractive in the first place.

Acknowledgments. The CA20139 - Holistic design of taller timber buildings (HELEN) is gratefully acknowledged for networking support.

References

1. Rainieri, C., Fabbrocino, G.: Output-only modal identification. In: Operational Modal Analysis of Civil Engineering Structures, pp. 103–210. Springer, New York (2014). https://doi.org/10.1007/978-1-4939-0767-0_4
2. Pasca, D.P., Aloisio, A., Rosso, M.M., Sotiropoulos, S.: Pyoma and pyoma_gui: a python module and software for operational modal analysis. SoftwareX **20**, 101216 (2022)
3. Pasca, D.P., Margoni, D.F., Rosso, M.M., Aloisio, A.: Pyoma2: an open-source python software for operational modal analysis. In: International Operational Modal Analysis Conference, pp. 423–434, Springer, Cham (2024)
4. Popovski, M., Chen, Z., Gafner, B.: Structural behaviour of point-supported CLT floor systems. In: 4th World Conference on Timber Engineering (WCTE2016) (2016)
5. Flores, E.S., Saavedra, K., Hinojosa, J., Chandra, Y., Das, R.: Multi-scale modelling of rolling shear failure in cross-laminated timber structures by homogenisation and cohesive zone models. Int. J. Solids Struct. **81**, 219–232 (2016)
6. Aicher, S., Reinhardt, H.-W., Garrecht, H.: Materials and joints in timber structures. Recent Dev. Technol. (2014)
7. Kellgren, L.V., Seeberg, A.C., Hagle, D., Øvregård, J., Rasmussen, J.: a moisture management strategy for CLT using sensor technology to create a robust Norwegian school. In: World Conference on Timber Engineering (WCTE 2023) (2023)
8. Strang, M., Leardini, P., Shirmohammadi, M.: Validating moisture-safe energy efficient CLT assemblies in hot and humid climates using experimental testing. In : World Conference on Timber Engineering (WCTE 2023) (2023)
9. Time, B., Andenæs, E., Karlsen, T., Geving, S., Kvande, T.: Moisture safety strategy for construction of CLT structures in a coastal nordic climate. In: Journal of Physics: Conference Series, vol. 2654, 13th Nordic Symposium on Building Physics (NSB-2023) (2023)

10. Tamagnone, G., Hairstans, R., Martin, J., Edmondson, V.: Method for monitoring the moisture response of a cross laminated timber (CLT) panel buildings. In: 2nd International Conference on Moisture in Buildings 2023 (ICMB23) (2023)
11. Ranta-Maunus, A.: Effects of Climate and Climate Variations on Strength, pp. 153–168. Wiley, Hoboken (2003)
12. Aloisio, A., et al.: Vibration analyses of an hybrid concrete and cross-laminated timber building case study. In: Italian Workshop on Shell and Spatial Structures, pp. 309–318, Springer, Cham (2023)
13. Ellis, B., Bougard, A.: Dynamic testing and stiffness evaluation of a six-storey timber framed building during construction. Eng. Struct. **23**(10), 1232–1242 (2001)
14. Reynolds, T., Harris, R., Chan, W.: Dynamic response of tall timber buildings to wind load. In: 35th Annual Symposium of IABSE/52nd Annual Symposium of IASS/6th International Conference on Space Structures, London (2011)
15. Reynolds, T., Feldmann, A., Ramage, M., Chang, W., Harris, R., Dietsch, P.: Design parameters for lateral vibration of multi-storey timber buildings. In: International Network on Timber Engineering Research Proceedings (INTER 2016) (2016)
16. Feldmann, A., Huang, H., Chang, W., Harris, R., Dietsch, P., Gräfe, M., Hein, C.: Dynamic properties of tall timber structures under wind-induced vibration. In: World Conference on Timber Engineering (WCTE 2016) (2016)
17. Mugabo, I., Barbosa, A., Riggio, M.: Dynamic characterization and vibration analysis of a four-story mass timber building. Front. Built Environ. **5** (2019)
18. Hafeez, G., Doudak, G., McClure, G.: Effect of nonstructural components on the dynamic characteristics of light-frame wood buildings. Can. J. Civ. Eng. **47**(3), 257–271 (2020)
19. Hafeez, G., Doudak, G., McClure, G.: Dynamic characteristics of light-frame wood buildings. Can. J. Civ. Eng. **46**(1), 1–12 (2019)
20. Hafeez, G., Doudak, G., McClure, G.: Establishing the fundamental period of light-frame wood buildings on the basis of ambient vibration tests. Can. J. Civ. Eng. **45**(9), 752–765 (2018)
21. Kim, Y.-M., Kim, Y.-M., Kim, H.-J., Ha, T.-U., Shin, E.-M., Kim, W.-J.: Analysis of roof load and dynamic characteristics of traditional timber building considering column axial force and ambient vibration measurement. J. Architectural Inst. Korea **37**(11), 271–280 (2021)
22. Tulebekova, S., Malo, K.A., Rønnquist, A., Nåvik, P.: Modeling stiffness of connections and non-structural elements for dynamic response of taller glulam timber frame buildings. Eng. Struct. **261**, 114209 (2022)
23. Aloisio, A., Pasca, D., Tomasi, R., Fragiacomo, M.: Dynamic identification and model updating of an eight-storey clt building. Eng. Struct. **213**, 110593 (2020)
24. Kurent, B., Brank, B., Ao, W.: Model updating of seven-storey cross-laminated timber building designed on frequency-response-functions-based modal testing. Struct. Infrastruct. Eng. **19**(2), 178–196 (2023). cited By 12
25. Kurent, B., Ao, W.K., Pavic, A., Pérez, F., Brank, B.: Modal testing and finite element model updating of full-scale hybrid timber-concrete building. Eng. Struct. **289**, 116250 (2023)
26. Kurent, B., Friedman, N., Aloisio, A., Pasca, D., Tomasi, R., Brank, B.: Bayesian model updating of eight-storey CLT building using modal data. Probab. Eng. Mech. **77**, 103642 (2024)
27. Peeters, B., De Roeck, G.: One-year monitoring of the z24-bridge: environmental effects versus damage events. Earthq. Eng. Struct. Dyn. **30**(2), 149–171 (2001)

28. Farrar, C.R., Worden, K.: An introduction to structural health monitoring. Philos. Trans. R. Soc. A: Math. Phys. Eng. Sci. **365**(1851), 303–315 (2007)
29. Rytter, A.: Vibrational based inspection of civil engineering structures (1993)
30. Larsson, C., Abdeljaber, O., Bolmsvik, A., Dorn, M.: Long-term analysis of the environmental effects on the global dynamic properties of a hybrid timber-concrete building. Eng. Struct. **268** (2022)
31. Granello, G., Palermo, A.: Monitoring dynamic properties of a pres-LAM structure: trimble navigation office. J. Perform. Constr. Facil. **34**(1), 04019087 (2020)
32. Aloisio, A., Pasca, D.P., Kurent, B., Tomasi, R.: Long-term continuous dynamic monitoring of an eight-story CLT building. Mech. Syst. Signal Process. **224**, 112094 (2025)
33. Schmidt, E., Riggio, M.: Monitoring moisture performance of cross-laminated timber building elements during construction. Buildings, **9**(6) (2019)
34. Tannert, T., Berger, R., Müller, A.: Remote moisture monitoring of timber bridges: a case study (2011)
35. Franke, B., Franke, S., Müller, A.: Case studies: long-term monitoring of timber bridges. J. Civ. Struct. Health Monit. **5** (2015)
36. Dietsch, P., Gamper, A., Merk, M., Winter, S.: Monitoring building climate and timber moisture gradient in large-span timber structures. J. Civ. Struct. Health Monit. **5** (2015)
37. Brischke, C., Rapp, A.O., Bayerbach, R.: Measurement system for long-term recording of wood moisture content with internal conductively glued electrodes. Build. Environ. **43**(10), 1566–1574 (2008)
38. Brischke, C., Rapp, A.O., Bayerbach, R., Morsing, N., Fynholm, P., Welzbacher, C.R.: Monitoring the "material climate" of wood to predict the potential for decay: Results from in situ measurements on buildings. Build. Environ. **43**(10), 1575–1582 (2008)
39. Fojtík, R.: Moisture content analysis of wooden bridges. Wood Res. **64** (2019)
40. Phares, B.M., Pence, T.L., Wacker, J.P., Hosteng, T.K.: Development of a smart timber bridge (phase iii): moisture and strain sensor investigation for historic covered bridges (2019)
41. Mahnert, K.-C., Hundhausen, U.: A review on the protection of timber bridges. Wood Mater. Sci. Eng. **13**(3), 152–158 (2018)
42. Seppälä, M., Ilgın, H.E., Karjalainen, M., Pajunen, S.: An analysis on finnish wooden bridge practices. Appl. Sci. **13**(7) (2023)
43. Björngrim, N.: Monitoring of a Timber Footbridge. PhD dissertation (2015)
44. Baas, E.J., Riggio, M., Barbosa, A.R.: A methodological approach for structural health monitoring of mass-timber buildings under construction. Constr. Build. Mater. **268**, 121153 (2021)
45. Granello, G., Leyder, C., Frangi, A., Palermo, A., Chatzi, E.: Long-term performance assessment of an operative post-tensioned timber frame structure. J. Struct. Eng. **145**(5), 04019034 (2019)
46. Kaya, Y., Ventura, C.E., Taale, A.: An ambient vibration test of an R/C wall of an 18-story wood building at the UBC campus. In: Caicedo, J., Pakzad, S. (eds.) Dynamics of Civil Structures, Volume 2. CPSEMS, pp. 315–320. Springer, Cham (2017). https://doi.org/10.1007/978-3-319-54777-0_39
47. Palma, P., Steiger, R.: Structural health monitoring of timber structures - review of available methods and case studies. Constr. Build. Mater. **248**, 118528 (2020)
48. Suzuki, K., Ito, T., Koike, K., Kawahara, T., Ke, M., Mori, K.: Improvement of generalization performance for timber health monitoring using machine learning (2020)

49. Chunyu, Q., Xiao, D., Jianguo, Z., Kai, Z., Wei, W., Junlian, L.: Timber structure ancient building vibration monitoring vibration pickup installation device (2016)
50. Zhibin, S., Shengnan, S., Zhaobo, M.: Healthy condition monitoring device under vibration for ancient wooden building (2019)
51. Oiwa, R., Ito, T., Kawahara, T.: Timber health monitoring using piezoelectric sensor and machine learning (2017)

Wind Design for Tall Timber Buildings

Martina Sciomenta[1](✉), Saule Tulebekova[2], and Haris Stamatopoulos[2]

[1] Department of Civil, Architectural and Building and Environmental Engineering, University of L'Aquila, L'Aquila, Italy
martina.sciomenta@univaq.it

[2] Department of Structural Engineering, Norwegian University of Science and Technology, NTNU, Trondheim, Norway

Abstract. The design of tall timber buildings (TTBs) is often governed by wind-induced vibrations at the serviceability limit state (SLS) due to the low density and stiffness which make these structures more flexible and susceptible to wind effects. Controlling sway-induced deformations and accelerations is critical to ensure occupant comfort and structural performance. This chapter provides an overview of the key dynamic properties, such as natural frequency and damping ratio, that significantly influence wind response, then explores the role of lateral load-resisting systems (LLRS) in mitigating the deformations and vibrations in TTBs under wind loads. Additionally, advanced methods for evaluating wind-induced responses, such as computational simulations and experimental wind tunnel testing, are discussed. Performance-based wind design (PBWD) is presented as a promising approach to address the limitations of traditional codes. The study highlights the importance of accurately predicting deformations and dynamic parameters while optimizing structural systems to meet SLS criteria and ensure safety and comfort.

Keywords: Wind design · Dynamic parameters · Wind performance · Serviceability Limit State Design · Performance-based wind design

1 Introduction

Tall timber buildings (TTBs) have, in principle, adequate strength capacity to withstand lateral loads (e.g. wind, earthquake) at the ultimate limit state (ULS). However, the inherently low density and stiffness properties of wood make tall-timber buildings lighter and more flexible and thus, more sensitive to wind-induced effects [1]. As a result, their design is predominantly governed by the vibration serviceability limit state (SLS), leading to restrict the wind-induced sway vibrations within certain limits [2–4].

The performance of TTBs under wind loads is largely determined by some dynamic parameters: mass, stiffness, damping ratio, natural frequency. While the natural frequency, dependent on the distribution of modal stiffness and mass within the structure, can be determined with confidence, the damping ratio is subject to significant uncertainty due to its reliance on different factors such as material properties, structural systems,

G. Fink et al. (Eds.): *Holistic Design of Taller Timber Buildings*, SPRTRCIENG, pp. 104–114, 2025.
https://doi.org/10.1007/978-3-032-02098-7_9

and connection types. Those latter, in fact, play a key role, since their stiffness and damping features, as their variation, resulting from the action of previous loads, could led to different response to a given dynamic load at different stages of building's life. Non-bearing components also positively influence the overall stiffness [5] and damping [6].

In practice, damping ratio and natural frequency are regarded to be independent upon the structural amplitude of buildings, however, results from field measurements suggest that their values may demonstrate typical amplitude-dependent features, especially for high-rise buildings [7].

The issue of assuring acceptable wind performance of TTBs is still an ongoing research topic. In particular, the key aspects, which will be hereafter discussed, concern the role and the prediction of dynamic parameters, the selection of the appropriate type of lateral load-resisting system and determining optimal strategies to meet the SLS comfort criteria limits.

2 Dynamic Properties of Tall Timber Buildings

2.1 Mass

The mass magnitude and distribution along the building have a large impact on the dynamic properties of the building. The total mass of the building is given by the sum of the dead and live loads.

In terms of dead loads, [8] analyzed two building types with the same floor layout and different lateral load-resisting systems: post-and-beam with diagonal bracings and CLT panels. They observed that post-and-beam buildings have a greater actual weight compared to CLT buildings. However, the equivalent mass, which is more effective in reducing acceleration levels, was also higher in post-and-beam buildings, making them more suitable for TTBs.

The increase of mass at floor level to reduce wind-induced accelerations on the upper floors was a successfully strategy applied in the design of both Mjøstårnet and Treet buildings, characterized by an all-timber lateral load resisting system with glulam mega-truss on the building's perimeter. In those case, the use of concrete layers on floors, or as fully concrete slabs (Mjøstårnet and Treet) or as additional slab on the top of CLT (Mjøstårnet) allowed to reduce the building's vibrations.

For live loads, in the Eurocodes, the recommended value for residential buildings is 2 kN/m^2, which is the upper limit load to achieve the most unfavorable condition at the ULS. When designing structure for SLS, the reduction factor $\Psi_2 = 0.3$ is introduced to account for the fact that the larger mass is often favorable in design for comfort criteria. Several studies on variable load estimation demonstrated that this design load can over-estimate the actual live load in the building. A study by Kurent et al. [9] on a 7-storey CLT building estimated the actual live load as only 30% of the Eurocode-recommended value. Similarly, a study of an 18-storey glulam frame building by Tulebekova et al. [10] estimated a lower value of the actual live load: less than 30% of the Eurocode recommendations.

2.2 Stiffness

The estimation of the stiffness properties of TTBs could be particularly challenging since it depends on connections and structural elements stiffness, as on the possible contribution from non-structural elements which was observed to be important in timber buildings.

For glulam frame building, the impact of variability of connection stiffness on the modal properties of the building was studied by carrying out a parametric analysis by Utne [11] Malo et al. [12] analysed the flexibility of the joints connecting beams and diagonals to the columns in the external truss for the Treet building highlighting that reducing the stiffness at the connections may also give an increase of the acceleration at the top of the structure. Tulebekova et al. [10], used model updating to demonstrate that the stiffness in the diagonal and beam connections as well as non-structural elements inside the building (partitions and facade) can influence the natural frequencies and the mode shapes of the glulam TTB. Reynolds et al. [13] demonstrated the considerable contribution of dowel-type connections on the structure global stiffness. Additionally, large-scale experimental study by Malo et al. [12] on the large glulam connections demonstrated that the stiffness of the connections is dependent on the group effect of dowels and the angle-to-grain.

Concerning CLT tall buildings, Tulebekova et al. [14] showed that variability in stiffness of connections between the CLT panels affect the natural frequencies and mode shapes of the building. Reynolds et al. [15] highlighted the role of non-structural elements on the modal properties for a 7-story CLT building during construction. Similarly, Mugabo et al. [16] found that stiffness of non-structural elements (exterior walls and glazing) significantly affected the modal response of the building. Kurent et al. [9] showed that the in-plane shear modulus of CLT walls has significant impact on the modal properties by using the FRF-based model updating.

Concerning timber-only building with core, Zhao et al. [17] conducted a numerical analysis on serviceability of a 30-storey glulam frame building with CLT core and determined that increasing the timber grade for glulam and CLT members can reduce the peak acceleration of the building by16.7%. A similar trend was observed in their parametric study of a 30-storey CLT building [17].

2.3 Damping

Damping is a crucial factor in tall timber buildings subjected to wind-induced excitations since it reflects the ability to auto-decrease a dynamical response by dissipation of energy in each component [18]. High damping values for buildings are thus important to prevent the discomfort of the occupant in case of strong winds and for safety issues [19].Compared to other dynamic properties, such as mass and stiffness, damping does not relate to a single physical phenomenon, since it depends on different factors that make the building dissipate energy and reduce the motion (i.e. the friction and impact between different moving parts (structural damping in the code), the interaction between airflow and the external part of the building (aerodynamic damping), the molecular and atomic friction in elastically strained material (referred as material damping) [20].

Therefore, there is no theoretical calculation process which can be performed to determine damping values for an individual structure. Damping estimation of a structure relies on empirical data from previous full-scale monitoring campaigns of buildings with many design codes offering values based on type of construction and building materials [21].

The EC1-1-4 Annex F [4] does not provide damping values for TTBs and the older Swedish building code, Boverket BSV 97 [22], provided just two generic damping values, one for timber structures with mechanical connections (about 1.5%) and another (1.0%) for timber structures without mechanical connections [20].

Despite advances in technology allowing for better modeling and scaled wind tunnel tests, the only accurate method for obtaining the damping ratio of a structure is from full-scale monitoring of the completed structure [21]. Damping values can be identified from ambient vibration tests (AVT) in both small and long-term conditions as by forced vibration tests (FVT) with known excitation (i.e. mass inertia shakers). Damping values identified from short-term AVT are more uncertain than ones achieved by FVT, while long-term AVT covering a large amount of strong wind events can yield accurate statistical damping values [23, 24].

The techniques used to identify damping factors and other modal properties of a structure based on vibration data collected under operating conditions, without initial or known artificial excitation, are generally referred to as Operational Modal Analysis or Output-Only Modal Analysis (OMA). Methods of OMA can be broadly classified by two aspects: *i)* the domain of analysis (frequency or time), and *ii)* the statistical approach (Bayesian or non-Bayesian). The most widely used frequency domain methods include the Frequency Domain Decomposition (FDD) and the Enhanced Frequency Domain Decomposition (EFDD). Examples of time domain methods include the Stochastic Subspace Identification (SSI) and the Random Decrement Method (RDM). The most widespread Bayesian method is the Bayesian Modal Identification, while, the aforementioned FDD, EFDD and the Peak Picking (PP) belong to the non-Bayesian methods.

In the presence of long-term ambient vibration tests, the dependence of damping on the amplitude of excitation can be evaluated using OMA [20]. Using the RDM to determine damping at various levels of excitation, Reynolds et al. [15] observed that non-structural elements contributed to the global damping properties of a CLT building. Tulebekova et al. [25] applied the same method and observed an amplitude-dependent behaviour in glulam frame building.

A large experimental campaign by Feldmann et al. [26] measured and analysed 12 tall timber structures, such as residential buildings and towers ranging in height between 24 m and 100 m. While the findings demonstrated no variation in fundamental frequencies, the damping ratios were scattered, indicating nonlinear behaviour.

Several forced vibration tests were also conducted to investigate the amplitude-dependence of damping properties in timber buildings. Ellis and Bougard [27] performed one of the first full-scale forced vibration tests on a 6-storey timber frame building using electrodynamic shaker. They observed amplitude dependence of dynamic properties: the natural frequency decreased with amplitude of excitation, while the damping ratio increased. A 3-storey light-frame building was experimentally tested by Steiger et al.

[28], where they observed increase in damping ratio with increase of excitation amplitude. More recently, a study of Tulebekova et al., [25] on the Mjøstårnet, demonstrated the significant variation and amplitude-dependence of damping in the building using the free vibration tests. The variation of damping was in the range between 0.5% and 2.0%.

3 The Influence of the Lateral Load Resisting Systems

According to [19] the most widespread types of Lateral Load Resisting Systems (LLRS) for TTBs used to resist wind loads or seismic ground displacements are:

- Bracing or truss systems with GLT diagonal members, often placed in the external walls.
- CLT panels in exterior and/or interior walls (core).
- Hybrid-timber structures with, e.g., concrete cores or steel trusses.

The comparison of different lateral load-resisting systems under the same wind conditions is still poorly investigated. Academic studies of timber buildings often rely on existing designs, leading to frequent analysis of "SOM timber tower-like" structures with CLT cores and shear walls, as seen in studies by Bezabeh et al. [29], Abeysekera and Málaga-Chuquitaype [30] and Chen and Chui [31]. In contrast, designers with access to multiple design options often lack the time for in-depth dynamic response analyses.

Ussher et al. [32] evaluated the performance of three lateral resisting systems—frame, shear wall, and diagrid—for tall timber buildings, using a reference structure in Ås, Norway. Wind loads and serviceability limit states were modeled following Eurocode standards, with the buildings' responses assessed based on peak acceleration, top displacement, and inter-story drift. The diagrid system emerged as the superior option due to its high natural frequency and lower peak accelerations, though all systems demonstrated acceptable performance. However, the diagrid system poses challenges such as increased structural complexity, precise fabrication, and higher construction costs compared to shear wall and frame systems.

Angelucci et al. [33], investigated the dual timber-concrete systems with the aim to analyze the contribution of a reinforced concrete core coupled to different stability systems made of timber (namely a GLT frame, a CLT shear walled system and a GLT diagrid) for the effective control of lateral drifts in multi-storey buildings subjected to severe loading scenarios. The building models show high lateral stiffness in withstanding seismic and wind-induced loads. This result is mainly attributable to the introduction of the rigid concrete core, which nearly supplies the demand for shear and bending stresses alone.

Compared to a typical cross laminated core, the concrete tube results in a stiffness increment of 68% for the frame variant, 45% for the wall variant and 23% for the diagrid variant. Therefore, the serviceability requirements, both in terms of top displacements and inter-story drifts, are inherently satisfied and kept well below the prescriptive limitations. The results confirm the excellent behavior of the diagrid systems, for which any variations of the inner core have almost insignificant impact on the global building performance.

The previous works indicate that the sizing of timber members is primarily driven by stiffness requirements, typically assessed through monitoring lateral sway, while strength demands are generally considered to be inherently satisfied. Furthermore, they highlight that both strength and stiffness demands can hold equal importance during the sizing of dual systems.

4 Serviceability Limit State Design for Wind-Induced Comfort in Tall Timber Buildings

The objective of building design for serviceability limit state is to ensure the comfort of the occupants by limiting excessive horizontal deformations and accelerations in case of wind-induced vibrations. Excessive swaying of the buildings can lead to significant discomfort for the occupants, where they can experience nausea, dizziness and fear [34].

Currently, there are no generally accepted international standards for the wind-induced comfort criteria in tall buildings. The common practice is to assess the comfort criteria on the human perception of horizontal acceleration and displacement. The acceleration limits vary between the standards, countries, and building occupancy types. The standardized base curves limiting the peak and root mean square of acceleration can be found in ISO 10137:2008 [35] and ISO 6897:2009 [36] with the former being recommended by the Eurocodes. ISO 10137:2008 [35] provides recommendations for a 1-year peak acceleration limit for buildings and offices as a function of the fundamental frequency.

The estimation in standards of wind-induced acceleration in TTBs is dependent on both dynamic properties of the building and the properties of the incoming wind. The estimation of wind-induced acceleration for comfort criteria varies around the world. In Europe, the Eurocode on wind load calculation [4] provides two methods for estimating peak acceleration in buildings for human comfort: Annex B and Annex C. Sweden has its own comfort criteria guidelines, which are presented in the Swedish National Annex [22]. In North America, the ASCE 7-2016 [37] standard is used for wind-induced acceleration estimation. While the procedure may vary between the codes, the fundamental approach is similar. The calculation wind loading on the structure is usually simplified as a conservative static load and the subsequent building response is evaluated with a simplified single-degree-of-freedom system. Additionally, the gust effect and the turbulence are accounted for in the calculations. Landel [19] conducted a comparison between different standards for the estimation of a long-wind acceleration and found that the resulting accelerations varied greatly between each other, demonstrating that different assumptions on the wind load and building properties lead to a large scatter and increased design uncertainty of wind-induced response of buildings.

Alternative methods to estimate the wind-induced dynamic response of the building exist. The finite element approach using computational fluid dynamics (CFD) can be performed in the frequency domain using wind spectral density function or time domain where the structure response is assessed using time series from the wind measurements. Another method for assessing the wind response is to conduct physical small-scale testing in the wind tunnel. The method can provide detailed results but requires significant planning and execution effort.

5 Performance-Based Wind Design

The wind load design is mainly considering the first mode vibration and serviceability limit state (e.g., cladding failure, occupant comfort) [38–40]. However, the previously mentioned code-based design approaches have certain limitations, including: (i) the predominantly descriptive nature of the procedures, (ii) the neglection of the plastic capacity of structural systems which results in a strictly linear wind analysis [29], (iii) insufficient consideration of the role of non-structural elements, which are highly susceptible to damage under extreme drifts of the lateral load-resisting system (LLRS), with such drifts often being difficult to predict, (iv) the lack of a clear evaluation of the risk related to the loss mechanism for non-structural elements [41].

Rather than being an alternative to prescription-based codes, Performance Based Design (PBWD) is an approach that does not exclude prescriptions but is aimed directly at the achievement of well specified performance objectives and/or their optimization [42].

In the last decades, many research were carried out with the aim to develop a PBWD framework that takes into consideration various sources of uncertainties and adopts structural reliability to predict and design for the performance of timber buildings and investigate the performance under serviceability conditions [38, 42, 43] and Bezabeh et al. [29, 39, 44, 45], Micheli et al. [46], specifically for timber structures.

Those works led to account for the PBWD procedure as alternative approach for wind design by some standards such as ASCE 7-22 [37] and Korean Design Standard [47], as to publish one of the few guidelines that provide a detailed procedures to utilize PBWD: the pre-standard for PBWD by ASCE/SEI [48].

This latter is one of the few available guidelines available for the design process of PBWD of the Main Wind Force Resisting System (MWFRS) and the building envelope. It has defined basic objectives for a system performance when subjected to a wide-ranging hazard intensity and the clear incorporation of acceptance criteria related to the envelope system performance. The main steps are summarised in Fig. 1.

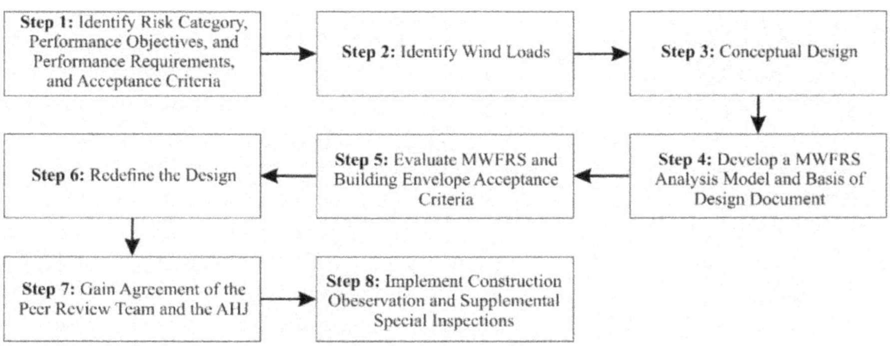

Fig. 1. PBWD procedure according to ASCE/SEI 2023.

Three performance objectives (POs) are specified in the ASCE/SEI Prestandard [48]: occupant comfort (OC), operational (OP), and continuous occupancy (CO). For the latter two POs, the hazard level is defined in terms of return periods based on the building risk category. The ASCE/SEI Prestandard [48] states that the MWFRS of a building shall remain elastic for OC and OP but allows inelastic deformation in specific elements or components of MWFRSs for the CO PO. To ensure that the POs are met, guidelines for evaluating the performance of a structure at global and component levels are also provided. Globally, acceleration, roof drift, residual drift, and deformation damage index (DDI) limits are provided to protect non-structural elements. At a component level, demand-to-capacity ratio limits for force- and deformation-controlled elements are given. A limit on the number of excursions beyond 1.5 times the yield capacity of deformation-controlled elements is also provided to prevent low-cycle fatigue failure [49, 50].

References

1. Foster, R.M., Reynolds, T.P.S., Ramage, M.H.: Proposal for defining a tall timber building. J. Struct. Eng. **142**(12) (2016). https://doi.org/10.1061/(asce)st.1943-541x.0001615
2. Abrahamsen, R., et al.: Dynamic response of tall timber buildings under service load - results from the dynattb research program. In: 13th World Conference on Timber Engineering, WCTE 2023, vol. 5, pp. 2907–2914 (2023). https://doi.org/10.52202/069179-0380
3. I. O. for S.- ISO: ISO 10137:2008, Basis for Design of Structures—Serviceability of Buildings and Walkways against Vibration (2008)
4. E. C. F. S. CEN: EN 1991-1-4: Actions on structures – Part 1–4: General Actions – Wind actions (2002)
5. Lazzarini, E., Frison, G., Trutalli, D., Marchi, L., Scotta, R.: Comfort assessment of high-rise timber buildings exposed to wind-induced vibrations. Struct. Des. Tall Spec. Build. **30**(12) (2021). https://doi.org/10.1002/tal.1882
6. Bjertnes, M.A., Malo, K.A.: Wind induced motion of "Treet"- a 14-storey residential building in Norway. In: *Proceedings of World Conference on Timber Engineering* (2014)
7. He, Y., Liu, Y., Wu, M., Fu, J., He, Y.: Amplitude dependence of natural frequency and damping ratio for 5 supertall buildings with moderate-to-strong typhoon-induced vibrations. J. Build. Eng. **78** (2023). https://doi.org/10.1016/j.jobe.2023.107589
8. Edskär, I., Lidelöw, H.: Dynamic properties of cross-laminated timber and timber truss building systems. Eng. Struct. **186**, 525–535 (2019). https://doi.org/10.1016/j.engstruct.2019.01.136
9. Kurent, B., Brank, B., Ao, W.K.: Model updating of seven-storey cross-laminated timber building designed on frequency-response-functions-based modal testing. Struct. Infrastruct. Eng. **19**, 1–19 (2021). https://doi.org/10.1080/15732479.2021.1931893
10. Tulebekova, S., Malo, K.A., Rønnquist, A., Nåvik, P.: Modeling stiffness of connections and non-structural elements for dynamic response of taller glulam timber frame buildings. Eng. Struct. **261** (2022). https://doi.org/10.1016/j.engstruct.2022.114209
11. Utne, I.: Numerical Models for Dynamic Properties of a 14 Storey Timber Building. NTNU Norwegian University of Science and Technology (2013)
12. Malo, K.A., Abrahamsen, R.B., Bjertnæs, M.A.: Some structural design issues of the 14-storey timber framed building "Treet" in Norway. Eur. J. Wood Wood Prod. **74**(3), 407–424 (2016). https://doi.org/10.1007/s00107-016-1022-5
13. Reynolds, T., Harris, R., Chang, W.S.: Stiffness of dowel-type timber connections under pre-yield oscillating loads. Eng. Struct. **65**, 21–29 (2014). https://doi.org/10.1016/j.engstruct.2014.01.024

14. Tulebekova, S., Malo, K.A., Rønnquist, A.: Dynamic identification and model calibration of connection stiffness in multi-storey cross-laminated timber buildings. J. Build. Eng. **72** (2023). https://doi.org/10.1016/j.jobe.2023.106607
15. Reynolds, T., Harris, R., Chang, W.S., Bregulla, J., Bawcombe, J.: Ambient vibration tests of a cross-laminated timber building. Proc. Inst. Civ. Eng. **168**(3), 121–131 (2015). https://doi.org/10.1680/coma.14.00047
16. Mugabo, I., Barbosa, A.R., Riggio, M.: Dynamiccharacterization and vibration analysis of a four-story mass timberbuilding. Front. Built Environ. **5**(86) (2019). https://doi.org/10.3389/fbuil.2019.00086
17. Zhao, X., Zhang, B., Kilpatrick, T., Sanderson, I., Liu, D.: Numerical analysis on global serviceability behaviours of tall glulam frame buildings to the Eurocodes and UK National Annexes. J. Civ. Eng. Constr. **10**(3), 109–122 (2021). https://doi.org/10.32732/jcec.2021.10.3.109
18. Reynolds, T., Chang, W., Harris, R.: Wind-Induced Vibration of Tall Timber Buildings. In: IABSE-IASS 2011: Taller, Longer, Lighter (2011) Available: http://opus.bath.ac.uk/. This version is made available in accordance with publisher policies. See http://opus.bath.ac.uk/forusagepolicie
19. Landel, P.: Wind-Induced Vibrations in Tall Timber Buildings: Design Standards, Experimental and Numerical Modal Analyses. Linnaeus University (2022)
20. Tamura, Y.: Amplitude dependency of damping in buildings and critical tip drift ratio. Int. J. High-Rise Build. **1**(1), 1–13 (2012). https://doi.org/10.1061/41000(315)39
21. Zhou, K., Li, Q.S.: Reliability analysis of damping estimation by random decrement technique for high-rise buildings. Earthq. Eng. Struct. Dyn. **50**(5), 1251–1270 (2021). https://doi.org/10.1002/eqe.3396
22. Boverket: *Boverkets handbok om snö- och vindlast, utgåva 2*, vol. 1, p. 1. Boverket (1997)
23. Magalhães, F., Brincker, R., Cunha, Á.: Damping estimation using free decays and ambient vibration tests. In: Proceedings of the 2nd International Operational Modal Analysis Conference, IOMAC 2007 (2007)
24. Macdonald, J.H.G., Daniell, W.E.: Variation of modal parameters of a cable-stayed bridge identified from ambient vibration measurements and FE modelling. Eng. Struct. **27**(13), 1916–1930 (2005). https://doi.org/10.1016/j.engstruct.2005.06.007
25. Tulebekova, S., Ao, W.K., Pavic, A., Malo, K.A., Rønnquist, A.: Identification of modal properties of a tall glue-laminated timber frame building under long-term ambient vibrations and forced vibrations. J. Struct. Eng. **150**(10) (2024). https://doi.org/10.1061/jsendh.steng-12810
26. Feldmann, A., et al.: Dynamic properties of tall timber structures under wind-induced vibration. In: WCTE 2016 - World Conference on Timber Engineering (2016)
27. Ellis, B.R., Bougard, A.J.: Dynamic testing and stiffness evaluation of a six-storey timber framed building during construction. Eng. Struct. **23**(10), 1232–1242 (2001). https://doi.org/10.1016/S0141-0296(01)00033-5
28. Steiger, R., Feltrin, G., Weber, F., Nerbano, S., Motavalli, M.: Experimental modal analysis of a multi-storey light-frame timber building. Bull. Earthq. Eng. **15**(8), 3265–3291 (2017). https://doi.org/10.1007/s10518-015-9828-9
29. Bezabeh, M.A., Bitsuamlak, G.T., Tesfamariam, S., Bezabeh, M., Bitsuamlak, G., Tesfamariam, S.: Performance-based wind design of tall buildings: concepts, frameworks, and opportunities. Wind Struct. **31**(2), 103–142 (2020)
30. Abeysekera, I., Málaga-Chuquitaype, C.: Dynamic response of tall timber buildings. In: *SECED 2015 Conference: Earthquake Risk and Engineering towards a Resilient World* (2015)
31. Chen, Z., Chui, Y.H.: Lateral load-resisting system using mass timber panel for high-rise buildings. Front. Built Environ. **3** (2017). https://doi.org/10.3389/fbuil.2017.00040

32. Ussher, E., Aloisio, A., Rathy, S.: Effect of lateral resisting systems on the wind-induced serviceability response of tall timber buildings. Case Stud. Constr. Mater. **19** (2023). https://doi.org/10.1016/j.cscm.2023.e02540
33. Angelucci, G., Mollaioli, F., Molle, M., Paris, S.: Performance assessment of timber high-rise buildings: structural and technological considerations. Open Constr. Build. Technol. J. **16**(1) (2022). https://doi.org/10.2174/18748368-v16-e2206270
34. Walton, D., Lamb, S., Kwok, K.C.S.: A review of two theories of motion sickness and their implications for tall building motion sway. Wind Struct. Int. J. **14**(6), 499–515 (2011). https://doi.org/10.12989/was.2011.14.6.499
35. International Organization for Standardization (ISO): ISO 10137:2007 - Bases for Design of Structures - Servicabiliy of Buildings and Walkways Against Vibrations (2007)
36. I. O. for S.- ISO: ISO 6897:1984, Guidelines for the evaluation of the response of occupants of fixed structures, especially buildings and offshore structures, to low-frequency horizontal motion (0.063 to 1.0 Hz) (1984)
37. A. S. of C. E. ASCE/SEI: Minimum design loads and associated criteria for buildings and other structures'. In: Minimum Design Loads and Associated Criteria for Buildings and Other Structures. Reston, VA, pp. 1–889 (2017). https://doi.org/10.1061/9780784414248
38. Ouyang, Z., Spence, S.M.J.: Performance-based wind-induced structural and envelope damage assessment of engineered buildings through nonlinear dynamic analysis. J. Wind Eng. Ind. Aerodyn. **208**, 104452 (2021). https://doi.org/10.1016/j.jweia.2020.104452
39. Bezabeh, M.A., Bitsuamlak, G.T., Popovski, M., Tesfamariam, S.: Probabilistic serviceability-performance assessment of tall mass-timber buildings subjected to stochastic wind loads. In: WCTE 2018, World Conference on Timber Engineering, 2018-January (2018)
40. Bernardini, E., Spence, S.M.J., Gioffrè, M., Kareem, A.: A reliability approach for the wind-induced response assessment of tall buildings using the high frequency force balance. In: Proceeding of Seventh International Colloquium on Bluff Body Aerodynamics & Applications (BBAA), pp. 825–834 (2012)
41. Abdelwahab, M., Ghazal, T., Nadeem, K., Aboshosha, H., Elshaer, A.: Performance-based wind design for tall buildings: review and comparative study. J. Build. Eng. **68**, 106103 (2023). https://doi.org/10.1016/j.jobe.2023.106103
42. Augusti, G., Paulotto, C., Ciampoli, M.: Some proposals for a first step towards a performance based wind engineering. Artic. Int. J. **10**(10), 1 (2004) Available: https://www.researchgate.net/publication/228826913
43. Spence, S.M.J., Gioffrè, M., Kareem, A.: An efficient framework for the reliability-based design optimization of large-scale uncertain and stochastic linear systems. Prob. Eng. Mech. **44**, 174–182 (2016). https://doi.org/10.1016/j.probengmech.2015.09.014
44. Bezabeh, M.A., Gairola, A., Bitsuamlak, G.T., Popovski, M., Tesfamariam, S.: Structural performance of multi-story mass-timber buildings under tornado-like wind field. Eng. Struct. **177**, 519–539 (2018). https://doi.org/10.1016/j.engstruct.2018.07.079
45. Bezabeh, M., Bitsuamlak, G., Solomon Tesfamariam, P.M.: Dynamic response of tall buildings to wind loads. J. Struct. Eng. **146**(10) (2020)
46. Micheli, L., Alipour, A., Laflamme, S., Sarkar, P.: Performance-based design with life-cycle cost assessment for damping systems integrated in wind excited tall buildings. Eng. Struct. **195**, 438–451 (2019). https://doi.org/10.1016/j.engstruct.2019.04.009
47. I. and T. of K. Ministry of Land: Korean design standard. [In Korean.] KDS 41-22
48. American Society of Civil Engineers ASCE/SEI (ASCE/Structural Engineering Institute): Pre-standard for Performance-Based Wind Design – V1.1. Reston, VA, (2023)

49. N. Kelemu, M. Bezabeh, C. Dickof, and M. Shahnewaz, Across-wind and torsional response of tall mass timber buildings: semi-empirical equations for structural design. Canadian Society for Civil Engineering (CSCE), 2024
50. Berile, N.K., Bezabeh, M.A.: Performance-based wind design of tall mass timber buildings with coupled post-tensioned cross-laminated timber shear walls. J. Wind Eng. Ind. Aerodyn. **257** (2025). https://doi.org/10.1016/j.jweia.2024.105981

Vibration-Based Wind Design Provisions for Tall Timber Buildings

Angelo Aloisio[1,2(✉)], Matiyas A. Bezabeh[3], Dag Pasquale Pasca[4], Marco Martino Rosso[6], Pier Francesco Giordano[5], Thomas Reynolds[7], Chiara Bedon[8], Kemal Hacıefendioğlu[9], Saule Tulebekova[10], and Blaž Kurent[11]

[1] Department of Civil, Construction-Architectural and Environmental Engineering, Università degli Studi dell'Aquila, L'Aquila, Italy
angelo.aloisio@univaq.it

[2] Faculty of Science and Technology, Norwegian University of Life Sciences, Ås, Norway

[3] Department of Civil Engineering, McGill University, Montreal, Quebec, Canada

[4] Norsk Treteknisk Institutt (Norwegian Institute of Wood Technology), Oslo, Norway

[5] Department of Architecture, Built Environment and Construction Engineering, Politecnico di Milano, Milan, Italy

[6] Department of Structural, Geotechnical and Building Engineering, Politecnico di Torino, Turin, Italy

[7] School of Engineering, University of Edinburgh, Edinburgh, UK

[8] Department of Engineering and Architecture, Università degli Studi di Trieste, Trieste, Italy

[9] Department of Civil Engineering, Faculty of Engineering, Karadeniz Technical University, Trabzon, Türkiye

[10] Department of Structural Engineering, Norwegian University of Science and Technology, Trondheim, Norway

[11] Faculty of Civil and Geodetic Engineering, University of Ljubljana, Ljubljana, Slovenia

1 Introduction

It is well acknowledged that as the height of a building increases, wind forces tend to become the controlling design loads. In tall, slender and flexible structures, wind-induced vibration serviceability is more critical than in low-rise buildings, where strength is usually the governing design criterion. Using engineered timber panels, tall mass timber buildings have reached heights comparable to those of mid-rise concrete and steel buildings. Therefore, wind-induced vibration needs attention even in mid-rise buildings when timber is used as a structural material for lateral load resistance. It has been shown that in timber buildings as low as seven stories wind-induced vibrations can be a governing design criterion [1]. Due to the relatively high strength-to-weight ratio of wood, timber buildings tend to be lighter and more flexible compared to their reinforced concrete counterparts. Typically, timber structures possess one-third the stiffness of concrete buildings and weigh one-fourth as much. This, in turn, means that they are more prone to

G. Fink et al. (Eds.): *Holistic Design of Taller Timber Buildings*, SPRTRCIENG, pp. 115–124, 2025.
https://doi.org/10.1007/978-3-032-02098-7_10

wind-induced vibrations that might cause discomfort to the occupants. Excessive wind-induced vibrations impair their long-term safety and functionality, which lowers their market value and may have an impact on the comfort and health of occupants. For these reasons, serviceability issues related to wind-induced vibration are considered a hindering factor for the design of tall timber buildings.

Compared to the ultimate limit states that define collapse or other forms of catastrophic failure of the structures, serviceability limit states define a loss of comfort or functional performance of the building. They establish a level of quality of the building and are often negotiated by the investor and the contractor to meet the expected functionality of the building. Its limitation may be set by the occupants' perception or technical requirements of the installed building elements or machines. Since serviceability limit states are not directly related to safety, professional committees and code bodies seem reluctant to codify serviceability issues rigidly. This reluctance is probably due to the different opinions on the purpose of building codes: protection for life safety or the establishment of complete minimum design standards. However, the fact that serviceability limit states are usually not codified should not diminish their importance. While safety is usually not an issue in this topic, the economic consequences can be substantial, especially when they are a part of contractual requirements set by the investors.

Whereas the research on the seismic response of timber buildings has progressed at a high pace, wind performance of mid and high-rise timber buildings has been studied more scarcely. There is no specific reported cases on wind-related human discomfort in tall timber buildings, mainly because there is no significant sample of such buildings for a meaningful statistical evaluation. Therefore, the wind modelling and the comfort criteria originate from past studies on other structural systems, see [2]. However, the lack of specific studies on timber buildings does not affect comfort criteria limits, which depend on the interaction between human perception and building dynamics and are not directly related to the building material.

2 Vibration Response Analysis Under Wind Excitation

Comfort criteria are based on the human perception of vibration, which is, due to substantial variation in individual physiological and psychological responses, very difficult to quantify. It is not fully understood how different quantities of motion (such as displacement, velocity, acceleration, and their derivatives) contribute to the perception of motion, however, the current comfort criteria are assessed through acceleration curves. These base curves represent either the threshold of perception of motion of some percentage of people or the limit for probable adverse comments by the occupants [3,4]. They depend on the vibration frequency of the structural system and the orientation of the vibration relative to the human body axes. Firstly, a performance indicator (e.g. running root-mean-square, peak acceleration) is computed based on the building's properties and the assumed wind loads. This indicator is then compared to the base perception curves to evaluate the system's performance. Some of the most relevant standards containing serviceability criteria related to motion perception of the occupants are hereby listed and reported in Fig. 1:

- ISO 10137 (2007) [3]. The standard recommends the serviceability criteria against the building's vibrations. Annex D provides a method for evaluating the human perception of wind-induced motions in buildings, giving acceptable limits in terms of peak acceleration, at the natural frequency, in the principal direction (along-wind, cross-wind, torsion) of the building. The peak acceleration should be calculated for wind speed with a one-year return period. The criteria is based on assuming peak = 3.5*RMS
- ISO 6897 (1984) [4]. The standard, based on [5]. covers building, whose frequency is 0.063 Hz-1 Hz, gives limit values of root mean square (RMS) accelerations for buildings used for general purposes. The limits are based on the wind levels with a five-year return period.
- AIJ (2004) [6]. The Japanese guidelines provide five curves: H-10, H-30, H-50, H-70 and H-90, where the number of each curve indicates the perception probability, expressed as a percentage of people who can perceive the given vibration level. The guidelines recommend no specific limit, which is to be decided by the owner and the designer.
- RWDI/BLWTL industry criterion [7]. A widely used industry practice, especially in North America, limits the 10-year return period peak acceleration to below 18 milli-g. While not formalized in a standard, this criterion is often adopted in practice.

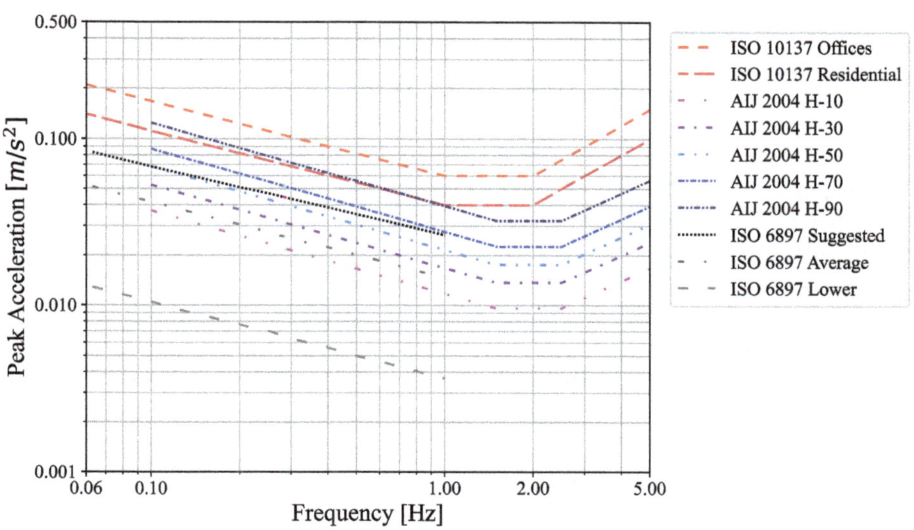

Fig. 1. Comparison between different perception curves (RMS values from ISO 6897 are multiplied by 3.5 to be comparable with peak acceleration limits).

The dynamic response of tall buildings to a wind load is a complex phenomenon, as the wind significantly varies in velocity and direction over time and space. To achieve simplification, building codes treat wind loads as quasi-static loads. The gust effect is also included as a turbulence factor added to the quasi-static component for high slender structures. Furthermore, the effect of the wind on a building will be affected by its exposure, the roughness of the terrain around it as well as the shape and height of the building. The gust load factor approach originates from the work of Davenport [8] and is a simplified frequency-domain method in which the wind load's standard deviation and the dynamic response's amplitude are multiplied by a peak factor to obtain the peak response. The standard deviation of the response and a peak factor are obtained based on random vibrations theory. Because of its simplicity, the gust factor method has received widespread acceptance worldwide and is employed in wind loading codes and standards in almost all major countries. According to the method in Annex B of EN 1991-1-4, for instance, the characteristic wind-induced acceleration, in the along-wind direction, for a point at height z, is calculated as:

$$a_{\text{peak}}(z) = K_p \, b_b \, h_b \, q_{z,\text{ref}} \, c_{fw} \, K \, \Phi_{1,x}(z) \, \frac{1}{M_1} \, R \tag{1}$$

$a_{\text{peak}}(z)$ Peak acceleration response (m/s^2) at height z
K_p Peak factor
b_b Building width
h_b Building height
$q_{z,\text{ref}}$ Reference wind load (N/m^2) at height z_{ref}
c_{fw} Force coefficient
K Dimensionless coefficient that scales turbulence effects
$\Phi_{1,x}(z)$ First-mode shape in the wind (cross-wind) direction, evaluated at height z
M_1 Modal mass of the first mode
R Resonance response factor

The peak response obtained can then be compared with peak acceleration criteria, such as ISO 10137 [9]. However, the response of buildings subjected to wind loads can be estimated more precisely with time-domain analysis using in-situ wind measurements or wind tunnel experiments, but also with methods combining spectral analysis and time-domain analysis [10–12]. The wind is described in a frequency-domain wind spectrum, and then a time series is generated from the spectrum. The dynamic response of the structure can be computed through numerical integration of the modal equation of motion. It should be mentioned that wind tunnel testing is recommended for mass-timber structures that are more than 20 stories tall, flexible, and have complex shapes, and are subject to wake buffeting from upwind buildings or channelling effects, and are susceptible to aeroelastic phenomena.

For calculating acceleration according to EN 1991-1-4 and checking it against the base curves, estimation of the natural frequency is necessary. Annex F of EN 1991-1-4 offers a simplified empirical equation for the fundamental frequency in Hertz

$$f_1 = 46/h \tag{2}$$

where h is the height of the building in meters. This equation can give a quick rough estimate, however, it can not be used for improving the design of the building, since the only variable is the height of the building. Reynolds et al. [13] compare the equation to ambient vibration test results for eleven multi-storey timber buildings in Europe, and find that it is a reasonable predictor of the fundamental frequency, although $f_1 = 55/h$ is a better fit to the measured data.

Designers of mid- and high-rise timber buildings often develop FE models to estimate their modal properties. Several research works have been carried out to find better modelling techniques for accurate estimation of modal properties of timber buildings (e.g. [14–19]). The main findings are that the nonstructural elements (such as façade, partition walls, plasterboards) can significantly increase the stiffness of the building (although this may not outweigh the effect of their added mass on the fundamental frequency) and that the type of construction (platform frame or balloon frame) importantly defines the stiffness properties of the building. The research is yet inconclusive about the influence of the connections in CLT buildings, however, it seems that soundproofing bedding under cross-laminated walls is an important factor in determining the stiffness of the connections. In glulam frame buildings, the stiffness and damping of the glulam connections can influence the modal properties [17,20]. It is important to note that modal properties of timber buildings vary seasonally, mainly due to environmental factors such as moisture content [21].

Most of the research work carried out during recent years deals with numerical studies where different structural systems (e.g. post-and-beams, CLT, etc.) are analysed throughout case studies and parametric analyses [12,22–32]. Specifically, Johansson et al. [22] studied the response of two archetypes (i.e. CLT buildings, glulam post-and-beam with a concrete shaft) to wind-induced acceleration according to EN 1991-1-4 and compared the results of the simplified analytical calculations with the limits of ISO 10137. They extended the analysis to a 48 m (16 storeys) high building where they studied the effect of doubling and tripling the mass, stiffness and damping ratio. In [23], the authors analysed 22-storey structures having an internal CLT core and a post-and-beam structure at the perimeter. They modelled the structure with FE software to get more reliable results regarding natural frequencies and mode shapes. Edskär and Lindelöw [24] performed a parametric analysis on the FE model of a CLT structure varying the footprint, height, damping ratio, wall stiffness, wall density and additional surface loads and studied the influence of these parameters on the dynamic response of the building. In [25], the authors extended the parametric analyses to a post and beam type of structure and compare it to the response of a

CLT one. In the same direction, Zhao et al. [26,27] performed parametric studies on the peak accelerations of CLT and glulam frames buildings, respectively. The structures were assumed to be located in Glasgow and have 30 storeys. Furthermore, the varied parameters were the timber material properties and building masses. In these papers, the accelerations were also calculated according to the Eurocodes, and the response is evaluated concerning ISO 10137 limits. Landel et al. [28] compared four procedures to evaluate the along wind accelerations on four existing tall timber buildings, highlighting high variations between the different codes. Cao and Stamatopoulos [12] delivered a numerical investigation on the response of moment-resisting frames subject to wind loads. They performed over one million simulations on planar frames where several parameters (e.g. floor height, floor number, beam stiffness) were varied. The response of the planar frames was evaluated with the simplified gust approach, but also performing time-domain analyses explicitly considering the time series of the wind force. A quite interesting finding was that the gust factor approach might underestimate the response for frames with up to 10–12 floors while overestimating the accelerations for frames with more than 10–12 storeys [12]. Bezabeh et al. [29] examined the dynamic response and serviceability performance of five case study tall mass-timber buildings varying in height (10-, 15-, 20-, 30-, and 40-story). Bezabeh et al. [10,11] proposed a probabilistic procedure to assess the serviceability performance of tall mass-timber buildings, applying the complete framework to a case study consisting of a 102-m tall building. The framework incorporated uncertainties at each step of the wind-loading chain. The design process consisted of a preliminary strength design using building code provisions. Then serviceability checks were performed using wind loads obtained from aerodynamic wind tunnel tests. Finally, the detailed probabilistic performance assessment was performed with structural reliability analysis using Monte Carlo sampling to propagate the uncertainties through the wind loading chain. The results from reliability analysis were used to develop fragility curves for wind vulnerability estimations. Lazzarini et al. [30] studied the comfort assessment of the 18-storey "Mjøstårnet" building in Norway, applying computational fluid dynamic (CFD) analyses to simulate the wind flow around the building. The results of the CFD analysis are then used to extrapolate detailed pressure data, which is applied to a generalized model and a reduced model to obtain accurate evaluations of wind-induced motions. Wind-induced vibrations in the across-wind directions were particularly strong, which is not captured by the current standard, indicating the importance of applying fluid dynamic analyses. Kurent et al. [1] performed a serviceability check of two timber and one hybrid timber-concrete buildings and showed that checking only the first mode of vibration is not enough, since the second mode can sometimes be more critical.

3 Conclusions

Wind-induced vibration governs the serviceability design of mid-and high-rise timber buildings at lower heights than it does for comparable concrete structures because the high strength-to-weight ratio of wood leads to lower mass and greater flexibility. Although comfort criteria such as ISO 10137, ISO 6897 and the AIJ guidelines are material-agnostic, their verification still hinges on an accurate prediction of peak accelerations; the review presented here shows that the traditional gust factor approach embedded in EN 1991-1-4 can be adequate only when its inputs, the first two natural frequencies, modal masses and damping ratios, are themselves reliable. Empirical height-frequency formulas (e.g. $f_1 = 46/h$) have yet to be validated for timber and therefore carry large uncertainty, while finite-element predictions remain sensitive to the stiffness contribution of non-structural components, connection bedding and seasonal moisture variation. Case-study simulations confirm that doubling mass or stiffness can halve the peak acceleration, but also reveal that higher modes, particularly the second bending mode, may control comfort more than the fundamental one [1]. Parametric studies further indicate that the gust-factor method may underestimate accelerations for 10- to 12-storey frames and overestimate them for taller systems, suggesting that time-domain analyses or hybrid spectral/time techniques should become routine for slender timber towers [12]. The scarcity of full-scale monitoring data remains the principal obstacle to refining both simplified and sophisticated models; without such data it is impossible to quantify the true damping achievable through façade friction, interior fit-out or supplemental devices, or to calibrate probabilistic fragility curves that capture the combined effect of wind variability and modelling uncertainty.

In practical terms, designers should (i) develop FE models that include sheathing, partitions and realistic connection stiffness, (ii) evaluate at least the first two lateral modes in both along- and cross-wind directions, (iii) supplement code-based gust calculations with time-domain checks when predicted peak accelerations approach 80% of the relevant perception limits, iv) consider wind tunnel testing for mass-timber buildings that are more than 20 stories tall, flexible, and have complex shapes, and are subject to wake buffeting from upwind buildings or channelling effects, and are susceptible to aeroelastic phenomena, and (iv) consider tuned mass or viscous dampers early in the design when mass or stiffness increases conflict with sustainability targets. On the research side, priority should be given to long-term dynamic monitoring of completed mass-timber towers coupled with in-situ wind measurements, to wind-tunnel databases that cover the distinctive geometries of timber bracing and cores, and to probabilistic frameworks that propagate material, connection, and environmental uncertainties through to serviceability risk metrics.

Acknowledgments. The CA20139 - Holistic design of taller timber buildings (HELEN) is gratefully acknowledged for networking support.

References

1. Kurent, B., Friedman, N., Brank, B.: Modelling of multi-storey cross-laminated timber buildings for vibration serviceability. Buildings **14**, 689 (2024)
2. Lamb, S., Kwok, K.C., Walton, D.: A longitudinal field study of the effects of wind-induced building motion on occupant wellbeing and work performance. J. Wind Eng. Ind. Aerodyn. **133**, 39–51 (2014)
3. I. O. for Standardization, Bases for Design of Structures–serviceability of Buildings and Walkways Against Vibrations. ISO (2007)
4. I. 6897, Guidelines for the evaluation of the response of occupants of fixed structures, especially buildings and off-shore structures, to low-frequency horizontal motion (0Âů 063 to 1hz) (1984)
5. Irwin, A.: Perception, comfort and performance criteria for human beings exposed to whole body pure yaw vibration and vibration containing yaw and translational components. J. Sound Vib. **76**(4), 481–497 (1981)
6. Aij, A.: Guidelines for the evaluation of habitability to building vibration, AIJ (2004)
7. RWDI, Structural performance: wind-induced motion of tall buildings (2017). RWDI Fact Sheet
8. Davenport, A.: How can we simplify and generalize wind loads? J. Wind Eng. Ind. Aerodyn. **54**, 657–669 (1995)
9. Kareem, A.: Synthesis of fluctuating along wind loads on buildings. J. Eng. Mech. **112**(1), 121–125 (1986)
10. Bezabeh, M., Bitsuamlak, G., Popovski, M., Tesfamariam, S.: Probabilistic serviceability-performance assessment of tall mass-timber buildings subjected to stochastic wind loads: Part i - structural design and wind tunnel testing. J. Wind Eng. Ind. Aerodyn. **181**, 85–103 (2018)
11. Bezabeh, M., Bitsuamlak, G., Popovski, M., Tesfamariam, S.: Probabilistic serviceability-performance assessment of tall mass-timber buildings subjected to stochastic wind loads: Part ii - structural reliability analysis. J. Wind Eng. Ind. Aerodyn. **181**, 112–125 (2018)
12. Cao, A., Stamatopoulos, H.: A theoretical study of the dynamic response of planar timber frames with semi-rigid moment-resisting connections subjected to wind loads, Engineering Structures, vol. 240 (2021)
13. Reynolds, T., Feldmann, A., Ramage, M., Chang, W., Harris, R., Dietsch, P.: Design parameters for lateral vibration of multi-storey timber buildings. In: International Network on Timber Engineering Research Proceedings (INTER 2016) (2016)
14. Kurent, B., Brank, B., Ao, W.: Model updating of seven-storey cross-laminated timber building designed on frequency-response-functions-based modal testing, Structure and Infrastructure Engineering, vol. 19, no. 2, pp. 178–196, (2023). cited By 12
15. Kurent, B., Friedman, N., Aloisio, A., Pasca, D.P., Tomasi, R., Brank, B.: Bayesian model updating of eight-storey CLT building using modal data. Probab. Eng. Mech. **19**(2), 178–196 (2024)
16. Kurent, B., Ao, W.K., Pavic, A., Pérez, F., Brank, B.: Modal testing and finite element model updating of full-scale hybrid timber-concrete building. Eng. Struct. **289**, 116250 (2023)
17. Tulebekova, S., Malo, K., RÃ, nnquist, A., Nåvik, P.: Modeling stiffness of connections and non-structural elements for dynamic response of taller glulam timber frame buildings, Engineering Structures, vol. 261 (2022)

18. Tulebekova, S., Malo, K.A., Rønnquist, A.: Dynamic identification and model calibration of connection stiffness in multi-storey cross-laminated timber buildings. J. Build. Eng. **72**, 106607 (2023)
19. Reynolds, T., Harris, R., Chang, W.-S., Bregulla, J., Bawcombe, J.: Ambient vibration tests of a cross-laminated timber building. Proc. Inst. Civil Eng. Constr. Mater. **168**(3), 121–131 (2015)
20. Tulebekova, S., Stamatopoulos, S., Malo, K.: A framework for the estimation of damping ratio of glued—laminated buildings by use of analysis in the time domain, Materials, vol. 18 (2025)
21. Larsson, C., Abdeljaber, O., Bolmsvik, Dorn, M.: Long-term analysis of the environmental effects on the global dynamic properties of a hybrid timber-concrete building, Engineering Structures, vol. 268 (2022)
22. Johansson, M., Linderholt, A., Bolmsvik, Å., Jarnerö, K., Olsson, J., Reynolds, T.: Building higher with light-weight timber structures-the effect of wind induced vibrations. In: 44th International Congress and Exposition on Noise Control Engineering, INTER-NOISE 2015 (2015)
23. Johansson, M., Linderholt, A., Jarnerö, K., Landel, P.: Tall timber buildings—a preliminary study of wind-induced vibrations of a 22-storey building, Proceedings of the WCTE (2016)
24. Edskar, I., Lidelow, H.: Wind-induced vibrations in timber buildings-parameter study of cross-laminated timber residential structures. Struct. Eng. Int. **27**(2), 205–216 (2017)
25. Edskär, I., Lidelöw, H.: Dynamic properties of cross-laminated timber and timber truss building systems. Eng. Struct. **186**, 525–535 (2019)
26. Zhao, X., Zhang, B., Kilpatrick, T., Sanderson, I.: Numerical analysis on global serviceability behaviours of tall clt buildings to the eurocodes and uk national annexes, Buildings, **11**(3) (2021)
27. Zhao, X., Zhang, B., Kilpatrick, T., Sanderson, I., Liu, D.: Numerical analysis on global serviceability behaviours of tall glulam frame buildings to the Eurocodes and UK national annexes. Journal of Civil Engineering and Construction **10**(3), 109–122 (2021)
28. Landel, P., Johansson, M., Linderholt, A.: "Comparative study of wind-induced accelerations in tall timber buildings according to four methods. In: WCTE 2021, World Conference on Timber Engineering, Santiago, Chile, 9-12 August (2021)
29. Bezabeh, M., Bitsuamlak, G., Popovski, M., Tesfamariam, S.: Dynamic response of tall mass-timber buildings to wind excitation. J. Struct. Eng. (United States) **146**(10) (2020)
30. Lazzarini, E., Frison, G., Trutalli, D., Marchi, L., Scotta, R.: Comfort assessment of high-rise timber buildings exposed to wind-induced vibrations, Struct. Design Tall Special Build. **30**(12) (2021)
31. Ussher, E., Aloisio, A., Rathy, S.: Effect of lateral resisting systems on the wind-induced serviceability response of tall timber buildings. Case Stud. Constr. Mater. **19**, e02540 (2023)
32. Reynolds, T., Casagrande, D., Tomasi, R.: Comparison of multi-storey cross-laminated timber and timber frame buildings by in situ modal analysis. Constr. Build. Mater. **102**, 1009–1017 (2016)

Performance of Taller Timber Buildings

Moisture Management During Construction

Matt Caldwell[1(✉)], Lucas Viereck[2], Patrik Aondio[3], and Nina Flexeder[4,5]

[1] Department of Architecture & Civil Engineering, Buro Happold, Bath, UK
matthew.caldwell@burohappold.com
[2] School of Engineering and Design, Technical University of Munich, Munich, Germany
[3] HBC Biberach University of Applied Sciences, Biberach, Germany
[4] Technical University of Munich, Munich, Germany
[5] Turner & Townsend, Munich, Germany

Abstract. Effective moisture management during construction is essential to achieve a successful, long-lived and sustainable outcome for timber structures. Several incidences of serious damage, originating in construction-stage moisture ingress or entrapment, have been documented. There are many aspects to consider and key decisions to be made throughout the design, planning and execution stages. Each of the key areas is identified and explored with best practice and guidance presented.

Keywords: Mass Timber · Wood · Glulam · GLT · Cross Laminated Timber · CLT · Moisture Content · Construction · Damage · Rot · Monitoring

1 Introduction: Why Moisture Management?

Timber building structures are getting larger, taller and more complex. This means a greater area of wooden surfaces and connections exposed to the outside environment during construction. The execution period, and thus duration of moisture exposure, also generally increases.

The result is a higher probability of the incidence and ingress of water. Since timber readily absorbs water, good planning and execution measures are necessary to avoid damage. Excess moisture exposure can lead to mould, fungal decay, dimensional changes, and reductions in structural strength and stiffness.

The basic principles of moisture control do not differ significantly between different forms of timber structures such as mass timber or timber framing. However, geometry and detailing can have a significant effect.

During construction, the risk that the timber elements are exposed to water is generally much higher than during the utilization phase for which they are ultimately designed. Construction-stage water can come from different sources, the most important in temperate climates, such as Europe, is precipitation in the form of rain. But other sources like snowmelt, condensation, water within fresh concrete or defects in plumbing should also be considered.

G. Fink et al. (Eds.): *Holistic Design of Taller Timber Buildings*, SPRTRCIENG, pp. 127–142, 2025.
https://doi.org/10.1007/978-3-032-02098-7_11

On-site measures for moisture control can require additional steps on-site as well as additional monetary resources, which should be accounted for during early-stage scheduling and the call for tenders. The costs of such measures are, however, small compared to the potential time and monetary costs of active drying or of rectifying moisture damage.

For example, a four-storey residential building supervised by one of the authors, has shown the protective covers made of tear-resistant mesh foil fabric, including all fastenings, anchoring, and weighting, cost around € 3.80/m^2 (price from 2022). If active drying was required to the CLT roof slab, the cost of temporary enclosure and drying could many hundreds of €/m^2. In the event of moisture damage to the timber, the cost of a repair can be several thousand €/m^2.

Moreover, design-stage moisture control measures also generally have beneficial side effects for the rest of the construction, such as a more robust design, a more efficient erection, as well as less moisture inside the building which is also good for other materials like masonry, gypsum boards and other finishes and fittings.

2 Moisture and Wood: Basics

Timber shows hygroscopic behavior by adapting its moisture content to the surrounding ambient conditions. The resulting timber moisture content is referred to as the equilibrium moisture content and can be estimated as a function of temperature and relative humidity. These models mathematically approximate the sorption history and thus the sorption hysteresis. As a more precise alternative to the well-known diagram by Keylwerth [1] for the drying of sawn timber, a more recent approach [2] provides models specific to wood species for calculating typical the equilibrium moisture content during construction and usage, see Fig. 1.

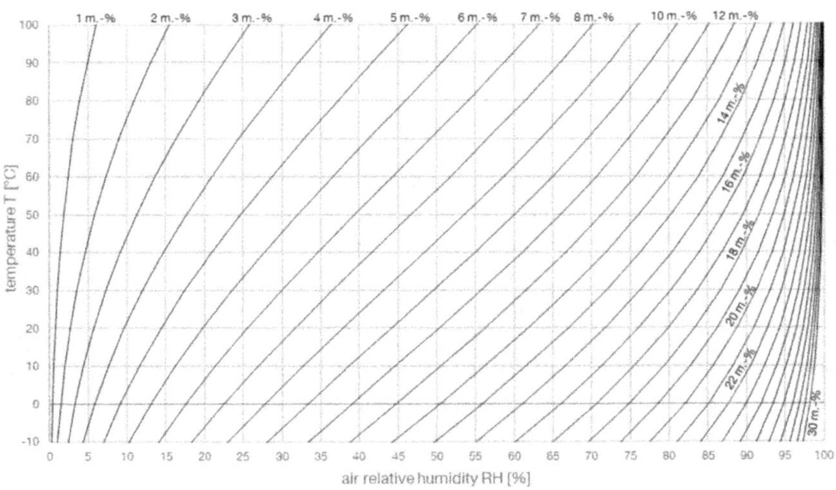

Fig. 1 Graph for estimating equivalent wood moisture content of Norway spruce (*Picea abies Karst.)* based on air temperature and relative humidity, developed by [3]

During production in the factory, transport, storage on the construction site, or during assembly, possible moisture gradients can arise, which can manifest themselves as moisture damage after the building has been put into operation. If the moisture content is too high, wood components can be severely damaged by moisture penetration and the associated deformation caused by swelling. In due course, it can also lead to rot, infestation by wood-destroying fungi, discoloration, infestation by dry rot, maceration, and mold growth. If a component is dried out soon enough after it has reached high moisture content, the swelling process is reversed, and the component can lose volume again as a result of shrinkage. However, if drying is too fast, differential moisture and shrinkage gradients can occur, leading to surface cracks, as statistical evaluations of damage cases show [3, 4].

In the range between 5% and 25% timber moisture content, the ratio of the swelling dimensions for spruce timber longitudinal: radial: tangential is approximately 0.01: 0.16: 0.32, which describes the percentage change in dimension per 1% change in timber moisture content. An average shrinkage and swelling dimension perpendicular to the fibre direction can be assumed for softwood, which is approximately 0.25% per 1% change in wood moisture content with unhindered swelling and shrinkage [5].

When the timber moisture content varies across the cross-section, deformations will occur under unrestrained conditions. Since wood products are typically kiln-dried, it can generally be assumed that the moisture content across the entire cross-sectional height is constant at the time of manufacture and supply. If moisture affects the top side of a wood cross-section, the resulting deformation will produce centerline strains and curvatures within the component. If a timber cross-section's deformation is constrained, e.g. by connectors like screws, nails or dowels, internal stresses will occur. The resulting internal forces can be very high and lead to the failure of the connectors and/or the timber local to them.

3 Organisation: Reducing Risks Through Management Processes

The suggested way to deal with the moisture during construction is the development of a project-specific Construction Stage Moisture Control Plan (CSMCP). The general requirements for a CSMCP will be stated in part 3 of the next generation of Eurocode 5 [6], however more detailed recommendations are found elsewhere [7, 8].

The construction stage starts with the transport of the timber elements and includes the storage on-site and the erection, as well as all necessary works after completion of the timber construction due to damages that were caused during the erection.

A comprehensive CSMCP should include technical procedures but also definition of the project, its functional requirements, the relevant stakeholders and clear allocation of responsibilities.

- Sections of the plan should include.
- Description of the project
- Definition of stakeholders and their various responsibilities
- Contact details of the responsible persons
- A description of the materials and products used
- A description of the relevant project-specific sources of moisture

- Identification of performance requirements of different timber elements
- Identification of high-risk locations within the construction
- A detailed description of the planned installation process
- Identification of failure criteria and triggers for actions
- Description of the chosen measures for the prevention of damage
- Measures for inspection, including moisture measurements
- A pre-defined system of recording and evidencing the full execution process

To maximise the success of the project execution, good communication at the construction site is crucial, specifically in cooperation with other construction trades who may lack the necessary experience with timber constructions. Moreover, irrespective of the CSMCP, everyone involved in the project must of course adhere to the relevant design standards.

A successful moisture-control outcome to the construction process depends on operational and organizational measures in addition to technical design solutions. There are multiple stakeholders who are required to contribute. The timing of their input will vary according to the design, procurement, and execution schedule. Typical example inputs by, and responsibilities of, the main project stakeholders are summarised below.

3.1 Client

- Carries out due diligence in appointing a consultant team which is suitably experienced and qualified to deliver the relevant scale and form of timber-structured project.
- Contributes to an unambiguous project brief. Especially regarding any performance requirements that have a direct effect on the moisture management of timber. For example, visual quality of exposed timber elements or considerations of asset and business protection which may exceed those of life-safety.
- Is active in reviewing the design and execution proposals and makes decisions based on the advice of the consultant team. An example would be considering whether a Contractor submitting a lower Tender price offer is doing so by omitting certain construction-stage moisture mitigations or quality assurance controls.

3.2 Project Manager

- Should have experience of delivering timber constructions of the relevant scale and performance requirements.
- Advises the Client on each of their responsibilities, including those above.
- Gives further assessment of the design and construction programme and critically reviews whether the right timescales are being targeted. Specific examples of poor practice include design-periods that are too short to allow proper review, construction schedules that are either too short to ensure good workmanship, or incorrectly phased such that timber is exposed to moisture for an excessively long period.
- Coordinates a clear and unambiguous allocation of design responsibility and identifies any scope-gaps or missing areas of expertise.

3.3 Principal Designer

- Should have experience of delivering timber constructions of the relevant scale and performance requirements
- Provides further review of the experience and qualification of the consultant team in relation to moisture control
- Ensures that the design process is coordinated and joined-up, such that each specialist consultant makes decisions cognizant of potential effects on moisture control
- Makes holistic design decisions, balancing and compromising between requirements of different consultants and stakeholders, whilst maintaining the required moisture control measures

3.4 Consultant Design Team

- Should have experience of delivering timber constructions of the relevant scale and performance requirements.
- Is honest about limitations of any one consultants' knowledge and experience and alerts the Principal Designer and Project Manager of any additional expertise that may be required.
- Considers and enacts advice from all consultants in order that effects on moisture management are evaluated as part of all relevant decision-making.
- Critically reviews and re-visits design assumptions and decisions that may lead to potentially high hazard outcomes.

3.5 Principal Contractor

- Honestly reviews and communicates their own expertise to deliver a project with the relevant scale and type of structural timber
- Appoints a suitably competent and experienced timber sub-contractor
- Critically and rigorously reviews the design and in particular from the specific perspective of moisture control during execution
- Carefully plans the overall execution schedule with moisture-control of the timber as a key factor, including operations immediately before and after the execution of the timber
- Ensures that moisture-control plans cover the full extent of execution, from manufacture to hand-over. Including hand-over between subcontractors and trades
- Ensures that moisture-control plans are enacted, reviewed, and modified where required.
- Ensures that moisture control data and processes are recorded and evidenced, covering the full period of execution, up to handover

3.6 Timber Subcontractor

- Honestly reviews their own expertise to deliver the particular project.
- Critically and rigorously reviews the design from the specific perspective of moisture control during execution

- Carefully plans the timber execution schedule with moisture-control of the timber as a key factor. Including consideration of operations immediately before and after the execution of the timber
- Ensures that moisture-control plans cover the full extent of execution, from manufacture to hand-over
- Ensures that moisture-control plans are enacted, reviewed, and modified where required
- Ensures that moisture control data and processes are recorded and communicated

PrEN1995 [6] invokes the requirement for preparing a 'Moisture Control Plan', but it does not expand in specific terms. The UK National Structural Timber Specification (NSTS) [9] also contains reference to a Moisture Control Plan and gives an outline of its requirements, without allocating responsibility precisely nor the document's timing and ongoing development.

Traditional practice has tended to consider planning for moisture control during construction to be the sole responsibility of the contractor. In this way, moisture control plans have often only been drawn up after award of the contract, at which stage the majority of the design decisions have already been made. However, certain design decisions have a very strong influence on whether the timber structure can be practically executed without moisture issues arising. They therefore need to be made in consideration of the execution stage and thus the CSMCP needs to be drafted during the design stages.

In the UK there have now been several incidences of significant disputes and legal claims in relation to moisture damage to mass timber buildings, these can be found in the construction press [10–15] and court records [16, 17]. It is noteworthy that Consultants have been the subject of the claims as well as Contractors. In some of these cases, the court specifically deemed that the Consultant had not properly exercised their duties to design appropriately for timber and moisture. In other cases, claims have been brought against the main contractor for failure to safeguard the structural timber elements after the timber subcontractor had handed over.

4 Planning: Reducing Hazards Through Building Design

Effective implementation of moisture control measures during construction requires careful planning and design from the outset. For large construction operations, it is advisable to use rapidly deployable coverings or durable EPDM sealing membranes designed for heavy-duty applications.

The internal structure should be enclosed as quickly as possible, ideally using the permanent façade and roof envelope. If this is not achievable, alternative coverings can be installed to act as a temporary enclosure. It is important to note that the permissible UV and weather exposure duration for any temporary covering is typically specified by the manufacturer and must be strictly adhered to.

Good moisture control measures require the following points:

- Early involvement of planners experienced in timber construction.
- High level of prefabrication
- Simple, rational geometry for the structure at large

- Connections that are efficient to assemble on site
- Clear construction stages
- Design & implementation of temporary water collection systems and outlets
- Placement of timber walls and columns on raised non-absorptive plinths
- Prevention of capillary moisture absorption, especially via end-grain
- Integration of expansion joints to accommodate moisture-related movements
- Vertical wet services distribution grouped in dedicated shafts
- Holistic moisture management strategy informed by project-specific risk analysis

Some construction products require special consideration in design and execution. For example, uniaxially laminated timber slabs, e.g. nail laminated timber or dowel laminated timber, are prone to significant swelling perpendicular to the grain and should not be installed without expansion/contraction joints. When uni-axially laminated timber planes undergo moisture movement, very significant forces develop if they are restrained. This can lead to crushing or splitting of the timber or damage to adjacent elements (Fig. 2).

Fig. 2 From left to right: temporarily sealed facade; heavy diffusion-tight EPDM membrane (sd-value >1400 m); lightweight, diffusion-open membrane (sd-value = 3.5 m) which is suitable for up to 12 weeks of direct weather exposure; Timber frame wall that was displaced by 5.2 cm due to a glued laminated timber slab with elevated moisture content.

Example The following section provides an example calculation of the forces that arise when an h = 180 mm thick glued uni-axially laminated timber roof plane (GL 24 h) has its moisture content increased by $\delta = 8\%$ on the top side, assuming a linear moisture gradient over the height and an average swelling and shrinkage rate of α_u 0.25%/% (Fig. 3).

Normal force [n] for restrained deformation:

(1)

Bending moment [m] for restrained deformation:

(2)

If a 1-metre-wide element is considered, the centre of gravity extension with unrestrained deformation can be calculated as follows.

(3)

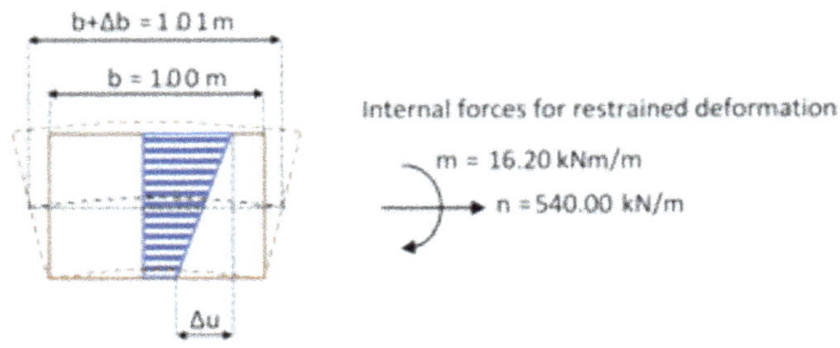

Fig. 3 Illustration of internal forces for restrained deformation

5 Manufacture, Transport and Storage: Protection of Timber Prior to Assembly

It is always relevant to consider and query the manufacture, transport and storage stages of the timber's supply chain. This is particularly the case if the timber has visual requirements, if the timber is travelling long distances or travelling through different climatic regions. Consideration should be given to the equilibrium moisture content of the timber and potential for condensation and mould growth inside packaging at all stages of the journey. Packaging and controls might need to be adapted in response, or expectations of surface finishes may need to be managed.

5.1 Manufacture

Control and monitoring of moisture content during manufacture is generally very good. Robust quality assurance procedures are required for the factory to meet codes and CE or equivalent marking. Therefore, the timber components should be at a reliable, certified moisture content when leaving the production facility.

5.2 Transport

Moisture risks during haulage are, in principle, simple to manage. Fabrication lots can be pre-determined and planned, and each wrapped individually in the factory. Use of curtain-side lorries is preferable to further reduce exposure to moisture and to mitigate any damage to wrapping. Panels are often loaded with any visible-quality faces orientated downwards (Fig. 4).

Where timber elements are to be transported over greater distances, they may be placed in shipping containers. These have potential to undergo large changes in temperature and humidity depending on where they are stacked within a ship or upon a train or lorry. It is in principle possible to integrate climate-control measures into containerized transport, however it comes with a significant cost increase.

5.3 Site Storage

The construction schedule should be planned, monitored, and adjusted to minimize the time timber is stored on site. Where timber is stored on site it should be protected from

Fig. 4 Left – pre-wrapped erection-lot of glulam beams on bespoke transport skid. Right – pre-wrapped erection-lot of CLT panels as delivered to site.

precipitation, rain-splash and condensation. It should also be ventilated to ensure that any water that contacts the timber can escape passively.

Timber should be separated on bearers to further facilitate ventilation and removal of water. Where the ground below or the shape of the elements leads to an inclined upper surface to any timber, bearers should be aligned down the slope as opposed to across it, to allow free passage of any liquid water (Fig. 5).

Fig. 5 Left – pre-wrapped erection-lots of timber elements stored on site, elevated and ventilated. Right – Pre-wrapped glulam beams stored on site, elevated and ventilated.

Particular attention should be made to condensation and humidity within any sheeted enclosures or non-breathable wrapping. Significant quantities of water and/or vapor can accumulate and impact the timber. The timber should have its moisture content checked at delivery and monitored during storage.

In certain climatic conditions there could be concern about excessively elevated temperatures and/or low humidity. This can lead to rapid moisture content reduction in the surface of mass timber elements, development of shrinkage differential to the centre of the element and consequent surface splitting. It can also lead to smaller-section elements attaining an excessively low moisture content throughout, leading to dimensional issues during installation and possible later dimensional changes relative to other building elements.

6 Erection: Reducing Risks During Site Execution

6.1 Atmospheric Moisture

The main external source of construction-stage moisture in temperate climates is likely to be liquid precipitation. However, snow-melt, frost, mist and high humidity can also be external sources, see Sect. 1.

Intercepting water and moisture before contact with the timber is an efficient form of protection. Mitigations include the following.

- Temporary scaffold roofs
- Sheeted scaffold enclosures
- Elevating bases of timber elements above foundation slab surfaces

Reduced dig or gravel margins adjacent to building perimeters (Figs. 6 and 7).

Fig. 6 Temporary sheeted weather protection to timber structure

Where moisture comes into direct contact with timber surfaces, mitigations are likely to focus on its efficient removal before it has time to penetrate.

Passive mitigations at the surface include surface-applied measures such as:

- Self-adhesive breathable membranes
- Application of hydrophobic compounds

Fig. 7 Glulam column (left) and CLT wall (right) elevated above concrete slab on galvanized and painted steel bases

Vapour-open protection layers allow passage of moisture out of the timber. This is useful for situations where the equilibrium moisture content of the timber on site is lower than at fabrication, or that excess moisture enters the timber by some other means, such as a penetration through, or flanking-path around the protective surface (Fig. 8).

Fig. 8 Left – Factory-applied tapes and membranes for end-grain protection. Right – Factory-applied self-adhesive membrane for CLT face protection.

Temporary vapour-closed layers are to be avoided except for short periods, as they tend to trap moisture against the timber surface or generate it through condensation. Vapour-closed layers can be effective where they are guaranteed to be applied to sufficiently dry timber and where there is no chance that water can track below it, i.e. that it is adhered directly to the timber surface. Examples are typically where one or more of the layers of the permanent roof waterproofing system are applied in the factory or immediately upon installation.

It is important to note that with significant temperature fluctuations between day and night, condensation can form on the underside of vapour-closed coverings. However, vapour-open membranes are generally more sensitive to mechanical impact and can quickly develop holes and damage, allowing water to penetrate.

Passive water removal via gravity could include laying the structural deck to a fall and minimising interruptions to the path of water such as inverts, valleys, upstands, and parapets. Water must be directed to planned locations for removal. These should be designed for the anticipated maximum flow rates, robustly detailed, and ideally piped, systems so that they are safe and can tolerate the actions of wind and site activities.

Active water removal methods include planned use of brushes, squeegees and wet-vac's. Where active removal is relied upon, consideration must be given to the required frequency of operation and how this is to be maintained over weekends and holidays (Fig. 9).

Fig. 9 Left – Fans used to promote air movement and passive drying in enclosed building. Right – Wet-vac for active water removal.

6.2 Escape of Water from Building Services

The principles of the above mitigations also apply to water originating internally from wet plumbing. As well as protection at the element surfaces, the path of any water, or other fluids, needs to be considered, especially as the quantity and flow rate from damaged pipework can be particularly large.

Particular attention must be paid to temporary protection and the capacity of temporary outlets in areas serviced by many wet services. It is common to specify more water-resistant structural surfaces to slabs in these areas, such as concrete toppings. Depending on its form and timing of installation, this could also assist in the temporary construction condition.

Sources of water internal to the building envelope generally originate from building services and are termed 'escape of water'. Forms of wet services distribution include incoming potable water, outgoing wastewater, wet fire risers, charged fire sprinklers and in some circumstances internal rain-water pipes. Other piped fluids can be associated with heating and cooling systems such as refrigerants and closed loop thermally activated building systems (TABS) and under-floor heating. Any of these sources can cause damage to the timber structure, and risks can be higher during construction as the systems are fitted, tested, and commissioned.

Mitigations against escape of water during construction can be part of design, planning execution or a combination. Several of the mitigations can also have a significant beneficial effect for the permanent in-service operation of the building. Specific mitigations include the following.

6.2.1 During Design

- Grouping of areas with wet services fixtures to localize risks and possible detection and interception measures. This includes alternative structural slab materials and finishes local to the higher-risk areas
- Minimising length of pipe runs and number of bends and joints
- Vertical routing of all wet services within dedicated risers, integrating access panels for inspection, maintenance, leak detection and repair.
- Horizontal routing of all wet services overhead, as opposed to within raised access floors, to facilitate inspection, maintenance, leak detection and repair.
- Selection of robust and durable pipe materials, jointing and fixtures.

6.2.2 During Execution

- Prevention of pipe freezing. Construction phase planning should ensure that the building envelope is complete and sufficient thermal controls are in place before filling wet services.
- Flow detection on incoming potable water. Zoned by area/floor to allow efficient location of leaks and shut-off of supply.

6.3 Moisture Measurement and Recording

The full duration of the construction process should be covered by a pre-determined programme of timber moisture content testing and recording. The testing should be tailored to the scale and geometry of the project. It will need to have sufficient testing points and frequency to identify the general trend in the timber structure at large. It will also need to have targeted locations and frequencies for areas that have been identified as having particular moisture hazards and/or requirements.

Examples are connection details where water entrapment is more likely and exposed end-grain are present, or an area of CLT wall that will be exposed in the final building and must not have any water staining.

The location of testing must be unambiguously recorded, and the data logged in such a way that trends can be identified and acted upon where required. The data should also be packaged and communicated to other stakeholders for record purposes. These include later evidencing of the moisture content at building completion and hand-over. The following chapter expands on the key times and periods for measurement of moisture content.

7 Quality Control: Monitoring Moisture and Control Measures on Site

To ensure the effectiveness of the planned measures, implementing quality control on a regular basis is recommended. This includes:

- Checking any protections such as foils or water outlets for defects
- Visually examining stored and installed timber elements
- Checking for condensation on elements beneath high diffusion-resistance foils
- Ensuring proper ventilation
- Documentation of moisture-related data, including moisture measurements

The wood moisture content should be determined during storage and in the part-completed structure, especially at critical locations such as connections or at exposed end grain. The wood moisture content should be determined at the following stages as a minimum:

- Immediately after the delivery of the timber products
- Prior to installation of any cladding or other kinds of enclosure of an element
- After any water ingress or a heavy rain
- Before handover to any subsequent trades
- During or immediately following the work of a trade that introduces moisture locally

Besides the moisture content of the timber, it is also sensible to plan for and determine the moisture content of adjacent materials, such as wet screeds or gypsum plaster before installation of impermeable coverings. The air humidity is also an important consideration. For example, the humidity can be extremely high in closed rooms with fresh concrete, screed or plaster. In these instances, temporary mechanical ventilation may be required.

Several methods can be used to determine wood moisture content, one of which is the electrical resistance method. The correct execution of the measurement, specific to the device, should be adhered to. This includes calibration of the instrument and numerical modifications for ambient conditions. For samples of wood or other materials like a fill, the moisture content can also be measured by oven-drying a sample. Active electronic monitoring systems can be used to gain large quantities of data, which can be used to quickly detect the degree and location of any elevated moisture contents and allow efficient interventions to address them. The planned locations and frequency of all required measurements should be stated clearly in the Construction Stage Moisture Control Plan (CSMCP).

If a wood moisture content above the specified limit is detected, measures for eliminating the water source and then re-drying should be taken. In most instances this critical moisture content will be set as 20% [18]. This value includes a safety margin to account for inaccuracies from the moisture measurement and is generalized across the typical timber species found in engineered timber. For wood products with a high amount of glue, like oriented strand board (OSB), the critical moisture content can be significantly lower.

The affected parts of the construction must be opened-up without delay to allow for drying to start, supported by a programme of carefully planned and controlled ventilation, humidity control and potentially low-level heating. The correct temperature, humidity and hence rate of drying depend heavily on the individual case and should be clarified with an expert and the manufacturer of the affected product. The whole process should be complemented by measurements of the moisture content of the timber at different depths through the section and rigorously documented.

After removal of the water source and drying, it should be checked if any parts must be replaced. For example, OSB can permanently lose its strength if it reached a certain critical moisture content threshold.

Many of the negative effects of moisture on wood are reversible, if the moisture is detected quickly enough and no damage to adhesives nor fungal decay has instigated, which in most cases, takes at least 1 month above fibre saturation point. Discolorations can be removed by sanding the surface or by treatment with a suitable agent such as oxalic acid. Surface mould can usually be removed with a mild detergent and a wet vac can be used to minimise further spreading of the mould spores. Whilst mould does not affect the structural capacity of timber, it can have negative implications for human health, along with some of the agents used for its removal. The safety of site operatives and future occupants must be considered.

8 Summary

Timber has vulnerabilities to moisture which are different to other common structural materials. The construction stage presents additional hazards associated with moisture exposure. However, the moisture issues can be adequately addressed through appropriate design decisions, planning and execution methods.

A Construction Stage Moisture Control Plan is an important tool for evaluating, planning, enacting, and recording the required measures. The process should be started at early design stages and recorded and re-visited throughout the design, procurement, and execution stages. Input will be required from multiple project stakeholders and should be tailored to the specific project.

It is in the best interests of all project stakeholders, and the timber construction industry at large, that timber buildings are carefully designed, planned, and executed with construction stage moisture in mind. Only then can timber structures reliably contribute their full potential to our built environment.

References

1. Keylwerth, R.: Grundlagen der Hochtemperaturtrocknung des Holzes (fundamentals of high-temperature drying of wood). Holz-Zentralblatt. **75**, 76 (1949)
2. Flexeder, N.: Monitoring wood moisture content. Doctoral Thesis, Chair of Timber Construction and Building Construction, TUM School of Engineering and Design, Munich, Germany (2024)
3. Flexeder, N.: Monitoring Wood Moisture Content. Technical University of Munich, Munich (2025)
4. Dietsch, P., Winter, S.: Structural failure in large-span timber structures: a comprehensive analysis of 230 cases. Struct. Saf. **71**, 41–46 (2018)
5. Niemz, P., Sonderegger, W.U.: Holzphysik: Eigenschaften, Prüfung und Kennwerte. Physik des Holzes und der Holzwerkstoffe. (Wood Physics: Properties, Testing and Characteristic Values): Hanser, (2021)
6. CEN/TC 250/SC5 CEN-CENELEC (ed.): prEN 1995-3:2023 Eurocode 5 - Design of Timber Structures - Part 3: Execution. CEN (2023)

7. RDH Building Science: Mass Timber Moisture Management Guide. RDH Building Science (2021)
8. Caldwell, M.: Timber Knowledge Sheet – Moisture Management During Construction. Timber Development UK (2024)
9. TRADA: National Structural Timber Specification. BM TRADA (2017)
10. Braidwood, E.: Exclusive: demolished eco school's design 'fell below expected level of competence'. Archit. J. (2017) http://www.architectsjournal.co.uk/news/exclusive-demolished-eco-schools-design-fell-below-expected-level-of-competence. Accessed 29 Oct 2023
11. Jessel, E.: Sarah Wigglesworth told to pay 1.3m over leaking Yorkshire eco school. Archit. J. (2022) https://www.architectsjournal.co.uk/news/sarah-wigglesworth-told-to-pay-1-3m-over-leaking-yorkshire-eco-school. Accessed 29 Oct 2023
12. W. Ing: Waugh Thistleton battles. £4m rotten roof legal claim. Archit. J. (2024). https://www.architectsjournal.co.uk/news/waugh-thistleton-in-4m-rotten-roof-row-set-for-january-trial. Accessed 18 Nov 2024
13. Thomas, D., Smith, I.: UK High Court backs Sky's multimillion-pound claim over leaky roof (2023) https://www.ft.com/content/2b45d15b-48e5-4079-b259-c5ccff4df85c. Accessed 29 Oct 2024
14. Ing, W.: Waugh Thistleton wins court case over rotted timber roof. Archit. J. (2025) https://www.architectsjournal.co.uk/news/waugh-thistleton-wins-court-case-over-rotted-timber-roof. Accessed 11 Apr 2025
15. Rogers, D.: Contractor called back in to fix leaking roof on AL_A's flagship Sky building (2021). https://www.bdonline.co.uk/news/contractor-called-back-in-to-fix-leaking-roof-on-alas-flagship-sky-building/5113097.article, Accessed 29 Oct 2024
16. Vitsoe v Waugh: Neutral Citation Number: [2025] EWHC 850 (TCC). (The High Court of Justice. King's Bench Division. Business and Property Courts of England and Wales. Technology and Construction Court) (2025) https://caselaw.nationalarchives.gov.uk/ewhc/tcc/2025/850
17. Sky UK Ltd. and another v. Riverstone Managing Agency Ltd. and others. Neutral Citation Number: [2023] EWHC 1207. (The High Court of Justice. Business and Property Courts of England and Wales. Commercial Court (KBD)) (2023) https://caselaw.nationalarchives.gov.uk/ewhc/comm/2023/1207
18. Deutsches Institut für Normung: DIN 18334: 2023-09: German construction contract procedures (VOB) - Part C: General technical specifications in construction contracts (ATV) - Carpentry and timber construction works (2023)

Property Protection in the Event of a Fire

Carl Pettersson[1(✉)], Leo Willem Menzemer[2,3], Petr Hejtmanek[4], and Daniel Brandon[3]

[1] RED Fire Engineers Sweden AB, Stockholm, Sweden
carl.pettersson@redfireengineers.se
[2] Dansk Brand- og Sikringteknisk Institut, Hvidovre, Denmark
lwm@dbigroup.dk
[3] Lund University, Lund, Sweden
{leo_willem.menzemer,daniel.brandon}@brand.lth.se
[4] Czech Technical University in Prague, Prague, Czech Republic
petr.hejtmanek@cvut.cz

Abstract. Timber buildings present unique fire safety challenges compared to non-combustible constructions. Fire damage in timber structures includes smoke, heat-related impacts and firefighting water. Effective fire mitigation strategies, such as smoke detection systems, automatic suppression systems, and compartmentation, can be crucial for enabling timely interventions and minimizing damages. Post-fire restoration focuses on water extraction, removal of smoke damage and replacing contaminated material as well as structural repairs. Modern timber buildings will more often than not require expert input to safely and effectively open or remove structural elements to allow necessary restoration. This chapter highlights the need for collaboration, innovative fire safety measures, and efficient restoration processes to improve the property protection in timber buildings. This is a field where further research is essential to address the evolving challenges of timber construction and enhance its sustainability.

1 Introduction

The use of combustible materials, particularly timber, increases potential fire damage. Timber buildings may experience different types of damage compared to non-combustible constructions like concrete. The consequences of fire in timber buildings include smoke, water and heat damage.

It is important to distinguish between mass timber buildings (e.g., CLT and engineered timber) and timber-framed construction. The repairability and restoration of timber buildings post-fire often depends on the type of construction. Modular timber-framed construction, for instance which connects complete modules into one building, requires unique approaches to dismantling and replacing damaged components, as these differ from traditional timber-framed structures where reverse assembly is possible.

While occupant safety remains the paramount goal of fire safety design as mandated by building codes, property protection is important for stakeholders such as insurers, business operators, and building owners. Balancing these objectives becomes increasingly complex for taller buildings or critical facilities like healthcare centres, where the

G. Fink et al. (Eds.): *Holistic Design of Taller Timber Buildings*, SPRTRCIENG, pp. 143–149, 2025.
https://doi.org/10.1007/978-3-032-02098-7_12

consequences of fire extend beyond physical damage to include operational disruptions [1]. Introducing combustible materials into such buildings can amplify property protection risks, underscoring the need for early stakeholder engagement to align fire safety objectives with risk tolerance.

2 Damages from a Fire

Damages caused by a fire impact multiple stakeholders, including building owners, tenants, and insurers. For insurers, the extent of damage—referred to as Estimated Maximum Loss (EML)—is a key metric in determining coverage and premiums [1]. Timber buildings present novel challenges in this context due to increased susceptibility to fire spread [2] and fire or water induced material degradation [3].

Types of Damage. In the event of a fire in a timber building the following damages can occur:

- Flame and heat damage: Structural weakening due to high temperatures or combustion leading to charring and reduced material strength.
- Smoke damage: Smoke spread leaving odours, stains and toxic combustion products, both internally and externally.
- Water damage: Resulting from firefighting operations, water suppression systems, or damaged water and wastewater pipes. Water exposure to wood can result in staining, mould and bacterial growth and movement of structural elements. Firefighting water can also spread smoke related damage in a building.

Key Factors Influencing Damage. The consequence of damage in a building will mainly be affected by the following factors:

- Fire size and spread: Larger and more widespread fires will increase the scale of damages.
- Fire duration: Extended exposure to heat and smoke increases the intensity of damages.
- Firefighting operations: Extinguishing water and demolishing measures (e.g., cutting into cavities) can increase the damage.
- Effectiveness of restoration: Timely intervention post-fire can significantly reduce long-term damage, particularly related to water exposure.

Timber structures are particularly sensitive to water damage. Excessive water use during firefighting can lead to structural saturation, which, if not promptly dried (within days), can result in staining, mould and bacterial growth and, if left saturated for a longer period, potential structural movement and eventual decay [3], requiring significant restoration measures. The degree of microbial growth over time is also influenced by the surrounding temperature. Rapid and effective restoration efforts [4] are critical to minimizing such damage. The key measure necessary to mitigate water damage risk after a fire event is efficient drying. The microbial growth phase following critical moisture levels occurs within the timeframes indicated in Fig. 1.

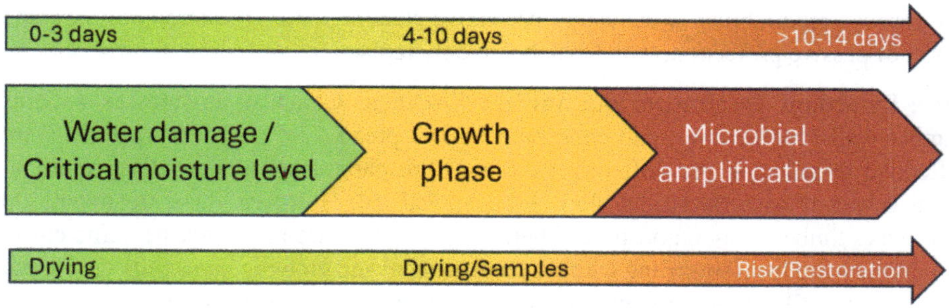

Fig. 1. Timeline of water damage progression and the critical phases of microbial growth following a fire event.

As shown in Fig. 1, microbial amplification typically begins after 10–14 days if the moisture level remains critical. This emphasizes the importance of prompt drying and restoration to prevent long-term structural and microbial damage.

3 Protection Against Fire and Smoke Damage

Damages specifically connected to fires are odours, stains and contamination from smoke spread, and charring and loss of structural strength from heating or by combustion of timber structures. During a fire, smoke can be expected to spread rapidly. However, smoke propagation within a building is often hard to predict with accuracy, especially at greater distances from the room where the fire originated [5]. The spread of smoke is influenced by the buoyancy of hot smoke, air currents resulting from pressure differences in the building, and ventilation systems. Cavities and gaps between construction elements, typically found in timber or modular construction buildings, can create passageways for smoke spread. Therefore, the use of appropriate fire sealants and fire stops—particularly around penetrations, connections, and within cavities—is crucial to protect against excessive smoke spread in such structures.

Fire Compartmentation. Compartmentation is a key fire safety strategy, designed to contain fire and smoke within specific areas. This reduces the potential fire spread and minimizes damage to occupants, firefighters, and property. In timber buildings, airtight construction is crucial to prevent smoke spread through cavities, as timber (or other combustible materials) in cavities can facilitate extended fire spread to large parts of the building.

Fire Resistance and Fire Protection Coverings. If designed adequately, fire may not cause a collapse of the structure, however, due to the large deformations of the structures affected by the fire, it may affect the usability of the building. Depending on the structural design of a building, enhancing fire resistance beyond the regulatory minimum can improve property protection. This may include the design of structural members to higher fire resistance than what is legally required. Restoring charred timber surfaces to its original (or usable) condition can incur significant time and financial cost. A solution to improve repairability is the use of protective claddings (i.e., plasterboard), or some form of passive protection, which can be quickly replaced.

Fire Retardant Treatments and Surface Coatings. Combustible surfaces, such as timber walls and ceilings, can contribute to fire spread. Fire retardant treatments are effective in reducing the different reaction-to-fire parameters as they delay ignition, reduce heat release rates, and slow flame spread [6]. However, these treatments do not make timber non-combustible. Therefore, this measure might not be sufficient in hindering a fire from spreading within a compartment and growing into a fully developed fire in which all exposed timber in the compartment will get involved.

Detection and Alarm Systems. Smoke detection and alarm systems provide a critical advantage by enabling early identification of a growing fire. This early detection allows for swift intervention, significantly reducing the potential for the development of fire and associated damages. As a complement to conventional detectors in the room interior, sensors inside of structures can also be installed. Sensors that are able to detect early phases of combustion or sensors that are able to detect smouldering are under development.

Suppression Systems. Water-based systems, such as sprinklers and water mist systems, are effective in containing fires and reducing fire damage. Research has shown that fire suppression lowers extent of fire damages compared to non-suppressed fire scenarios [7]. Oxygen reduction systems (ORSs) or gas extinguishing systems could be considered as an alternative to water-based suppression systems, where appropriate, to reduce water exposure to the structure. While ORSs aim to constantly lower oxygen levels within a protected area to prevent ignition and onset of a fire, suppression systems using gaseous agents aim to extinguish a detected fire through suffocation (oxygen displacement) or cooling (heat extraction). However, such systems need to be commissioned with care as lowering oxygen concentrations, constantly or momentarily, can impose health and life risks for people in the area [8, p. 2].

Exhaust/Ventilation Smoke Control. Effective smoke management systems, including exhaust and ventilation strategies, reduce hazards of limited visibility and toxicity, aiding both evacuation and firefighting efforts. These systems can also reduce smoke related damages to the building.

Fire Service Intervention. Effective fire service operations are a critical factor in mitigating damage. Building owners can facilitate effective intervention by:

- Providing clear signage and firefighting plans.
- Ensuring accessibility to hydrants, stairs, and fire safety features.
- Hosting familiarisation visits for local fire services.

4 Protection Against Water Damage

Water exposure within the building can be caused by firefighting operations using water, activated water suppression systems or damaged and leaking water and wastewater pipes. Water damage is challenging to prevent but can be mitigated through design considerations and efficient response operations. Vulnerable construction details in timber buildings, such as penetrations through slabs and exposed end-grains, should be sealed to prevent water ingress (e.g. see Fig. 2). Some sensitive construction details in timber buildings have been identified to be prone to water damage [3]:

- Water entrapment below concrete toppings,
- Joints in which water can be trapped, wetting the end-grain,
- Penetrations through slabs where the end-grain is exposed,
- Curtain wall to floor slab connections where the fire seal is located below the line of the slab allowing water to pool against the exposed CLT or timber edge,
- Glulam beam to column concealed connections that include cut-outs with cavities and allow water to collect within the concealed cavity.

An effective measure to protect against water damage is incorporating inbuilt drains (e.g., in bathrooms) that allow excess or extinguishing water to be drained quickly. Successful protection against water damage in a fire scenario will rely on effective responding operations, starting with firefighters on-site and followed by salvage teams. By employing well-informed tactics and methods, firefighters can minimize damage to the building—for example, by using water efficiently, understanding fire spread, and effectively extinguishing smouldering fires within structures or cavities.

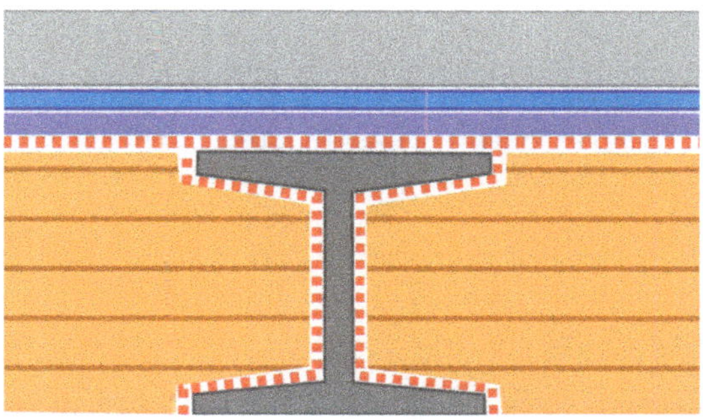

Fig. 2. CLT slab with built-up steel beam, potential water entrapment areas indicated by the red dots.

5 Post-fire Repair and Restoration

Post-fire damage can be categorized and ranked to prioritize restoration efforts:

- Heat Damage: Repair or replacement of burnt (charred or otherwise thermally decomposed) structural elements (e.g., see [9, 10]).

– Smoke Damage: Removal of odours and stains from structures, preferably starting from the areas damaged by fire.
– Water Damage: Requires immediate action to prevent biodeterioration, like mould and bacterial growth.

Fire and Smoke. Once the fire has been extinguished, it is important to establish a pressure differential using fans within the building. This helps hinder the further spread of smoke and odours beyond the fire-exposed rooms. Additionally, measures should be taken to prevent further damage and the spread of odours that may occur during the removal of damaged materials or the demolition and refurbishment of affected structures.

Smoke-stained materials typically need to be identified and replaced. The same applies to materials contaminated by odours. However, addressing odour contamination can be more challenging, as such contamination is not always visible.

Water. Once the fire is extinguished, the first responders on site can change focus from firefighting to salvaging, which in relation to water damage means "drying". The objective should be focused on extracting any water as soon as possible. Visible water should be extracted out of the building, through workable drains or redirected from spreading into the structure [4]. The next crucial step is to start the drying out process, which must be started within the first couple of days.

Restoration Process. Timely and well-coordinated restoration efforts are essential. Key steps include:

1. Damage Assessment: Identify and document affected areas.
2. Water Extraction and Drying: Initiate within the first couple of days to prevent mould and bacterial growth.
3. Structural Repairs: Replace damaged elements, ensuring compatibility with original construction.
4. Surface Cleaning and Restoration: Remove smoke stains and odours.

Restoration strategies should be tailored to the type of construction (e.g., modular vs. traditional timber framing or mass timber construction) and guided by best practices to ensure cost-effectiveness and durability. Extended exposure to fire or water may make demolition of structural elements necessary, requiring input from experts.

6 Conclusion

Property protection in timber buildings requires a multi-faceted approach encompassing fire prevention, damage mitigation, and efficient restoration. Stakeholder collaboration, advanced fire safety technologies, and robust post-fire recovery plans are critical to balancing life safety and property protection objectives. Further research and innovation are essential to address the unique challenges posed by timber construction and to enhance the resilience of sustainable building designs.

References

1. Corbey, S.: Mass timber insurance playbook – providing common ground in addressing insurance risks. Struct. Eng. **101**(5), 12–13 (2023). https://doi.org/10.56330/FXQJ5376

2. Andrew, B., Östman, B.: Fire Safe Use of Wood in Buildings: Global Design Guide. CRC Press, Boca Raton, FL (2022). https://doi.org/10.1201/9781003190318
3. Brandon, D., Su, J.: Mass timber structures post-fire: A gap analysis. (2024) Available: FPRF-2024-02
4. Lindsay, R.: Solutions for Upper Mid-Rise and High-Rise Mass Timber Construction - Rehabilitation of Mass Timber Following Fire and Sprinkler Activation., Project No. 301013070. FPInnovations (2019)
5. Ebus, J., et al.: Smoke Propagation in Residential Buildings. The Main Report on the Field Experiments Conducted in a Residential Building with Internal Corridors. Organisation of Dutch Fire Services (2020)
6. Klippel, M., Just, A.: Guidance on fire design of CLT including best practise. In: COST Action FP1404, Zürich, Switzerland, N223-07 (2018). https://doi.org/10.3929/ethz-b-000319542p
7. Brandon, D., Plathner, F.V., Mikkola, E.: SAFITS - statistical analysis of fires in timber structures. Report. **2021**, 97 (2021) Available: ISBN:978-91-89385-87-0
8. van Hees, P., Åström, J., Nilsson, M., Meacham, B.: Review of Oxygen Reduction Systems for Warehouse Storage Applications – Phase 2 (2024).
9. Brandon, D., Sjöström, J., Kahl, F.: Post-fire rehabilitation of CLT. RISE Rep. **67**, 2021 (2021)
10. Chorlton, B., Gales, J.: Structural repair of fire-damaged glulam timber. J. Archit. Eng. **27**(1), 04020043 (2021). https://doi.org/10.1061/(ASCE)AE.1943-5568.0000445

Damage and Repair of Tall Timber Buildings

Boris Forsthuber[1(✉)], Daniel Brandon[2], Eleni Toumpanaki[3], Jan Vcelak[4], and Luka Vojnovic[3]

[1] Holzforschung Austria, Vienna, Austria
B.Forsthuber@holzforschung.at
[2] Lund University, Lund, Sweden
daniel.brandon@brand.lth.se
[3] University of Bristol, Bristol, UK
{eleni.toumpanaki,nf21539}@bristol.ac.uk
[4] Czech Technical University in Prague UCEEB, Prague, Czech Republic
jan.vcelak@cvut.cz

Abstract. The chapter summarizes the different methods available for damage repair after water and/or fire events. Different drying methods as well as their advantages and disadvantages are discussed. This is done separately for timber framing and mass timber constructions. The section for repair after fire events points out the different steps that should be taken after a fire event. This includes methods for soot removal, cleaning but also for the removal of odour. Aside from damage repair, also methods for damage prevention, i.e. moisture monitoring methods are discussed. These sensors detect either water (leakage detection) or wood moisture content. In the last section, the importance of asset management and post-construction management is discussed.

Keyword: Repair of Timber constructions · Water Damage · Fire Damage · Moisture Monitoring

1 Introduction

Recent developments have led to a significantly increased use of mass timber in tall buildings. With the increasing height and value of mass timber buildings, the potential property damage is increased as well. It is, from an insurance perspective, essential that expected events do not lead to irreparable damages, as otherwise high costs are the consequence. These associated risks have led to high insurance costs for timber buildings up to 800% more than conventional materials. On a societal level, water and fire damages are recognized to be the costliest forms of damages in buildings [1]. As timber is sensitive to water and does not recover fully from significant heating, uncertainties arose regarding the repairability of timber structures. A project by the USA's National Fire Protection Association, indicated gaps of knowledge regarding repair of particularly mass timber structures. These knowledge gaps did not only concern the assessment and repair, procedures, but also how to deal with insurance policies and recertification of the building.

G. Fink et al. (Eds.): *Holistic Design of Taller Timber Buildings*, SPRTRCIENG, pp. 150–176, 2025.
https://doi.org/10.1007/978-3-032-02098-7_13

In general repair procedures after a fire event are kept confidential. Despite that a number of publications provide recommendations based on real post-fire repairs and experimental studies of post fire or water damage repair. This chapter provides guidance for damage assessment, repair and maintenance based on current knowledge.

2 Water Damage and Repair

Probably all building materials are to some extend prone to water damage due to vapour condensation, roof leaks, failures at building envelope penetrations such as doors or windows and wicking from wet foundations [2]. Therefore, moisture management is crucial for any building material. Nevertheless, wood is special in that sense, that wood is a hygroscopic and biodegradable material that tends to wet far more quickly than it dries. Furthermore, the dimensional changes connected to those changes in wood moisture content makes this especially critical in timber constructions.

Wood takes up water from the environment, both in gaseous form (from the humidity in the air) and liquid form. Without direct water contact, the highest wood moisture content that once dried wood can reach is marked by the fibre saturation point (FSP). This point is specific for each wood species and vary between 20% to 40%. For Spruce, the FSP is appr. 30%. It is important to note that up to the FSP, the water is mostly in the cell walls. Only with wood moisture content values above the FSP, free water is in the cell lumen *(i.e.* the hollow space inside the wood cells where water is not bound to the cell walls). This happens, when wood comes into contact with liquid water. Liquid water is easily accessible, and its occurrence strongly increases the probability of a fungal attack. Practices generally target a wood moisture content below 20% (e.g. for glulam production lamination, wood moisture content should be $<18\%$ for treated and $<15\%$ non-treated wood according to the EN 386) and practice [3] with no single measurement exceeding 24%. Eurocode 5 specifies that this moisture content limit is maintained in indoor environments, *i.e.* service classes 1 and 2. For outdoor applications, wood is often treated to enhance its durability.

During production, wood is dried to reach target moisture contents generally below 20%. Notably, removing water from the cell lumen requires significantly less energy than extracting water from within the wood cells themselves.

Water damages usually happen during the use phase of a building, e.g. due to construction defects, leaking pipework or even from extinguishing water after a fire event. However, moisture management is also crucial during the erection phase. Guidelines are available e.g. from Caldwell [4].

When water damage has occurred in timber constructions, the following procedure is recommended by Tscherne et al. [5]:

1. Wood moisture content measurement in different spots of the affected timber construction and adjacent construction elements.
2. In most cases opening the construction, removal of wet backfills and insulation materials
3. Visual examination for moisture related fungal infestation, especially mould and wood destroying fungi.

4. Removal of infested wood, depending on the type of fungi and degree of damage, restoration of statically necessary measures if required.
5. Use of technical drying methods, depending on the type of construction and the degree of moisture damage
6. Control of the drying success by measuring the wood moisture content especially on the most critical parts of the construction.

These steps are discussed in more detail in the following sections.

2.1 Drying of a Structure

Prior to all almost all drying methods, the construction has to be opened and inspected for fungal infestation. After that, the drying procedure can be started. Generally, the aim of the drying process is the reduction of wood moisture content below 20% over the cross section of all timber construction elements. Control of the moisture content during and after drying is crucial. It is also imperative not only to assess the wood moisture content on the wood surface, but also in deeper layers of the wood, as the drying on the surface is much faster than inside the bulk wood.

It is advisable to measure the wood moisture content especially in the parts of the construction, where drying is difficult. This is, for example, the case of the supports of a wooden-beam ceiling.

Various techniques are available for drying of timber constructions (and other materials). Most of them require a difference in partial pressure between outer and inner layers of wood. While higher moisture gradients will lead to faster drying, there are certain limits. When the moisture difference is too high, capillaries will be interrupted, and the drying time might get even longer. Furthermore, a high moisture gradient over the cross section can lead to damages such as cracking of wood. Therefore, the right balance of the severity of drying is recommended.

The duration of the drying procedure depends on the following factors:

1. Intensity of moisture penetration: What is the highest wood moisture content and how deep does it reach into the wood
2. Drying technique and drying program used
3. Accessibility to dry air

2.2 Drying Methods

If a dry environment with low relative humidity is available**, natural drying** in fresh air is a viable option, as it is usually very mild and cheap. For this method to work properly, the wet areas have to be fully exposed. Heated rooms are especially suitable but need constant ventilation to get rid of the moist air.

Condensation dryers (Fig. 1, left) are the most used drying systems, as they are rather cheap and give excellent drying results. The basic principle behind this technique is blowing heated (and therefore dried) air in in the environment to allow moisture desorption from wood. Subsequently, the resulting warm, humid air that results from the drying process is forced to flow across a cold surface, where the water condenses and is collected. The air is heated up again and is reused for drying. The condensed water has to be removed continuously. As a result, the relative humidity inside the room is continuously reduced. It is crucial that the dry air gets in contact with the wet surface areas of the construction elements. Ventilators are thereby excellent accessories, that, when positioned appropriately, are a great help getting the hot, dry air to all regions of the affected construction elements (Fig. 1, right).

Fig. 1. Left: Condensation dryer; right: Ventilators help to guide the air flow during drying

For efficient drying, it is furthermore important to seal of the room. Condensation dryers are ideal for closed and sealed rooms, that have a room temperature of at least 6 °C. An overview of advantages and disadvantages of condensation drying as a repair method are given in Table 1. An alternative to condensation dryers are **adsorption dryers**. With these devices, humid air is drawn in and routed through a rotor that contains drying materials, such as silica gel or zeolites. These minerals are immobilized, e.g. on a honeycomb-shaped and air-permeable glass fibre structure. The adsorbed water is removed continuously with a heated second air stream in counter current mode. The moist air is then led outside (e.g. through a window), while the heat is recovered by a heat exchanger. Advantages and disadvantages of adsorption drying as a repair method are given in Table 1.

For very difficult to access areas (e.g. soaked floor structures in new buildings), the above-mentioned drying techniques can be assisted by **over-** and **underpressure** drying methods. For these techniques, dry air is either pressed into or vacuumed from cavities of the construction. With the **overpressure drying**, dry air is blown in through injection openings that are arranged in a grid. The humid air escapes at outlet openings or at exposed edge joints (i.e. between floor and wall). The advantages and disadvantages of the overpressure method compared to the under pressure method is listed in Table 2.

In contrast to the overpressure method, humid air is sucked out with the **underpressure drying method**, using vacuum turbines. Air flows through openings of the floor or through exposed edge joints. For efficient drying, the air has to be dried simultaneously, using condensation or adsorption drying. One of the biggest advantages of the underpressure method is that the air can be filtered, and no mould spores or other contaminants are blown into the room. This makes this an especially suitable method for rooms with high hygienic requirements.

A currently rather exotic method is the **infrared drying technique** and is currently used only on rare occasions. Infrared irradiation does not penetrate deep into the wood

Table 1. Advantages and disadvantages of various drying methods condensation dryers

Drying method	Advantages	Disadvantages
Condensation dryer	Simple technique, robust and durable devices	Water from condensation has to be removed constantly
	Relatively cheap	Air turbulence can cause mould spores to swirl and distribute inside the room
	High efficiency	Minimal temperature of 6 °C
	Drying is not too fast and thereby prevention of drying damages such as extensive cracking	
Adsorption dryer	Wider application spectrum compared to condensation dryer	High energy consumption
	Also usable with air temperatures below 0 °C	Since wet air is continuously removed, supply air must be provided
	Humidity is removed via an air flow and hence, no water tank has to be emptied	Danger of overdrying and resulting damages (e.g. cracking of wood)
	Very fast drying, as the dried air has a humidity around 15–20%	

Table 2. Advantages and disadvantages of overpressure drying compared to underpressure drying

Advantages	Disadvantages
Higher efficiency and therefore faster drying than underpressure drying	Contamination of the room (e.g. mould spores or other materials like fibres. Underpressure drying is also feasible for hygienic-sensitive rooms,
Easy installation	Filtering of exhaust air is not possible and can therefore not be used in rooms with strong hygienic requirements
Air ventilation is easily done	

and hence, only the surface is dried efficiently. For efficient drying, another heat source and a (forced) air flow is necessary.

3 Drying of Ceiling Elements

3.1 Drying of a Wood-Beamed Ceiling

Figure 2 shows a typical wood-beam ceiling construction, typical for floors for intermediate storeys. In modern wood-beamed ceilings, a recessed formwork (6) and a gypsum plaster board (7) are also common. Between the beams there is usually also a sound insulation present. The top layers are typically a combination of several wood-based materials as well as a concrete screed, with/without backfills and with/without underfloor heating.

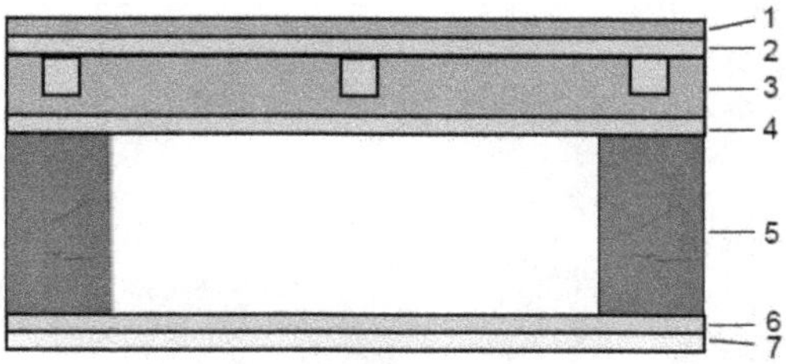

Fig. 2. Schematic of a traditional Austrian wood-beam ceiling construction; 1) Floor covering, e.g. parquet flooring; 2) blind floor, typically softwood; 3) backfill; 4) formwork; 5) wood beams with app. 80 cm distance; 6) recessed formwork; 7) gypsum plasterboard

If there is not insulation material between the beams (which is often the case in older buildings), the drying procedure is straightforward: in each cavity between the beams, openings are drilled and over- or under pressure drying is performed (Fig. 3). If present, the insulation material between the beams has to be removed prior to the drying process.

Drying from below is, however, not sufficient, if there is moisture ingress from the top, i.e. the parquet flooring and backfill. In this case, flooring and backfill have to be removed. The main reason why drying does not work properly through the backfill is the formation of drying paths, through which the dry air is blown. The surrounding material tends to stay wet.

3.2 Drying of Timber Frame Walls

Drying of timber frame walls is achieved by opening the construction and removing the insulation materials in the partitions. Gypsum boards and organic materials are prone to infestation with mould and should therefore be removed quickly. When removing the bracing internal planking, it is important to keep at least one side (e.g. the exterior side) intact. If that is not possible, it is crucial only to replace one board after the other not jeopardize the stiffening properties.

Drying is usually done with condensation or adsorption dryers, assisted by ventilators. Special attention should be paid to connection joints to avoid moisture islands. Important are also other hard-to-dry building elements, such as horizontal beams on the floor.

Before closing the construction again, wood moisture content has to be monitored conscientiously. Especially the wood moisture content in the deeper parts is of great

Fig. 3. Drying of wooden-beam ceiling with overpressure method

importance. For that, it is advisable to use stick-in electrodes with at least 40 mm in length. After drying, walls have to satisfy mechanical properties (stiffening function), but also building physics (fire, sound, heat insulation). Especially important is also moisture protection and restoring the vapour barrier.

3.3 Drying of Mass Timber

Drying of mass timber elements such as walls and ceilings is much more difficult compared to timber frame buildings, as they do not have any cavities, where dry air could be blown through. It is therefore important to expose the cross laminated timber elements (i.e. removal of e.g. gypsum boards) and start drying promptly (Fig. 4). This can be done by conventional dryers such as condensation and adsorption dryers, again strongly assisted by ventilators. Most critical areas are joints and end-grain areas.

The aim is again reducing the wood moisture content to less than 20%. Same as with the timber frame building, before closing the construction, wood moisture content has to be checked, especially on the hard-to-dry parts and not only superficially but also inside the CLT elements.

A study of 80 soaked wall-to-floor connections [6] compared three drying methods: air drying (9 months outdoors), mild drying [3] (500 h at 30–35 °C), and aggressive drying (200 h at 60–82 °C). While load capacity and stiffness remained largely unchanged, aggressive drying increased the brittleness of Douglas fir, unlike other wood species. The authors warn of biodeterioration risks with natural drying and recommend mild

Fig. 4. Mass timber elements before drying; all panels were removed and the CLT elements were exposed

drying, though insurance industry representatives indicated aggressive drying is the most common method applied in practice [8] (Fig. 5).

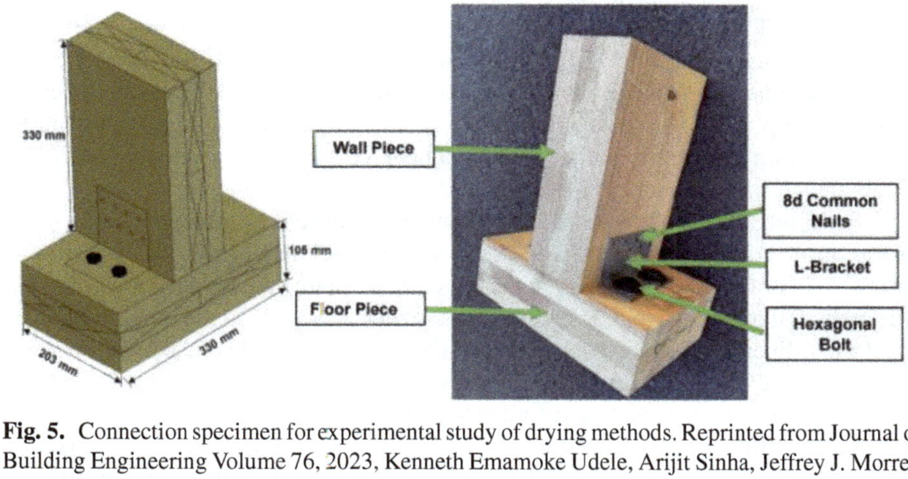

Fig. 5. Connection specimen for experimental study of drying methods. Reprinted from Journal of Building Engineering Volume 76, 2023, Kenneth Emamoke Udele, Arijit Sinha, Jeffrey J. Morrell Effects of Re-drying on properties of cross laminated timber (CLT) connections, Copyright (2023), with permission from Elsevier.

embedment strength, a 10–15% reduction in strength and a grea[…]rease of up to 20% were recorded due to artificial cracks simulating shrinkage effe[…]

To mitigate the effects of water damage in timber connections, the […] is to eliminate the root cause of water ingress (e.g., improve ventilation, repair […] ing membrane, protect exposed end grain faces). To mitigate the effects […] cracking in timber connections, repair methods can vary from reinforcement […] screws, bonded-in rods, and Fibre Reinforced Polymer (FRP) wrapping systems […] tionally, gap-filling adhesives or metal plates with screws can also be employed. […] latter option, although not lightweight, has been commonly used in rehabilitating […] timber buildings (e.g., Beaulieu Church in Hampshire [21]) to increase the rotational stiffness and load-carrying capacity of joints.

The use of screw reinforcement in moment connections with multiple dowels and artificial cracking in the middle row proved to increase the rotational capacity of joints showing limited crack growth upon loading in [22]. However, a slight improvement in the load-carrying capacity was observed compared to the uncracked, unreinforced specimens. The location of the artificial crack in the middle row rather than the side rows which are under extreme tension and compression could explain the slight difference in failure loads. An increase in the load-carrying capacity and ductility of timber joints with a single dowel has been reported in [23] and [24].

4 Fire Damage and Repair

A fire event can be considered an uncontrolled burning process. During a fire, wood forms a char layer on exposed surface, while the portion of the core remains relative cool [25]. That is the reason, why the structural strength of both mass timber but also timber frame buildings will mostly be retained, which is a huge benefit compared to other building materials.

However, important questions remain: How complex and costly is the restoration after a fire? How much of the construction can be repaired or replaced? What steps need to be taken to comply with insurance repair policies and to be able to recertify the building? Or is it even better to demolish the whole building and start all over? These are important considerations and affect building owners but also insurance companies. The following sections give an overview on the methods and considerations for damage repair after a fire. It also has to be kept in mind, that fire damage is in most cases combined with water damage from extinguishing water.

During a fire event, harmless materials can emit potentially hazardous combustion products. As an example, after a longer period of smouldering fire, also construction wood can emit toxic combustion products.

According to Matzinger and Polleres [26] general procedure for inspection and repair of fire damage can be summarized as follows:

1. Inspection and first measures after a fire event:
    - Determining the extent of damage and, if required, structural securing of the building to prevent building collapse.
    - Temporary sealing of wall and ceiling openings

Although glued-in rods are promising connections for their application in taller timber buildings due to their high axial load transition and stiffness, as well as greater connection efficiency, care should be paid in the case of water protection. The use of glued-in rods (GiRs) is not recommended in wet conditions and service class 3 due to risk of loss of interfacial bonding between the wood and the adhesive as a result of internal stresses building up from swelling and shrinkage effects. However, a decrease in pull-out load capacity of up to 23.8% was observed in wet conditions for steel rods in glulam specimens after wet and drying conditions [16], and the duration of load effect in GiRs with epoxy adhesive is similar to timber under varying humid climate conditions [17].

Corrosion of metal fasteners can take place at prolonged exposure to water and high humidity environments (e.g., swimming pools) despite corrosion-resistant measures being in place (e.g., use zinc coated screws or electrogalvanized carbon steel dowels). Visual inspection can be restricted due to the accessibility of timber connections and replacement of corroded fasteners can be limited due to either gluing/embedment in the timber or due to wedging of the fasteners from the expansive corrosive products. Corrosion can be accelerated in some instances due to the natural extractives of certain wood species in combination with high moisture content.

Varying relative humidity (RH) conditions, from wet to dry, can significantly affect the load-bearing performance of timber-steel connections. This is due to the restraining effects caused by rigid steel elements opposing the dimensional changes in timber caused by its hygroscopic nature. Sudden shifts from high to low humidity can occur during the construction sequence—especially in the absence of a moisture management plan—when timber is exposed to rainfall, followed by low humidity due to heating at operational conditions. Additionally, water ingress and trapped moisture caused by poor detailing can further contribute to moisture-related issues, including biodegradation. A decrease of up to 90% in the withdrawal capacity of screws in wet conditions (change in moisture content of 6%) has been recommended due to the confinement effects of thick metal plates, whereas additional detrimental effects of screw failures can take place when combined effects of high moisture and over-torquing of screws occur on-site. In timber connections with dowels, the change of moisture content from 12% to dry conditions results in the contraction of timber holes and high localised stresses due to the presence of the metal fasteners restraining this movement. These internal stresses, combined with stresses from the applied loads, can exceed the tensile strength perpendicular to the grain, leading to cracking and a reduction in the moment-resisting capacity and ductility of connections, which are critical for tall timber buildings [18]. A great propensity for cracking increases with the increase in the size of structural timber elements and the high moisture content gradients taking place with sudden changes in relative humidity. The effect of cracking on the load-bearing capacity of connections depends on the depth, length, and location of the crack. Longer and deeper cracks have an adverse effect on the load-carrying capacity of bolted timber connections when they are placed on the tension side of a moment connection [19]. Decreases in the mechanical performance of up to 31% were experimentally observed in tensile testing of timber connections with multiple dowels due to shrinkage cracking [18]. A decrease in fastener yielding and ductility was observed with a greater extent of cracking in timber connections [20]. In terms of dowel

embedment strength, a 10–15% reduction in strength and a greater decrease of up to 20% were recorded due to artificial cracks simulating shrinkage effects [15].

To mitigate the effects of water damage in timber connections, the first measure is to eliminate the root cause of water ingress (e.g., improve ventilation, repair of waterproofing membrane, protect exposed end grain faces). To mitigate the effects of extensive cracking in timber connections, repair methods can vary from reinforcement with metal screws, bonded-in rods, and Fibre Reinforced Polymer (FRP) wrapping systems. Additionally, gap-filling adhesives or metal plates with screws can also be employed. The latter option, although not lightweight, has been commonly used in rehabilitating old timber buildings (e.g., Beaulieu Church in Hampshire [21]) to increase the rotational stiffness and load-carrying capacity of joints.

The use of screw reinforcement in moment connections with multiple dowels and artificial cracking in the middle row proved to increase the rotational capacity of joints showing limited crack growth upon loading in [22]. However, a slight improvement in the load-carrying capacity was observed compared to the uncracked, unreinforced specimens. The location of the artificial crack in the middle row rather than the side rows which are under extreme tension and compression could explain the slight difference in failure loads. An increase in the load-carrying capacity and ductility of timber joints with a single dowel has been reported in [23] and [24].

4 Fire Damage and Repair

A fire event can be considered an uncontrolled burning process. During a fire, wood forms a char layer on exposed surface, while the portion of the core remains relative cool [25]. That is the reason, why the structural strength of both mass timber but also timber frame buildings will mostly be retained, which is a huge benefit compared to other building materials.

However, important questions remain: How complex and costly is the restoration after a fire? How much of the construction can be repaired or replaced? What steps need to be taken to comply with insurance repair policies and to be able to recertify the building? Or is it even better to demolish the whole building and start all over? These are important considerations and affect building owners but also insurance companies. The following sections give an overview on the methods and considerations for damage repair after a fire. It also has to be kept in mind, that fire damage is in most cases combined with water damage from extinguishing water.

During a fire event, harmless materials can emit potentially hazardous combustion products. As an example, after a longer period of smouldering fire, also construction wood can emit toxic combustion products.

According to Matzinger and Polleres [26] general procedure for inspection and repair of fire damage can be summarized as follows:

1. Inspection and first measures after a fire event:
   - Determining the extent of damage and, if required, structural securing of the building to prevent building collapse.
   - Temporary sealing of wall and ceiling openings

- If required, erection of a temporary roof
- Removal of inventory
- Compartmentalization of undamaged building components, e.g. to avoid the distribution of potentially toxic substance in other areas of the building
- Turn of ventilation systems, e.g. a controlled domestic ventilation system.

2. Damage Repair

   - Removal of:

     - damaged interior panels (gypsum boards, wood-based materials boards, etc.)
     - damaged construction elements after proper securing measures
     - sooty insulation materials
     - flooring
     - electric installations and cable ducts

   - Renovation of construction
   - Removal of odors (e.g. by ozonisation).

Some of these steps are described in more detail in the following sections.

4.1 Determining the Extent of Damage - Structural Safety

Visual inspection of the damage and residual carrying capacity of the construction is a crucial first task. After the first visual inspection, opening of the construction elements is the next step. If panels like gypsum boards are present, these have to be removed, as they are either burnt, full of soot and/or wet from the fire extinguishing action. Figure 6 shows a kitchen after a fire event. Although no water damage is present, strong sooting occurred due to the smoldering fire.

Figure 7 on the left shows the ceiling over the fire source after opening the construction. Especially around the panel joints and cracks that occurred during the fire, soot particles penetrate the deeper layers of the construction. Convection allows sooting also far away from the fire source. Typically, sooting can be seen in the vicinity of sockets or light outlets. Especially to remove odor, it is important to check installation levels and cable ducts for sooting.

Figure 7 on the right shows considerable sooting inside the insulation plane, although the actual fire happened only on the ground floor.

For fires with significant structural damage, it is recommended to involve structural fire engineers to assess the residual load bearing capacity of the structure. The most critical parts are load bearing structures and their residual load bearing capacity. If panel materials for stabilizing in timber frame buildings are affected, additional temporary measures have to be taken to sufficiently brace the structure.

The heat of a fire can reduce the mechanical properties of wood. If the heating is significant enough to cause char formation, the charred material has no significant strength and will not recover in strength. Scraping off the char layer and determining the thickness of the discolored layer of wood, therefore, can give a good indication of structural damage. It is, however, important to note that weakening of timber also takes

Fig. 6. Room after smoldering fire, with heavy sooting also in adjacent rooms

Fig. 7. Left: Ceiling over the fire source: Heavy sooting especially around the panel joints and cracks; Right: Sooting of the insulation panels in the roof, although the actual fire happened only on the ground floor.

place below the decomposition temperatures of 200 °C [27], which means that not all damage is visible. It is unknown to what extent uncharred wood recovers in strength.

Resistance drilling with a long thin drill using a, so named, resistograph can be used to not only assess the thickness of the char layer, but also the thickness of the damaged layer reduced properties [27]. A larger number of drill measurements will lower measurement uncertainties which are caused by the variability of wood properties and the impact of cracks and knots. Alternative methods include material testing of cut out samples of the fire exposed wood and modeling of the fire scenario based on fire dynamics, whereby information of the fire duration is used and results should be validated against other measurements such as resistance drilling.

4.2 Damage Repair - Cleaning of the Surface

Cleaning of the surfaces can be done in certain cases **by hand**. Special cleaning agents contain wetting agents that help cleaning the soot. The work itself is potentially dangerous due to the toxic substances inside the soot. Therefore, appropriate protective clothing have to be worn. The wastewater should be professionally disposed.

Another method is the **peel-off procedure** or **latex cleaning**. For this method, a highly viscous dispersion of y polymer is sprayed or brushed onto the surface. After 1 to 24 h, the dispersion is dried and can be peeled of (Fig. 8, left). While this method works very well on non-porous substrates, it works only very superficial on wood (Fig. 8, right).

Fig. 8. Left: Removal of dried soot removal film; Right: Cleaned and uncleaned surfaces.

For heavy contaminations, sandblasting or dry-ice blasting methods can be effective methods. Both methods have a strong abrasive effect on the surface and remove several millimeters of the wood surface. While with sand-blasting, quite a high amount of (contaminated) sand will remained, dry-ice blasting uses frozen CO_2 as abrasive medium. Therefore, dry ice blasting does not leave any secondary waste, as dry ice particles sublimate upon impact. However, both methods leave a heavily roughened surface and require in most cases a proper post-treatment (planing, sanding).

Another method is manual planing and sanding Fig. 9. It is highly advisable to remove at least 3 mm more than the discoloured surface to get rid of the burning smell.

4.3 Cleaning of Ventilation System

Modern buildings quite often contain housing ventilation and or controlled domestic ventilation. After a fire event, these are usually contaminated with soot and have to be either removed or carefully cleaned (Fig. 10). Cleaning is done, using a tube brush equipment as well as a powerful suction unit. Results from a recent research project (Brand-Wasser_Schaden, Holzforschung Austria) indicate good results were obtained using this method.

Fig. 9. Renovation of the wood surface using planing and sanding methods

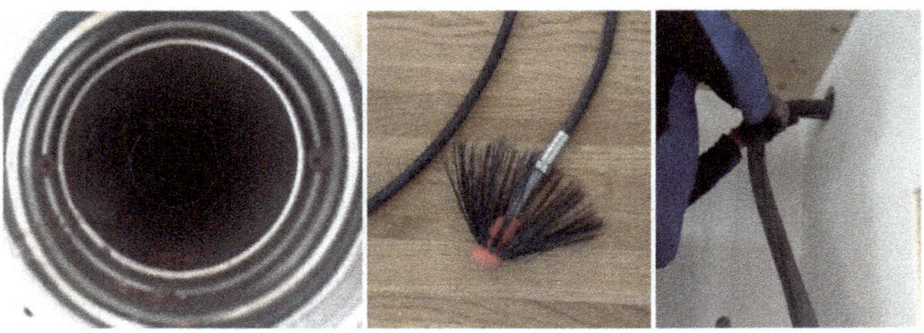

Fig. 10. Left: Soot inside air ventilation pipes; middle: Special tube cleaning brush; right: a powerful vacuum is needed

4.4 Renovation of the Wood Construction

After removal of the fire damage, the construction have to be restored to fulfill all the statical and building physics requirements. In addition, it is likely required, in agreement with an insurance policy, to obtain the same aesthetics as before the fire. Especially important is the stiffening function of the panelling and the building physics in terms of fire, sound and heat protection. An experimental case study [27], see Fig. 11, indicated that mass timber can be repaired to achieve the same structural capacity as it had before the fire. This was achieved by replacing damaged timber with timber that has a high strength grade, and by bonding the replacement material to the cleaned mass timber it with a gap-filling phenolic resin. The repairability of structurally damaged connections can, however, be significantly more challenging if metallic members conducted heat deeper into the timber material. It is to achieve better repairability, therefore, recommended to sufficiently embed metallic parts of structures that would be challenging to repair.

4.5 Removal of Smell

Even after thorough cleaning and exchange of burnt building materials, the smell might still be present. One possibility to get rid of the remaining smell is the use of ozone. This procedure destroys relevant VOCs (e.g. butylbenzene or cresole derivates). It has to be kept in mind that the removal of the burning smell is one of the most difficult parts of a renovation after a fire event. Furthermore, there is also a strong psychological component: Burning events can be a traumatic experience, and inhabitants are very sensitive for the burning smell.

5 Overview over Moisture Monitoring Technologies

Moisture related problems and limited service-life of timber elements are one of the most cited barriers of investors when selecting the base construction materials for new projects. Although fire might have fatal consequences for the timber buildings, statistically problems related to increased humidity or moisture have much higher incidence.

Fig. 11. Experimental setup for post-fire repair of a section of a compartment ceiling after a severe 4 h long fire [27]

Fire resistance of buildings is covered by international and national standards including fire detection and suppression systems while moisture monitoring and control is not required by any standard.

While in case of single floor family house the consequences of water damage are usually repairable in case of multistorey timber buildings the repairs might be extremely expensive and thus early detection or warning can save owners money, specifically when there are problems hidden in the structure and not visible from exterior or interior of the buildings. Therefore system to detect and localize increased moisture or humidity problems became part of tall timber projects and today there is multiple systems on the market to choose from. There are several typical sources of the increased humidity in the building structures:

- Water leakages (drinkable water, heating systems, waste water, HVAC condensate, rain water)
- Condensation of water vapours (damaged vapor barrier, high interior humidity, heat bridges)
- Ground water penetration thru foundation
- Embedded moisture (trapped water during construction, exposure of the construction site to rain water, insufficient drying of the timber elements, sealing of the structure with high level of moisture, wet construction processes screed)
- User behaviour (insufficient air exchange, high interior humidity, leaks of water, damage of hydro isolation layers)

Sensor system does not always need to measure directly moisture content in material (MC). In many cases relative humidity (RH) sensors located directly in the timber structure would detect the increasing trends of relative humidity. While moisture content is local variable, relative humidity spreads in breathable materials in the building structure and thus RH sensor can detect leakages located meters away from the sensor spots.

Measurement of **relative humidity** is applicable using cheap miniature sensors that can be easily integrated in the structure. The measurement of RH makes sense only when sensor is located in air or in breathable internal materials such as insulations from rockwool, mineral wool, cellulose or some wood fibre boards. The detection area is limited to borders of the space where the sensor is located.

The main advantage of these sensors is the thigh detection speed, which is much faster than in case of wood moisture content sensors (e.g. those based on the electrical resistance method). The disadvantage of RH sensors is the high temperature influence and measurement range: While 100% relative humidity is the maximum value, the wood moisture content can still further increase due to the presence of liquid water (Fig. 12). After reaching 100% RH it is not possible to judge if the structure is already drying out, stabilized, or getting even more wet.

Sensors for measurement **moisture content** MC or in case of wood often used WMC (**wood moisture content**) are more complicated and expensive. Handheld devices are available but they can not be integrated directly to timber structure and can measure only in accessible locations of the structure. Increase of moisture in the material when exposed to water or high humidity is a long-time process (days to months) and thus MC sensors are not optimal for immediate detection of sudden leaks. Moisture content is local variable and horizontal and depth gradient can be high (5–10% over 10 cm). The

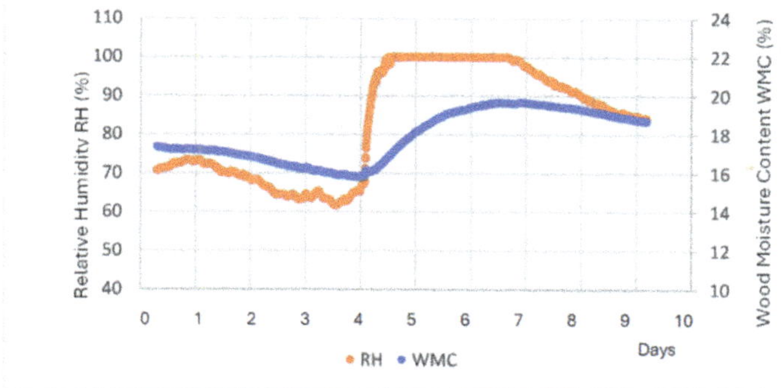

Fig. 12. Comparison of detected leak in relative humidity and moisture content measured in the same spot (Senzomatic.com)

most common method for measuring WMC is converting measured electrical resistance to wood moisture content value by available equations for specific wooden species. The method is invasive because of penetration of the wood with two rod electrodes. WMC measurement is defined in EN 13183-2. The advantage of this method is that the temperature effect on measured value is very small, measurement range is quite high 5–40% WMC (depending on the fibre saturation point of the respective wood species) and this method provides valid data even in situations where relative humidity shows constantly 100% (Fig. 13).

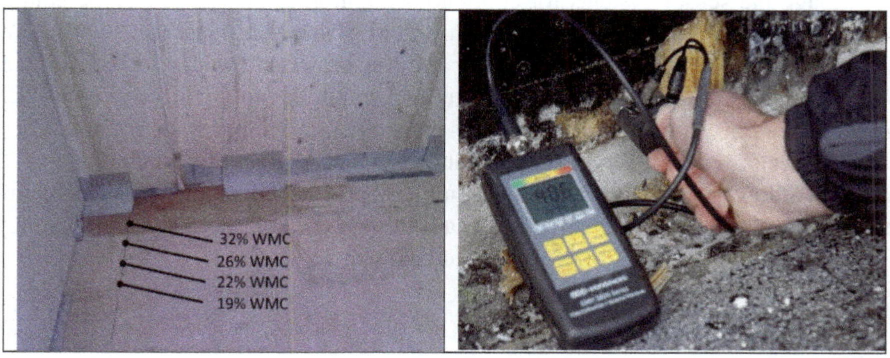

Fig. 13. Horizontal gradient of moisture content in CLT, example of handheld measurement device (Senzomatic.com)

In cases where we need to detect the **presence of liquid water** there is a possibility to use detectors with binary output (leakage detection). In such cases the users do not get the information if the moisture content is increasing or decreasing but rather if water is present in the structure or not. In most cases these detectors are based on flat tapes, flooding cables or spot detectors located in the parts of the structure that are exposed to high risk of water related problems. These detectors work on the principle that on the tape or cable there are two isolated electrical wires. Immersing of water will change the isolating resistance between the wires or pins. In case of tapes or flooding wires the sensing area is along the whole length of the tape or cable. Therefore the main advantage is the detection area, while disadvantage is absence of localization within the cable or tape length or limited information about the trend of the detected problem due to the binary information only (Fig. 14).

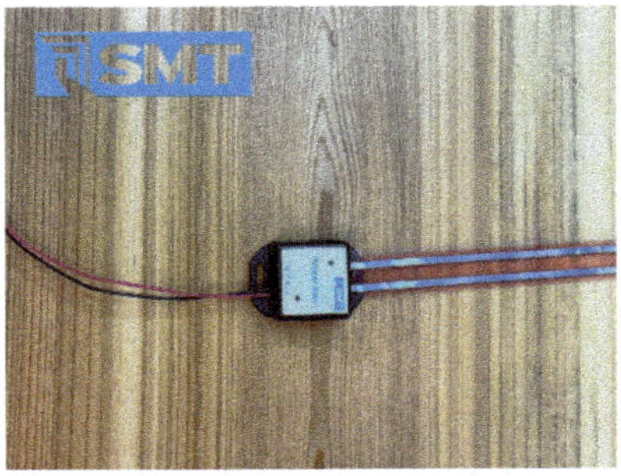

Fig. 14. Water sensitive tapes (SMT-Research)

5.1 Systems for Moisture Monitoring in Timber Buildings

The different methods for wood moisture content measurements were already described in Although the sensor systems for continuous monitoring of wood moisture content measurements timber structure monitoring are relatively new on the market, they became standard for most of the tall or large timber building projects nowadays. The advantage of early detection and localization of moisture related problems is attractive to builders and for owners-investors too. In some cases more advanced sensor system can track the moisture levels already from the prefabrication, thru storage and transportation, on-site storage, building erection, up to building commissioning, standard operation to structure decomposition and reuse-recycling. The main factor to be considered when selecting a proper system for specific projects are listed below in Table 3.

5.1.1 Wired Sensor Systems

The main advantage of wired sensors compared to their wireless counterparts is unlimited service-life which predetermines them to be used in inaccessible locations in the structure where it is not possible to change the batteries. The data could be sent with higher frequency e.g. 5 min. The main disadvantage is that the system requires cable infrastructure to be installed which makes installation more complicated and expensive. Wire usually provides not only communication line but also the power for sensors connected. The communication interface could one of industrial standard bus (e.g. RS485), and the communication protocol also could be open and standardized such as Modbus, M-Bus, UART. The topology of the network could be bus, star or combinations (Fig. 15).

Table 3. Systems for moisture monitoring overview

Functionality/parameter	Explanation
Type of the timber or hybrid structure	Wood resistance method is suitable for materials such as (Timber Frame, CLT, Glulam, LVL), but not applicable for OSB, wood fibre or chip boards
Investment & operational cost	Initial purchasing cost of the whole system and regular operational fees and maintenance cost
Number of measuring spots	Selection of spots for sensor placement, risk assessment vs investment cost
Speed of problem detection	Various measurement / detection methods have different response time
Service life of the structure and monitoring system	Expected service life of the structure and accessibility of the sensors should be considered in sensor technology selection
Sensor interface & communication protocol	Analog output, digital output, data communication protocols, wireless transmission
Wired or wireless	Main selection criteria considering cable infrastructure, cost of installation works and service life of the system
Detection or measurement	Water detection or real wood moisture content measurement or relative humidity measurement
Localization of problem	Precise localization of problem within the structure
Real time or walk over	Real-time monitoring can warn user immediately, other technologies require manual or walk-over readout to get data from the sensors (e.g. RFID sensors)
Data period	Frequency of data sampling is very different in case of wired or wireless systems
Service and maintenance of the system	Serviceability and regular fees related to operation of the system has to be considered when the system is selected

5.1.2 Wireless Sensor Systems

Wireless moisture monitoring systems do not need the cable infrastructure and therefore the installation procedure is much faster and cheaper. Furthermore, these systems can be easily installed in already built structures. The disadvantage is limited service life which is strongly dependent on the battery used and frequency of data transmissions. Usually, the frequency of data transmission for moisture values is e.g. 4 times per day, and for relative humidity 1x in an hour. With proper design of the power circuit and

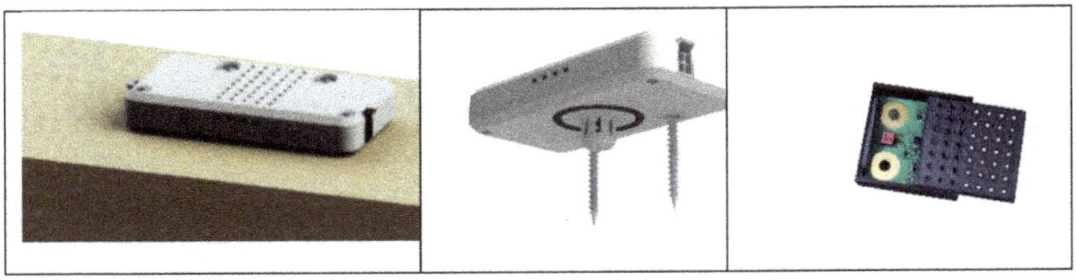

Fig. 15. Left and middle: Wired combined sensor MHT (Senzomatic.com). Right: Wired sensor (tagtron monitoring gmbh)

optimal communication protocol it is possible to achieve service life of the battery up to 10 years. There is a number of wireless technologies that can be used in licensed as well as in unlicensed radio bands. To name the most common LPWAN technologies it is LoRa/LoRaWAN, NB-IoT, Zigbee, IQRF and others. The range as well as the transmitting power depends on the wireless technology selected. Figure 16 shows several examples of sensor technologies currently available on the market.

5.1.3 RFID Sensors

A special group of wireless sensors are RFID sensors. Passive or battery assisted RFID sensors do not transmit data periodically. Operators with special RFID reader have to approach the integrated sensor and then the reader excite the sensor by electromagnetic radiation, the sensor uses the energy to wake up, take the measurement, transmit data to the reader and sensor falls back to sleep mode again. The sensors are usually very simple providing temperature data and/or binary information of detected water. Service life is not limited by the lifetime of a battery. Operators walk-over or drone fly-over readout with special RFID reader is necessary which makes the operation costs higher. This technology is not suitable for application where we need to have immediate warning when sudden increase of moisture or humidity happens.

5.1.4 Detection Systems with Tape or Flooding Cable

Water sensitive tape or cable is detector which can cover quite large areas of the structure such as flat roofs. The tape or cable consists of two or more isolated wires that are separated by immersible material. In case of water presence over the length of the tape the water cause drop for sensed resistance and system warns the user. The tape can be several meters long and equipped with special detection patterns of electrodes. The method of detection is reversible and after drying the resistance increases again. The tape and cable should be manufactured from durable material which is resistant to periodic exposure to water. The disadvantage might be that such system detect liquid water. Detection of increase humidity is problematic and dependant on other factors. Advantage is that tape is fully flat design and is easy to be integrated in the structure (Fig. 17).

 Wireless sensor (Tector.com)	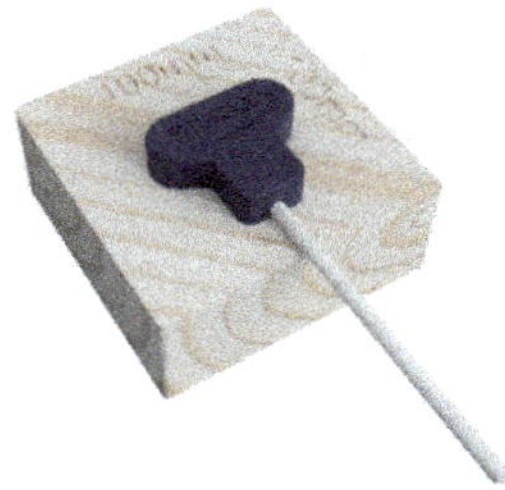 Wireless sensor – cable is connected to transmitter (tagtron monitoring gmbh)
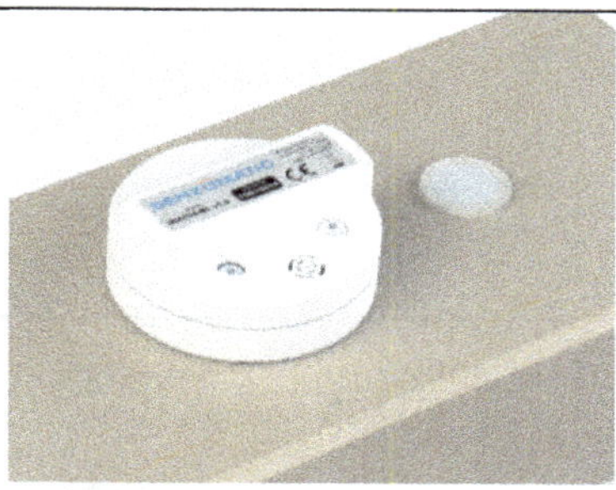 Wireless sensor MHT/WL (Senzomatic.com)	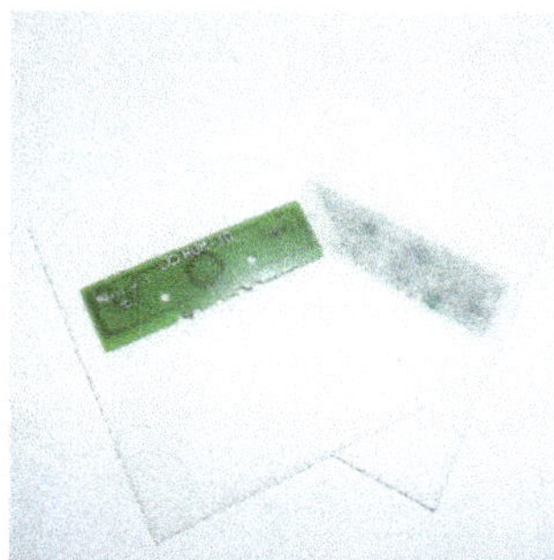 RFID Wireless sensor KD-1 (HUM-ID.com)

Fig. 16. Various commercial wireless moisture monitoring systems

In several research projects, also the possibility of integrating moisture sensors directly into wooden building components are investigated. Forsthuber et al. [28] developed a printed moisture sensors, that can be integrated directly into wooden construction elements, such as CLT or Glulam (Fig. 18). This allows continuous *in-situ* monitoring of wood moisture content. With the novel developed printed moisture sensors, it was possible to reliably monitor accumulation of moisture within CLT elements.

Fig. 17. Examples for detection systems with tape. Left: printed paper sensor, right: woven polyester monofilament fibres with increased durability (images: tagtron monitoring gmbh).

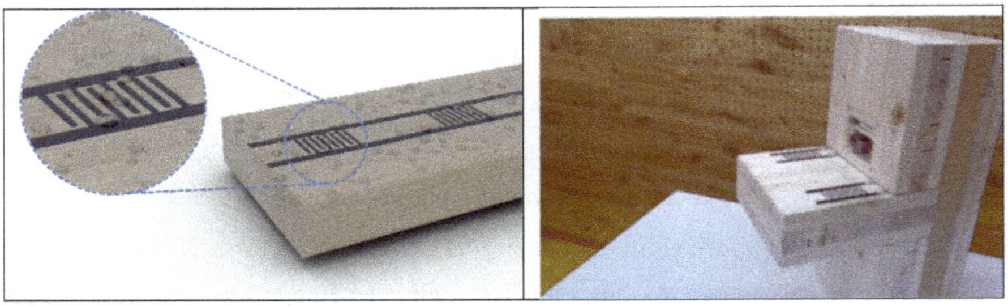

Fig. 18. Left: principle of printed moisture sensor on wood; Right: Demonstrator of a printed wood moisture sensor on a CLT floor (©Forsthuber/Holzforschung Austria)

6 Asset Management

6.1 Importance of Managing Tall Timber Buildings Post-construction

The condition of tall timber structures after the construction can be determined with structural inspections. Inspections and their outcomes, such as reports or a digital twin can inform the owner about the structural health (detection of defects, severity and extent of defects) and provide engineering recommendations, such as maintenance activities, to keep structures and users safe.

Ideally, the scope, requirements, and frequency of inspections should be defined in a standard. Frequency may be an annual visual inspection, detailed inspections every 5 years and additional inspections, such as monitoring inspections. The overall aim of managing tall timber structures is to have durable, resilient, robust building structures for long-term service life and at the lowest annual average cost possible. Similarly, competencies required for a person to undertake inspections shall be defined in a standard that can be widely used in the construction industry, i.e. how to obtain the competencies, its validity and process of renewal etc.

Monitoring is an important and efficient way of managing tall timber buildings. They are affordable, require fewer resources and service disruption. Monitoring equipment is readily available, usually non-intrusive and often operated remotely. The main aim is to detect change, predict trends of deterioration and support making an informed engineering decision based on data.

6.2 Managing a Building Within and Beyond Its Design Life

Design of tall timber buildings presents engineering challenge. Furthermore, a lack of experience in managing tall timber buildings makes engineering even more complex when it comes to preparing operations and maintenance manual on how to use and manage tall timber buildings. Current Eurocodes suggest design life of 50 years for new buildings with minimal maintenance activities [29] The latter in itself offers three options when a structural design team presents an operations and maintenance manual with identified longer-term operational expenditure (OPEX) to the owner and promoter of a tall timber building project.

The steps below describe an indicative level of involvement of the owner Depending on the use of the buildings use, importance and where in the world is these may be either voluntarily or legally driven:

In an ideal scenario from the owner's point of view, the owner may want to have little to no involvement and expense during the design life of that tall timber building. This may be due to the impeccable design and construction as well as use. Alternatively, as we see too many times, due to lack of finance or willing to invest in long-term prevention of deterioration of the tall timber buildings may dictate little to no involvement by the owner.

Next step up towards a good stewardship of the tall timber structure - follow the lowest annual average cost [30] to maintain tall timber building's structure, fabric and other elements, such as gutters, services, communal spaces, wet rooms, lift shafts etc.

The third step up would be a standardised and regular inspection regime to capture extent and severity of defects, identification of any change thus, condition of the structure. Furthermore, once a trend of change is captured, a risk-based approach of regular inspections may be developed.

Given that each tall timber building is bespoke and unique, it may be well worth exploring if its maintenance regime should be to some degree bespoke. Moreover, methods, techniques and procedures could be either fully standardised following known and tested solutions and best practices or include non-standardised applications of engineering and asset management principles.

Surveys (or stewardship actions) are meant to provide certainty around tall timber buildings performance by identifying any defects, their change in severity and extent and determine the condition of the tall timber building. Thus, preventing uncontrolled and unpredictable deterioration which may lead to a sudden failure of a part of a structure.

The next step may be to undertake structural assessment of the tall timber building [31]. This is usually done after an incident or deterioration beyond acceptable and could impede safety of the buildings and the users. An example of structural assessment is provided [32]. Furthermore, it may be useful to undertake a level of structural assessment even though an incident that would initiate a structural safety concern did not occur (Fig. 19).

Post-construction phase: owner of the building is ultimately accountable for managing and keeping the building safe and durable in order to keep the users and relevant stakeholders safe. They may employ manager, such as facility manager responsible for managing on the owner's behalf. Whereby other stakeholders need to be informed about the planned work, such as inspection or any remedial, refurbish or renewal programmes.

References

1. Anonymus: Damages caused by fire and lightning are the most expensive claims. Accessed 27 Feb 2025
2. Wang, Y.J.: Durability of mass timber structures: a review of the biological risks. Wood Fiber Sci. **50**, 110–127 (2018)
3. Anonymus: Moisture in Timber, Wood Information Sheet WIS 4-14

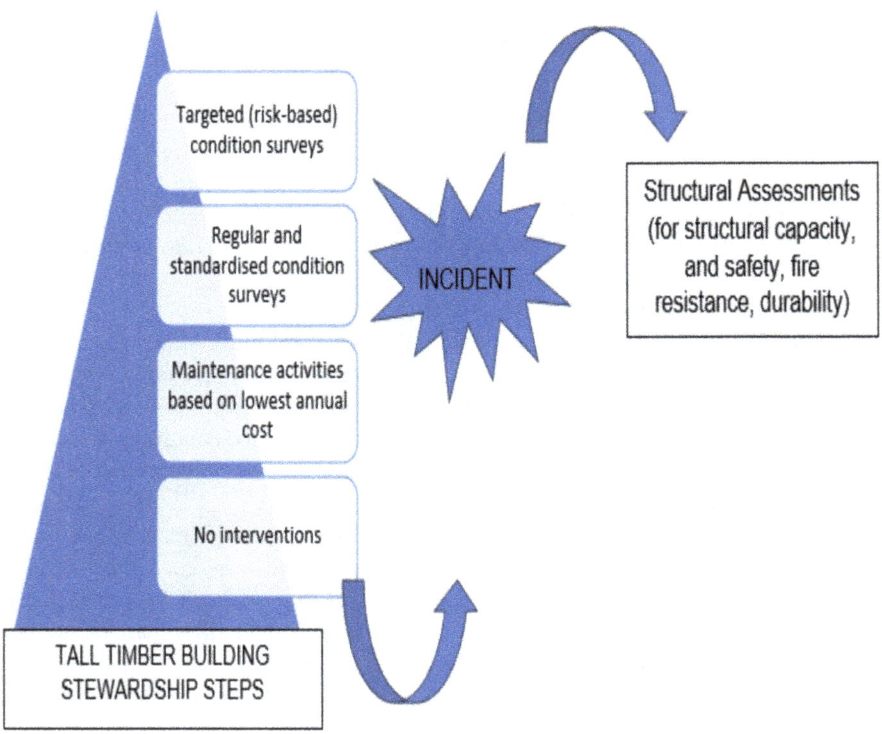

Fig. 19. Indicative approach to managing tall timber structures

4. Caldwell, M.: Moisture management during construction. Timber Development UK, UK (2024)
5. Tscherne, F., Polleres, S., Matzinger, I.: Leitfaden Trocknung nach Wasserschäden im Holzbau. Holzforschung Austria, Vienna, Austria (2021)
6. Udele, K.E., Sinha, A., Morrell, J.J.: Effects of re-drying on properties of cross laminated timber (CLT) connections. J. Build. Eng. **76**, 107298 (2023). https://doi.org/10.1016/j.jobe.2023.107298
7. Gutknecht, M.P., Macdougall, C.: Withdrawal resistance of structural self-tapping screws parallel-to-grain in common Canadian timber species. Can. J. Civ. Eng. **46**(10), 952–962 (2019). https://doi.org/10.1139/cjce-2018-0374
8. Brandon, D., Su, J.: Mass timber structures post-fire: a gap analysis. RISE Rapport; 2023:140 (2023)
9. Rammer, D.R., Winistorfer, S.G.: Effect of Moisture Content on Dowel-Bearing Strength. Wood Fiber Sci. (1) (2001)
10. Pirnbacher, G., Brandner, R., Schickhofer, G.: Base parameters of self-tapping screws. Dübendorf, Switzerland, International Council for Research and Innovation in Building and Construction CIB-W18/42-7-1 (2009)
11. Görlacher, R.: Untersuchungen an altem Konstruktionsholz: Die Ausziehwiderstandsmessung. Bauen mit Holz **92**(9), 904–909 (1990)
12. Silva, C., Branco, J.M., Ringhofer, A., Lourenço, P.B., Schickhofer, G.: The influences of moisture content variation, number and width of gaps on the withdrawal resistance of self tapping screws inserted in cross laminated timber. Constr. Build. Mater. **125**, 1205–1215 (2016). https://doi.org/10.1016/j.conbuildmat.2016.09.008
13. Toumpanaki, E., Gawne, A., Humphreys, R., Vojnovic, L.: Effect of moisture and rate of loading in the withdrawal capacity of screws in Cross Laminated Timber (CLT). Structures **69**(16), 107530 (2024). https://doi.org/10.1016/j.istruc.2024.107530
14. Yermán, L., Vargas, J., Ottenhaus, L.-M.: Effect of wetting and fungal degradation on the withdrawal capacity of stainless-steel screws in radiata pine. Eur. J. Wood Prod. **82**(1), 15–22 (2024). https://doi.org/10.1007/s00107-023-01988-7
15. Schweigler, M., Lemaître, R., Shehadeh, Z., Bader, T.K.: Moisture and assembly history effects on embedment properties of steel dowels in spruce and birch loaded in grain direction. In: World Conference on Timber Engineering (WCTE 2023), pp. 1146–1153 (2023)
16. Deng, J.X.: Strength of Epoxy Bonded Steel Connections in Glue Laminated Timber. Department of Civil Engineering, University of Canterbury (1997)
17. Aicher, S., Reinhardt, H.-W.: PRO 22: influence of moisture, temperature and load duration on performance of glued-in rods. RILEM Publications (2001)
18. Sjödin, J., Johansson, C.-J.: Influence of initial moisture induced stresses in multiple steel-to-timber dowel joints. Eur. J. Wood Wood Prod. **65**(1), 71–77 (2007). https://doi.org/10.1007/s00107-006-0136-6
19. Tian, P., Qiu, H., Wang, H.: Influence of shrinkage cracks on the mechanical properties of bolted-type glulam joint with slotted-in steel plate: experimental study. J. Build. Eng. **44**(4), 102632 (2021). https://doi.org/10.1016/j.jobe.2021.102632
20. Zhang, J., He, M.-J., Li, Z.: Numerical analysis on tensile performance of bolted glulam joints with initial local cracks. J. Wood Sci. **64**(4), 364–376 (2018). https://doi.org/10.1007/s10086-018-1709-9
21. Lawrence, A., Ross, P.: Appraisal and Repair of Timber Structures. ICE Publishing, London (2020)
22. Zhang, C., Guo, H., Jung, K., Harris, R., Chang, W.-S.: Screw reinforcement on dowel-type moment-resisting connections with cracks. Constr. Build. Mater. **215**(11), 59–72 (2019). https://doi.org/10.1016/j.conbuildmat.2019.04.160

23. Chen, C.J., Lee, T.L., Jeng, D.S.: Finite element modeling for the mechanical behavior of dowel-type timber joints. Comput. Struct. **81**(30–31), 2731–2738 (2003). https://doi.org/10.1016/S0045-7949(03)00338-9
24. Windorski, F.D., Soltis, A.L., Ross, J.R.: Feasibility of Fiberglass-Reinforced Bolted Wood Connections. United States Department of Agriculture, Forest Service, Forest Products Laboratory (1997)
25. McLain, R.P.: Repair of Fire-Damaged Mass Timber: A summary of assessment considerations and repair methods for post-fire conditions. WoodWorks - Wood Products Council (2023)
26. Matzinger, I., Polleres, S.: Leitfaden Sanierung nach Brandschäden im Holzbau. Holzforschung Austria, Vienna, Austria (2021)
27. Brandon, D., Sjöström, J., KAhl, F.: Post-Fire Rehabilitation of CLT. RISE, RISE Report 2021:67 (2021)
28. Forsthuber, B., et al.: Printed sensors for monitoring wood moisture content inside timber building elements. In: World Conference on Timber Engineering (WCTE 2023), pp. 3874–3879 (2023)
29. Eurocode - Basis of structural design, EN 1990 (2010)
30. Anonymus: Asset Management - an anatomy. The Institute of Asset Management (2024)
31. Dietsch, P., Köhler, J.: Assessment of timber structures. Shaker Verlag, Aachen (2010)
32. Smith, A., Hayes, P., Lawrence, A.: Risk and existing building assessments: understanding the approach. Struct. Eng. **102**(1), 12–16 (2024). https://doi.org/10.56330/KAQN3720

Lifecycle Impact Assessment for Tall Timber Building: Learning from HAUT

Shady Attia(✉)

Sustainable Building Design Lab, Department of UEE, Faculty of Applied Sciences, Université de Liège, Liege, Belgium
shady.attia@uliege.be

Abstract. The HAUT building in Amsterdam, the tallest timber structure in the Netherlands, redefines sustainable high-rise construction. This 21-story hybrid design combines mass timber with concrete, offering a benchmark for the Lifecycle Impact Assessment (LCIA) of tall timber buildings. The chapter addresses challenges in embodied carbon, biogenic storage, and end-of-life pathways, using dynamic LCIA modeling to assess environmental impacts. Research shows timber buildings often surpass a 150-year lifespan, exceeding the 50-year evaluation standard in regulations. However, findings suggest timber-only designs lose efficiency above 60 m, requiring hybrid solutions. Lessons from HAUT provide actionable strategies for advancing sustainable timber high-rises.

Keywords: environmental impact · GHG emission · carbon modelling · Cross-Laminated Timber · end-of-life

1 Introduction to the HAUT Building and LCIA

This chapter provides a comprehensive review of Lifecycle Impact Assessment (LCIA) approaches and challenges for tall timber and hybrid buildings, using the HAUT building in Amsterdam as a central case study. HAUT, a 21-story residential tower and the tallest timber building in the Netherlands, combines a concrete core with Cross-Laminated Timber (CLT) floors and walls [1]. By integrating hybrid structural systems and biogenic materials, HAUT exemplifies how design choices affect embodied carbon, service life, and end-of-life scenarios in high-rise timber construction. Through this case, the chapter aims to bridge practice and theory by examining real-world data while generalizing key LCIA considerations for future timber projects aligned with EU climate neutrality targets [2–3].

2 Goal and Scope Definition

Based on ISO 14044 and CEN 15804, the system boundary for HAUT's LCIA includes raw material acquisition (A1), transport (A2), manufacturing (A3), product delivery (A4), and end-of-life scenarios (C1–C4) [12]. Functional units are defined per cubic meter (m^3) of CLT or square meter (m^2) of structural components [4, 5]. This ensures comparability with other LCIA studies in the construction sector.

G. Fink et al. (Eds.): *Holistic Design of Taller Timber Buildings*, SPRTRCIENG, pp. 177–185, 2025.
https://doi.org/10.1007/978-3-032-02098-7_14

Assumptions include regional forestry practices, standard sawmill energy mixes, and typical transportation distances. Challenges arise from wood species variability, preservation treatments, and the hybrid nature of HAUT's structural systems in the Netherlands. In 2024, the Sustainable Building Design Lab conducted a comprehensive study on the lifespan of timber buildings and their components [5]. The study, based on both expert insights and data from historical timber buildings, concluded that timber structures typically exceed an average life expectancy of 150 years, particularly when well-maintained and properly treated. This finding reflects industry perceptions as well as case studies of existing timber buildings that have surpassed this lifespan.

Biogenic Carbon Storage in Timber: Biogenic carbon, captured during tree growth, constitutes approximately 50% of the dry weight of wood, making timber a significant temporary carbon sink. 1 kg of dry wood sequesters about 0.5 kg of carbon, equivalent to 1.83 kg of CO_2-equivalent (CO_2e) when oxidized. The timber used in the HAUT building was sourced from PEFC-certified forests in Austria, managed under sustainable forestry practices with continuous net growth.

End-of-Life Scenarios: Incineration vs. Landfilling: Timber's environmental impact at the end of its life hinges on its disposal method: (1) Incineration with Energy Recovery: Timber's biogenic carbon is released as CO_{2e}, but the process generates energy that offsets fossil fuel use, reducing net emissions. HAUT's prefabricated components, produced by Binderholz and processed by Brüninghoff, minimize waste throughout the lifecycle [6]. (2) Landfilling: Timber decomposition in landfills can release methane (CH4), a potent greenhouse gas.

3 Life Cycle Inventory (LCI) Analysis

The primary challenge in assessing the building's life cycle lies in integrating embodied, operational, and end-of-life GHG emissions calculations across all stages. Figure 1 illustrates the life cycle modules, covering stages such as A1 (raw material supply), A2 (transport), A3 (manufacturing), and A4 (transport to site). For timber products, these stages include forestry operations, sawmill activities, engineered wood processes, and distribution networks. This comprehensive view highlights the complexity of harmonizing emissions data across these interconnected phases for an accurate life cycle assessment.

Data collection prioritizes primary sources supplemented by established LCA databases (Ecoinvent, GaBi). When primary data prove difficult to obtain, validated secondary sources are used to fill information gaps. All data sets undergo consistency checks to align with ISO 14040/44 and are documented to ensure traceability. Primary data were collected from HAUT's project documentation, including mill-level production data, energy consumption records, and transport logistics. Secondary data were sourced from established LCA databases (e.g., ecoinvent, GaBi) to supplement gaps in the inventory.

Key LCI components include the origin of timber (certified forestry vs. non-certified sources), energy inputs for drying and cutting, by-product utilization (e.g., bark, sawdust), and transport logistics. Distinguishing among different product types—solid sawn lumber, cross-laminated timber (CLT), glued laminated timber (glulam)—is crucial, as

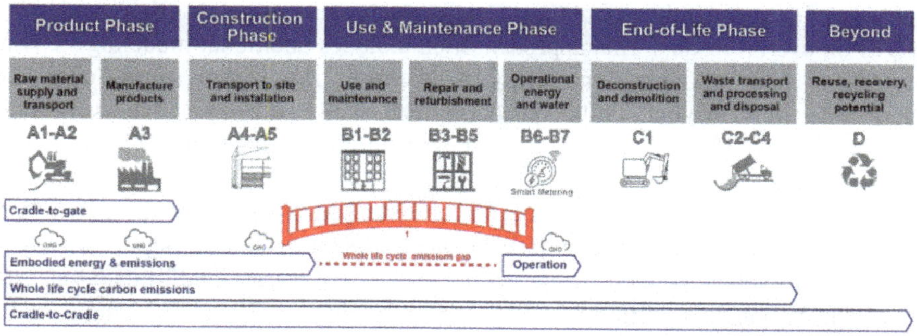

Fig. 1 Building Life Cycle stages and modules and system boundaries [2].

each manufacturing process exhibits unique energy and material flows. Detailed foreground data (e.g., kWh consumed per m^3 of lumber cut) feed into background processes (power generation, fuel production) drawn from recognized databases. HAUT's inventory includes the origin of certified timber, energy inputs for CLT production, and by-product utilization. Transport logistics were modeled to capture emissions from delivering prefabricated components to the construction site.

Service life considerations entail modeling maintenance intervals (e.g., protective coatings every 5–10 years), replacement cycles, and end-of-life scenarios of timber building components and finishes. The chapter specifically addresses how different use conditions (indoor vs. outdoor, high vs. low humidity) affect decay and insect vulnerability [7]. By capturing the frequency of repair or replacement within the 50-year reference period (or another chosen timeframe), the LCI can reflect realistic long-term environmental impacts of timber elements.

4 Life Cycle Impact Assessment (LCIA)

In line with ISO 14044, LCIA emphasizes key categories such as Global Warming Potential (GWP), Acidification, Eutrophication, Photochemical Ozone Creation, and Resource Depletion [4]. Timber's role as a temporary carbon sink is critical for GWP evaluations, as carbon sequestered during tree growth reduces net greenhouse gas emissions—sometimes yielding net-negative results when storage is included. However, common end-of-life scenarios like incineration or landfilling often release this stored carbon, making disposal methods and timeframes pivotal in determining climate benefits. Also, timber's insulation reduces operational energy, though extra layers may offset these gains.

LCIA methods can be categorized based on their impact scope, modeling approach, and geographical relevance. The taxonomy of these are classified by the impact scope including (1) Climate Change Focused (e.g., IPCC 2021, EF 3.0 Climate Change), (2) Midpoint-Oriented (e.g., CML, ReCiPe Midpoint) or (3) Endpoint-Oriented (Damage-Oriented) (e.g., ReCiPe Endpoint, IMPACT World+).

Comprehensive (Multi-Impact) (e.g., EF 3.0, ReCiPe, IMPACT 2002+). It is recommended for timber buildings LCIA to use the IPCC 2021 method because it incorporates biogenic carbon storage, delayed emissions, and cascading use scenarios. Sensitivity analysis evaluates the influence of transport distances, material substitutions, and hybrid designs on overall results. This approach often includes additional modeling steps for delayed emissions, allowing dynamic LCIA models to track carbon storage over a 50-year service life or beyond while considering end-of-life pathways, such as recycling into secondary wood products or energy recovery through incineration [8].

Moreover, a Material Flow Analysis (MFA) has to be conducted for timber buildings to track biogenic carbon uptake, storage, and emissions across the life cycle (A1–C4). At end-of-life (C1–C4), MFA differentiates between reuse in secondary applications (e.g., furniture, particleboard), incineration with energy recovery, or landfill disposal, each influencing carbon fluxes differently. This approach enhances Global Warming Potential (GWP) calculations by integrating cascading material use and disposal scenarios in LCIA.

Finally, sensitivity analysis is essential, as variations in local waste management practices, forestry conditions, and energy mixes can significantly affect outcomes. System boundaries must be clearly defined. Some studies focus on cradle-to-gate impacts (A1–A3 in CEN 15804 terminology) (CEN 15804 + A2), while others adopt cradle-to-grave or cradle-to-cradle perspectives for more comprehensive insights into disposal or reuse scenarios. The functional unit—typically one cubic meter of sawn timber—must align with the system boundary to ensure transparency and comparability across different timber products or building components.

Timber's service life can dramatically influence overall environmental impacts since more frequent replacements raise the total embodied carbon attributable to manufacturing, transport, and construction. Balancing these concerns, proper detailing and careful selection of protective layers can reduce the likelihood of structural damage and allow timber components to remain in use beyond the assumed standard lifespan.

LCIA results often reveal that the initial increase in certain impact categories (e.g., chemical production for treatments) can be more than offset by the avoided manufacturing and transport involved in multiple replacements. These practical implications underscore the importance of accurate service-life data within an LCIA model.

5 Results

Figures 2 and 3 illustrate the percentage increase in greenhouse gas (GHG) emissions between biobased-only and hybrid constructions (A1-A3, excluding under ground) over 50, 100, and 150 years. At 50 years, emissions rise from 5 kg CO_2e/m^3 (biobased) to 8 kg CO_2e/m^3 (hybrid), representing a 60% increase. At 100 years, emissions increase from 4 kg CO_2e/m^3 (biobased) to 5 kg CO_2e/m^3 (hybrid), a 25% rise. By 150 years, emissions grow from 3.4 kg CO_2e/m^3 (biobased) to 5.4 kg CO_2e/m^3 (hybrid), reflecting a 59% increase. These trends demonstrate that hybrid constructions consistently result in higher GHG emissions compared to biobased-only designs over the entire lifespan.

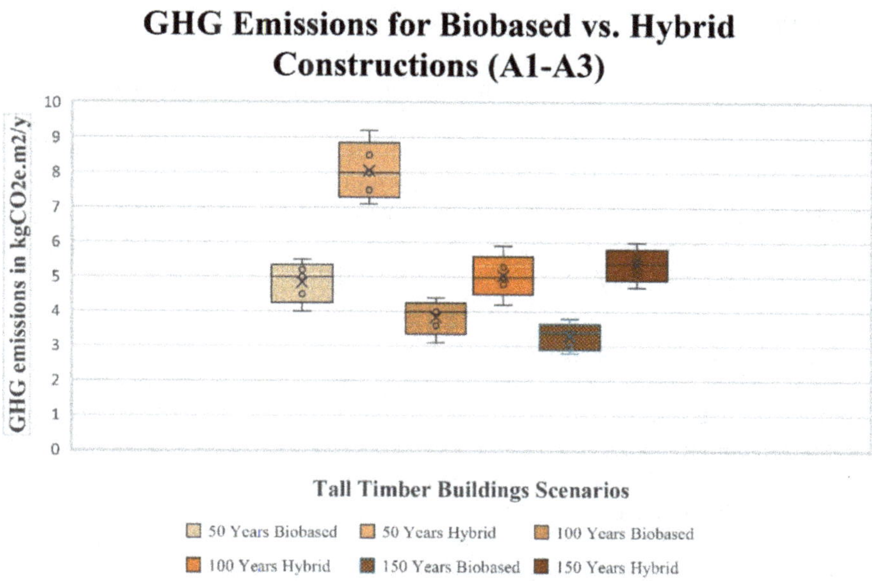

Fig. 2 Comparative bar chart illustrating the greenhouse gas (GHG) emissions for biobased-only construction versus hybrid construction with steel and concrete consolidation over 50, 100, and 150 years

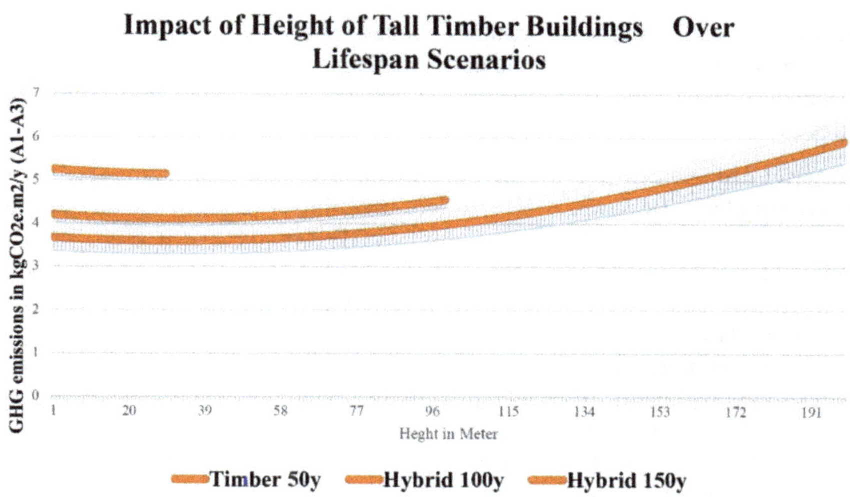

Fig. 3 The impact of GHG emissions across building heights in tall hybrid timber building, emphasizing the relationship between building height and emissions over lifespan scenarios.

6 Conclusions

The HAUT building serves as a benchmark for sustainable tall timber construction, illustrating the synergies between biogenic carbon storage, hybrid systems, and sustainable sourcing. By adopting dynamic LCA methods and prioritizing end-of-life management, HAUT sets a precedent for leveraging timber's environmental benefits while addressing its challenges [10]. As the construction industry strives toward net-zero emissions, projects like HAUT underscore the transformative potential of timber when used thoughtfully and responsibly.

6.1 Key Insights for LCIA in High-Rise Timber Buildings

- Impact Characterization: The IPCC 2021 method improves the accuracy of biogenic carbon accounting.
- System Integration: Hybrid systems, such as HAUT's concrete core and CLT walls, balance structural demands with sustainability goals.
- Lifecycle Durability: Maintenance schedules and protective measures significantly influence long-term environmental impacts.
- Cascading Use: Promoting reuse and recycling of components enhances the sustainability profile.

6.2 Specialized LCIA Considerations for Timber

When dealing with high-rise or complex timber applications, certain characteristics must be captured in the LCIA to avoid underestimating or overestimating impacts. Timber's dimensional changes, load-bearing requirements, fire safety rules, and end-of-life realities can each shift the environmental balance.

- **Material Properties and Biogenic Carbon Storage:** Timber sequesters carbon over its growth phase. However, if over 70% of used timber in Europe is incinerated or landfilled at the end of its life, the ultimate GWP profile can worsen. Shrinkage and deformation also introduce the need for additional connectors, raising resource depletion and embodied carbon, while necessary protective layers may temper any natural insulation benefit.
- **Structural Stability, Foundations, and Underground Features:** Excavation for underground parking or mechanical spaces further inflates environmental burdens, particularly regarding global warming and resource depletion.
- **Fire Safety and Protection:** Fire safety measures in high-rise timber can include chemical treatments or additional layers like gypsum board or concrete cladding, all of which boost embodied impacts and can complicate end-of-life recycling. Regulatory compliance tends to be stricter for timber, often necessitating more conservative design margins or redundant protective systems.
- **Hybrid and Composite Designs:** Many tall or complex buildings employ hybrid solutions: concrete cores for stiffness and elevator shafts, steel frames in critical load-bearing zones, and timber infills or façade elements for reduced weight. This mixing of materials makes LCIA more complex since each material has its own

production processes, maintenance needs, and EoL pathways. Dynamic or seismic loads can also push toward heavier or reinforced assemblies, thereby increasing the share of high-impact materials like steel or concrete. The use of a concrete core in HAUT (73 m) was crucial to ensure the structural integrity required for a building of its height, particularly for wind and seismic resistance. The need for concrete in cores and steel reinforcements increases the embodied carbon footprint. CLT can be effective for buildings up to 60 m, beyond this height, it becomes less efficient or sustainable to rely solely on timber [9]. This is due to the increased material intensity required to meet structural and fire safety standards at greater heights and the growing need for hybrid systems.

- **Lifecycle Durability, Maintenance, and Cascading Use:** Cascading uses strategies—reuse, recycling, or repurposing before disposal—can defer carbon release and reduce reliance on virgin resources. Prefabrication improves quality control and lowers on-site emissions but may raise transport impacts, especially with packaging or long-distance shipping. Tall structures often require significant steel reinforcement, increasing overall embodied emissions.
- **End-of-Life Scenarios and LCIA Adjustments:** If timber is disposed of via incineration or landfilling, biogenic carbon is eventually released. Cascading use offers an alternative: secondary products like panel boards, particleboard, or even new cross-laminated timber (CLT) layers can extend the timeframe over which carbon remains stored. By adopting dynamic LCIA models, practitioners can better capture the net effect of multiple use cycles on overall GWP. Aligning with circular economy policies that incentivize reuse, and recycling can further improve timber's environmental profile.
- **Transportation and Construction Logistics**: Prefabricated components can streamline on-site construction and reduce local emissions, yet greater protective packaging or specialized transport requirements can erode these benefits. Hybrid structures involving many different materials can increase overall transport demands if coordination is poor. Consequently, the net effect on GWP and resource depletion depends on an interplay of factors such as shipping distances, on-site waste management, the ability to integrate deconstruction-ready designs.

6.3 Recommendations for Future Projects

While timber is gaining traction in the construction industry, experts caution against its overuse in high-rise buildings due to technical limitations such as strength and combustibility. Nevertheless, HAUT demonstrates that these challenges can be addressed effectively with thoughtful design and the use of sustainably sourced timber. PEFC certification ensures responsible forest management, aligning with EU policies on sustainable construction. The project highlights timber's potential to reduce embodied carbon in urban densification strategies, aligning with the goals of the European Green Deal. Future LCIA studies should adopt dynamic modeling approaches, prioritize IPCC 2021 for characterization, and incorporate detailed service life data to capture the full environmental benefits of timber high-rise buildings. Studies from the Dutch Environmental Database emphasize the importance of accounting for biogenic carbon in lifecycle assessments [11]. For instance, temporary carbon storage in timber can provide measurable Global Warming Potential (GWP) benefits. However, these benefits must be credited appropriately to avoid double counting, particularly when materials are recycled or reused.

6.4 Limitations of Current LCIA Approaches

Finally, while LCIA methods provide a robust framework for assessing environmental impacts, they still face significant limitations. Most notably, standard LCIA models—particularly midpoint methods like ReCiPe or CML—tend to overlook local ecological impacts, such as biodiversity loss, soil degradation, or changes to local hydrology, especially relevant in forestry-based materials like timber. Current indicators often fail to distinguish between sourcing wood from monoculture plantations versus ecologically diverse forests, leading to an underestimation of ecosystem-level trade-offs. Additionally, social aspects such as labor practices in forestry, land use conflicts, or the displacement of communities are not typically captured in conventional LCA tools. These limitations underscore the need for more holistic, next-generation LCIA frameworks that integrate regionalized biodiversity metrics and social life cycle assessment (S-LCA) into mainstream building assessment workflows.

References

1. Verhaegh, R., Vola, M., de Jong, J.: Haut: a 21-storey tall timber residential building. Int. J. High-Rise Build. **9**(3), 213–220 (2020)
2. Attia, S.: Net Zero Energy Buildings (NZEB): Concepts, Frameworks and Roadmap for Project Analysis and Implementation, 2nd edn. Elsevier, Cambridge, MA (2025)
3. Attia, S., Hoxha, E., Piccardo, C., Schau, E., Novais Passarelli, R., Van Acker, J., Al-Obaidy, M.: Life Cycle Assessment for Taller Timber Buildings Summer School. Sustainable Building De sign Lab., Liege (2024)
4. CEN. 15804:2019+A2 sustainability of construction works—Environmental product declaration—Core rules for the product category of construction products. Brussels, Belgium (2019)
5. Verdier, T., Attia, S.: Lifespan of high-rise timber building components (Webinar 1) [Video recording (2024) https://www.youtube.com/watch?v=zFED83Z5w-A
6. Martien van den Oever, M., Gursel, I.V., Weterings, H., de Munck, E., van der Burgh, F., Verspeek, S., Drissen, J.: Bio-based Building Products in the Dutch Environmental Database (NMD): Proposal for Crediting Biogenic Carbon Storage. Wageningen Food & Biobased Research, Wageningen (2024). https://doi.org/10.18174/647711
7. Arup: Rethinking Timber Buildings: Seven Perspectives on the Use of Timber in Building Design and Construction, London (2019)
8. Hoxha, E., Passer, A., Mendes Saade, M.R., Trigaux, D., Shuttleworth, A., Pittau, F., Al Lacker, K., Habert, G.: Biogenic carbon in buildings: a critical overview of LCA methods. Build. Cit. **1**(1), 504–524 (2020). https://doi.org/10.5334/bc.46
9. Martien van den Oever, M., Telleman, Y., van Kampen, A., van der Voort, M., van der Weide, R., van Baren, S., Jacobs, S.: Biobased grondstoffen voor hoogbouw: Geïndustrialiseerde modulaire en lage emissie hoogbouw in de G4. Wageningen University (2024)
10. Martien van den Oever, M., Gursel, I.V., Elbersen, W., Kranendonk, R., Michels, R., Smits, M.J.: Regional Supply of Herbaceous Biomass for Local Circular Bio-based Industries in the Netherlands. Wageningen Food & Biobased Research, Wageningen (2023)

11. Wageningen Food & Biobased Research: 230607 Timber Myth booklet: A Dialogue Be Tween Our Ambitions and the Facts, Wageningen (2023)
12. ISO, 14040:2006: Environmental Management—Life Cycle Assessment—Principles and Framework. International Standardization Organization, Geneva (2006)

Estimation of Mechanical Properties and Assessment of Existing Timber Structures Using Non-Destructive Tools

Patrick J. McGetrick[1,2(✉)], Roohollah M. Pirooz[1], David Gil-Moreno[1,2], Rimjhim Kashyap[1,2], and Conan O'Ceallaigh[2,3]

[1] School of Engineering and Ryan Institute, Timber Engineering Research Group, University of Galway, Galway, Ireland
patrick.mcgetrick@universityofgalway.ie
[2] Construct Innovate, University of Galway, Galway, Ireland
[3] Atlantic Technological University, Galway, Ireland

Abstract. The structural integrity of existing buildings, particularly those of historical significance, is a critical concern, often complicated by limited knowledge of the original material properties and the effects of aging and changing use. This study explores the use of non-destructive testing (NDT) methods as practical, reliable alternatives to destructive techniques for assessing timber structures. In the first phase, 100 Irish-grown Sitka spruce specimens were tested in a laboratory setting to evaluate the correlation between the dynamic modulus of elasticity (E_{dyn}) - derived from density and Time-of-Flight acoustic measurements - and key mechanical properties such as bending strength, and the local and global bending moduli of elasticity. Bending tests were conducted in accordance with the EN 408 standard. Regression analysis confirmed that E_{dyn} is a reliable indicator of the mechanical properties of timber, supporting its use in the evaluation of existing timber structures. The second phase involved an on-site evaluation of a historic timber roof in Ireland, composed of queen post trusses. Acoustic NDT was used to estimate E_{dyn} and infer mechanical properties, while moisture content (MC) measurements helped adjust readings and identify potential decay. This assessment provided insights into the condition of the structure and highlighted practical limitations of in-situ testing while recommendations are provided for future inspections. Overall, the study demonstrates that the NDT methods can be effective for evaluating the mechanical performance of timber in both laboratory and real-world contexts, offering a viable approach to support the preservation and maintenance of timber structures.

Keywords: Acoustic Methods · Non-Destructive Testing · Time-of-Flight · Dynamic Modulus of Elasticity · In-Situ Measurement · Timber

1 Introduction

The in-situ assessment of existing timber structures is of critical importance as they age. Several factors can adversely affect the integrity and performance-among these factors are moisture content, fungal infestation, insect damage, exposure to chemicals, biological

G. Fink et al. (Eds.): *Holistic Design of Taller Timber Buildings*, SPRTRCIENG, pp. 186–204, 2025.
https://doi.org/10.1007/978-3-032-02098-7_15

growth, aging, poor maintenance and historical building usage. Consequently, inspecting existing structures to ensure their safety and serviceability is essential, particularly for older buildings lacking proper maintenance. Furthermore, modern timber buildings need reasonable assessment methods that are economically feasible due to the increase in mid and high-rise timber buildings, typically constructed using softwood mass timber. To ensure that structural integrity is not compromised, a thorough assessment of the structure may be necessary. This is particularly crucial in buildings of historical significance, highlighting the imperative to rehabilitate and preserve these buildings for future generations.

The assessment and conservation of timber structures requires steps to ensure their safety. Typically, a desk study is the first step, mainly aimed at establishing the history of the building and its previous uses. Next, a preliminary visual survey is carried out to identify decay, mechanical damage, etc. that may need to be investigated in more detail at a later stage. This can also include the determination of the measures required to ensure safe access to the structure. Measuring relative humidity also helps to focus efforts, since areas of high humidity can highlight areas susceptible to rotting and fungal attacks. The next important step is the identification of species. This allows a better understanding of the structural behaviour of timber elements, as well as a first approximation of density values. The next phase of the inspection, which is sometimes the last phase of the survey, must address the determination of mechanical properties; this is the primary focus of this study.

To gain insight into the mechanical properties of structural elements of an existing building, destructive testing methods are an option. In such cases, it is required to cut a sample from the element to destructively test it in bending, tension, or compression. However, this approach may not always be feasible, particularly if the building under consideration is of historical or architectural significance, or if access is restricted at occupied floors in a mid or high rise building. As a general rule, it is not possible to disassemble the timber members, which would allow inspection of all four sides, measurement of knots, stress testing, etc. Instead, a non-destructive assessment is generally the only valid option. Non-destructive techniques (NDT) are of greater interest to engineers, researchers, and stakeholders. Compared to destructive methods, they are typically more cost-effective, practically feasible, and allow for rapid evaluation.

A diverse array of techniques and devices exists for the non-destructive condition assessment of existing structures. In a comprehensive examination of a brick-timber architectural heritage building, Jiao et al. [1] utilized the ArborSonic 3D material tester to detect decay in timber columns. Mol et al. [2] utilised laser scanning to construct 3D models of existing timber structures, followed by the application of NDTs such as pin penetration, drilling resistance tests, and ultrasounds to assess their condition. The laser scanning and drilling resistance tests were also employed by Abraldes et al. [3] for condition assessment of historic timber structures.

The mechanical properties of timber structures are also commonly evaluated using acoustic wave measurements. Wood stiffness, similar to other materials, is linked to the acoustic behaviour of a wave passing through it. As a result, the dynamic modulus of elasticity (E_{dyn}) can be calculated based on the velocity of the acoustic stress wave combined with the density of the material. E_{dyn} has been used in several studies in order to estimate the modulus of elasticity (E_{m}) and strength (f_m) of timber specimens [4–10]. This method offers a cost-effective and straightforward means to measure stiffness and predict bending strength. The relationship of E_{dyn} with E_{m} is typically strong. However, caution must be exercised when estimating the strength of a structural member as the weakest cross-section may occur due to a local defect whose influence will vary depending on the position and the type of load subjected on the timber member.

The most common devices used for acoustic assessment are either based on resonance, which measures the resonant frequency obtained by impact excitation of the timber specimen, or Time-of-Flight (ToF), which calculates the speed of propagation of a sound wave by measuring the time delay between two points. The resonance method requires the piece to vibrate freely and therefore cannot be applied to a specimen incorporated within a structure. Arriaga et al. [11] utilized stress wave ToF and probing tests for the in-situ assessment of a 7-story timber structure in Madrid, Spain. Arriaga et al. [12] utilized drilling chip extraction and stress wave velocity to estimate the mechanical properties of an eighteenth-century structure. Percussion-based techniques have been shown to be applicable for accurate evaluation of wood elements' internal structure [13]. The application of NDTs (stress wave, micro drilling resistance) has also been demonstrated in development of a structural health monitoring system for timber structures by Wang et al. [14]. Riggio et al. [15] summarize the most common NDT techniques for in-situ assessment of structural timber, comparing advantages and limitations [15].

Building on scientific studies such as those cited above, this study investigates the use of NDT for the prediction of mechanical properties of the species on which the forest industry in Ireland is based, Sitka spruce (*Picea sitchensis* (Bong.) Carr.). This knowledge is critical given that mechanical properties and their relationships with predictor variables are often species and site specific among others. In particular, the first phase of this study seeks to delve deeper into the correlation between acoustic measurements (ToF), visual inspection, and the mechanical properties of Irish-grown Sitka spruce when subjected to bending under laboratory conditions.

ToF measurements obtained using acoustic assessment methods are typically conducted in controlled laboratory settings using an end-to-end approach. However, when assessing real-world structures, this method is impractical, particularly if the structural timber is hidden behind secondary fixings or finishes such as plasterboard. Instead, alternative sensor placements must be considered. Therefore, this paper also focuses on the challenges of an *in situ* inspection of timber structures, focusing on the assessment necessary to estimate mechanical properties for design purposes. The second part of the study outlines the tests and analysis conducted by members of the Timber Engineering Research Group at the University of Galway to evaluate the condition of a historic building in Ireland.

2 Materials and Method

2.1 Laboratory Materials and Equipment

In the experimental laboratory phase of this study, 100 pieces of Sitka spruce grown in Ireland was sourced from the regular production runs of an Irish sawmill. Each piece had a nominal cross-section of 100 × 44 mm and a length of 3.6 m. The FAKOPP

Microsecond Timer (Fig. 1) was employed to determine the Time-of-Flight (ToF) for the full length of each piece. This device measures the time delay for an acoustic wave to travel between a start and stop sensor. Given the length (L), mass (M), width (H) and thickness (B) of the section (measured and averaged on three points) and ToF, the E_{dyn} was computed as Eq. (1):

$$E_{dyn} = \rho V^2 = \frac{M}{LHB}\left(\frac{L}{ToF}\right)^2 \tag{1}$$

Where ρ represents the wood density and V denotes the acoustic velocity, or speed of sound calculated from distance measured divided by time delay.

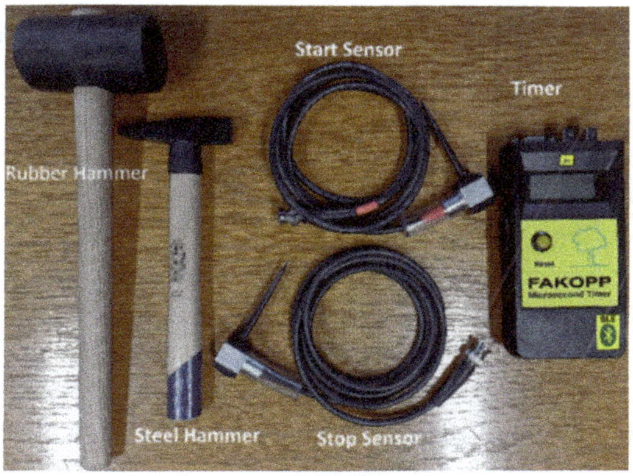

Fig. 1. FAKOPP microsecond timer

Figure 2 illustrates the experimental setup of the FAKOPP microsecond timer for determining the ToF of the timber boards. Arriaga et al. [11] and Íñiguez-González et al. [16] explored various sensor placements, suggesting an end-to-end configuration for sensors. As illustrated in Fig. 2, this study employed these proposed sensor positions to ascertain ToF. The sensors were placed at the ends of the timber, and a rubber hammer was used to drive them in. Upon striking the start sensor with a steel hammer, an elastic wave was generated, propagating along the piece. Subsequently, the timer recorded the elapsed time (ToF) for the wave to traverse from one end to the other in microseconds. Although accessing the entire length of elements in existing buildings is often impractical, this setup offers the most precise ToF estimation in the laboratory setting.

2.2 Methodology for Laboratory Study

The boards were subjected to X-ray scanning using the Goldeneye 702 system (by MiCROTEC s.r.l. – GmbH) at Murray Timber Group, Galway, Ireland. The Goldeneye 702 employs X-ray radiation attenuation to detect and locate knots and defects in timber. The critical section for destructive testing in bending was determined based on the X-ray images obtained and validated by visual inspection. The data obtained at the sawmill

Fig. 2. Experimental setup to determine ToF

was analysed to pinpoint the weakest positions for the bending tests. Additionally, the accuracy of the results from the X-ray images was confirmed through visual inspection in the laboratory.

In accordance with the guidelines specified in EN 408 [17], a length of 1800 mm from the boards was designated for destructive testing in a four-point bending configuration, with the weakest section positioned at the midpoint of the span.

The knots were measured around the weakest section of the piece, and the total knot area ratio (tKAR) and marginal knot area ratio (mKAR) were calculated using the online software Web Knot Calculator v2.2 by MiCROTEC. According to IS 127 [18], the tKAR represents the cumulative cross-sectional areas of all knots within a 150 mm length, divided by the cross-sectional area of the piece. mKAR was determined in the same manner as tKAR, but at a margin of 1/4 of the board's depth. Based on the results derived from the destructive bending tests, the local modulus of elasticity (E_l), global modulus of elasticity (E_g), and bending strength (f_m) were determined for each piece in accordance with EN 408.

To compute E_l, the relative deflection of the mid-section was measured relative to points positioned at a distance of 2.5 times the section height, where theoretically there is no shear stress. Additionally, E_g was determined by computing the deflection of the midpoint relative the supports. Figure 3 demonstrates the schematic of four-point bending tests carried out in this study. It is worth mentioning, since the structural elements of an existing timber building might possess any level of moisture content, the mechanical properties derived from destructive test results were not adjusted to any reference moisture content. Instead, they were utilized for further analysis based on the moisture content present at the time of testing.

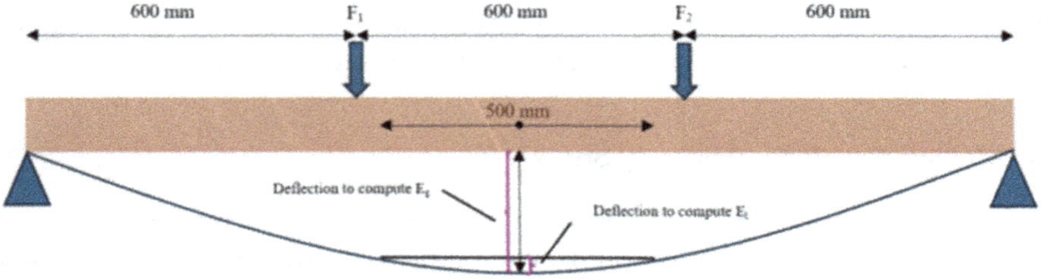

Fig. 3. Four-point bending test schematic

2.3 Methodology for the Case Study Assessment on an Existing Building

The case study assessment was carried on an existing structure that was built in the early nineteenth century whose roof has been subject to water ingress in recent years. The exact location of the building is not provided to protect sensitive information. The structure comprised of a series of roof trusses is depicted in Fig. 4. Some truss ends and the bearing ends of other components were embedded into the walls, while others were supported by corbels. As it was not possible to carry out destructive measurements, acoustic NDT measurements were carried out on a total of 77 timber members: beams, struts, queen posts, and rafters.

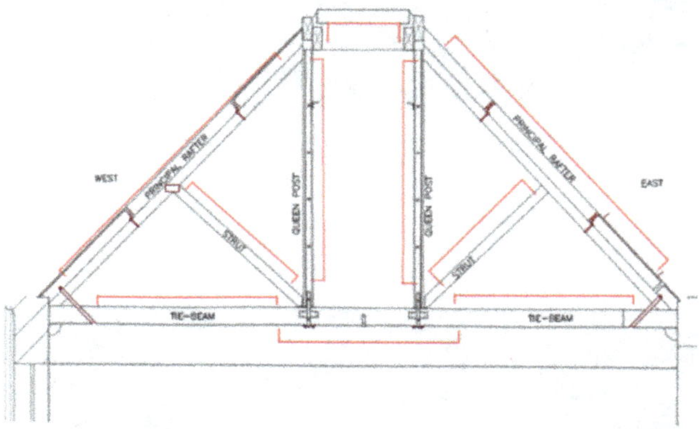

Fig. 4. Queen truss with position of ToF measurements carried out marked in red

Measurements were carried out using the Microsecond Timer, MST (Fakopp, Sopron, Hungary; Fig. 1), over a 5 day period. As presented earlier, the MST is an impact stress wave device that measures the time delay for a wave induced by a hammer to travel between two sensors inserted at a known distance apart in the timber member, thus enabling calculation of the wave velocity, V. This velocity can be used to calculate the dynamic modulus of elasticity as per Eq. (1).

Due to the impossibility of accessing the ends of the timbers, by default, both sensors were placed on the same face of the assessed member (Figs. 5 and 6). This was found to offer the best estimation for the prediction of the mechanical properties [19] in situ where an end-to-end configuration is not feasible. If the readings were judged to be defective, crossed measurements were carried out, placing a sensor on one face and the other on the opposite face. The measurements were judged to be reliable when three identical consecutive readings were obtained, or when a deviation of less than 2% was observed between readings. Two members, in particular two restraining beams, could not be assessed acoustically since the readings were judged to be unreliable. It must be noted that besides the ends of the timber members, sections near the connections (Fig. 5) could not be assessed due to space limitations for hammering in the positions of the sensors. These locations were inspected visually and assessed with a resistograph R650-SC (RinnTech, Heidelberg, Germany). This is a device that measures the resistance against the introduction of a small diameter drill at a constant speed [20]. However, this assessment is outside the scope of the study at hand.

Fig. 5. Measurements of moisture content on a queen post (left) and detail of a sensor near a connection (right)

Fig. 6. ToF measurements on the lower edge of a rafter; start and end probes marked in red and yellow measuring tape covering a distance of almost 6 m.

Due to the impossibility of accessing the ends of the timbers, both sensors were placed on the same face of the assessed member (Figs. 5 and 6). This was found to offer the best estimation for the prediction of the mechanical properties [19] in situ where an end-to-end configuration is not feasible. If the readings were judged to be defective, crossed measurements were carried out, placing a sensor on one face and the other on the opposite face. The measurements were judged to be reliable when three identical consecutive readings were obtained, or when a deviation of less than 2% was observed between readings. Two members, in particular two restraining beams, could not be assessed acoustically since the readings were judged to be unreliable. It must be noted that besides the ends of the timber members, sections near the connections (Fig. 5) could not be assessed due to space limitations for hammering in the positions of the sensors. These locations were inspected visually and assessed with a resistograph R650-SC (RinnTech, Heidelberg, Germany), which measured the resistance against the introduction of a small diameter drill at a constant speed [20], however, this assessment is outside the scope of the study.

The angle of insertion of the MST probes was approximately 45°, as recommended by the manufacturer, and always less than 60 degrees. The measurement distance between probes was extended over the maximum possible length of the truss members, and at least 2 m (Fig. 6). This obeys the velocity dependency on length, due to a time-lag in measured time [4, 5] that produces a tendency for velocity to decrease as the distance measured increases. Previous studies on Scots pine found that for each additional metre measured in the range from 1 to 4 m, the velocity decreased by approximately 83 m/s [21], whereas Arriaga et al. [22] quantified the difference as 70 m/s per metre, equivalent to a 1.4% decrease for each additional metre. A decrease in velocity of roughly 1.4% per metre was also observed in Llana et al. [5] for Scots pine. To reduce any bias of the shorter measurements below 4 m presented in this study, the measured velocities for these measurements were adjusted to the reference length of 4 m by applying a reduction of 1.4% for every metre shorter. No adjustments were applied to measurements over 4 m.

The tree species identified within the trusses was Scots pine (*Pinus sylvestris* L.), commercially known as red deal or redwood. Models published in scientific literature predicting the mechanical properties of this species from acoustic measurements were examined [4–6, 23]. The models determined by Montero García-Andrade [23] using the MST on Scots pine was chosen. The study included pieces with and without wane, which is deemed relevant for the current structure:

$$E_m = 942.177 + 0.665957 E_{dyn,end} \tag{2}$$

$$f_m = -6.9677 + 0.00362785 E_{dyn,end} \tag{3}$$

However, the models for Eqs. (2) and (3) are based on end-to-end measurements. Hence, the velocities, measured on the surface, were corrected to predict the end-to-end values according to [24] before applying Eq. (1) to obtain $E_{\mathrm{dyn,end}}$. Regarding the measurement of density ρ, before conducting the acoustic assessment, a preliminary inspection team used a Pressler borer to extract sample cores from various points throughout the structure, focusing on key areas that were potentially more susceptible to decay. These cores were used to calculate density, and the fifth percentile value (the characteristic value) was used for the calculation of E_{dyn}.

The moisture content (MC) was measured using a Hydromette HT 85 (GANN Mess-u.Regeltechnik GmbH, Germany; Fig. 5) with long pins that roughly penetrated a third of the thickness of the element assessed, as per the standard EN13183-2 [25]. These measurements were aimed at obtaining an overview of the moisture content, and also to adjust the acoustic measurements, influenced by the MC, to a reference 12% MC. Additionally, the MC measurements allowed for the detection of areas of high MC where decay may occur, although it was not an objective to specifically find and measure points that could be more prone to high moisture content.

3 Results of Laboratory Study

3.1 Acoustic Measurements and Moisture Content (MC)

Figure 7 illustrates the distribution of E_{dyn} for the 100 boards examined in this project. As shown, 52 boards exhibit E_{dyn} values between 10–12 kN/mm^2. The average E_{dyn} also falls withing this range (11.1 kN/mm^2). The minimum, maximum and standard deviation for E_{dyn} are 6.5, 15.3 and 1.7 kN/mm^2 respectively.

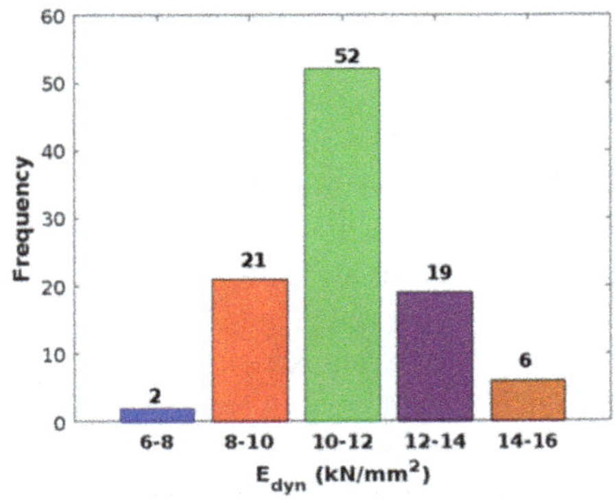

Fig. 7. Distribution of E_{dyn} from laboratory tests

After each test, a small piece free of knots and other defects was cut as close as possible to the point of failure, covering the entire cross-section of the board to determine the clear density. The samples were then dried in an oven and based on their dimensions and mass (before and after drying), their Moisture Content (MC) was calculated following the guidelines outlined in EN 13183-1 [26]. The average MC of all 100 tested samples was 20%, with 2 exceeding 25% and 10 with MC below 16%.

3.2 Relationship Between E_{dyn} and Mechanical Properties

After testing, the f_m, E_l and E_g properties for each board were determined according to the provisions outlined in EN 408 [17]. To gain an understanding of the distribution of these mechanical properties, Table 1 summarises statistical characteristics for f_m, E_l and E_g.

In their study, Gil-Moreno et al. [27] reported an average bending strength value of 30.8 N/mm^2 for British spruce (a blend of Sitka spruce and Norway spruce), which is 7.8% lower than the value found in our this research.

The relationship of mechanical properties (f_m, E_l and E_g) with E_{dyn} were examined and the strength of fitting a linear model was evaluated by calculating the coefficient of determination (R^2). In Fig. 8, E_{dyn} is plotted against E_g and the straight line fitted to the data is also being shown. The coefficient of determination (R^2) has been calculated as 0.62, which confirms a reasonable correlation between E_{dyn} and E_g.

Table 1. Statistical summary of mechanical properties

	Property (N/mm²)		
	E_g	E_l	f_m
Minimum	3501.4	4591.4	11.7
Maximum	13051.9	12214.4	54.1
Mean	7772.6	7945.9	33.4
Standard Deviation	1937.9	1611.4	7.6

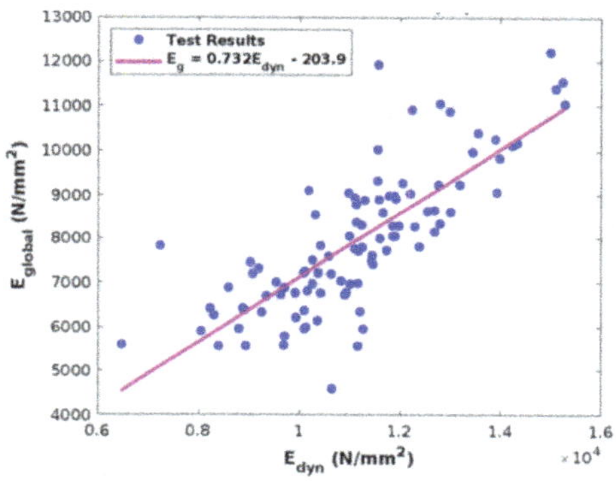

Fig. 8. E_{dyn} versus E_g ($R^2 = 0.62$)

Incorporating the knot indexes into the model for predicting E_g results in a slight increase in R^2. With tKAR and mKAR included, the R^2 values increase to 0.65 and 0.66, respectively. This outcome is within expectations. While E_g characterises the overall stiffness of the piece in bending along its entire length, the knot indexes were measured only locally around the critical section of the board (within 150 mm to the left and right). It is important to note that since E_{dyn} is determined for the full length of the piece, it inherently encompasses the influence of all defects. Consequently, incorporating knot indexes into the model is not expected to enhance the output significantly. The revised model equations are presented in Table 2 as Eqs. (4) and (5).

The relationship between E_l and E_{dyn} was explored in Fig. 9. The findings indicate that E_l exhibits a weaker dependence on E_{dyn} compared to E_g. Nevertheless, E_{dyn} can still be effectively utilized for predicting E_l. Since E_l solely characterizes the behaviour of the board within the pure bending region (between two-point loads in the bending test), it is reasonable to anticipate that incorporating the knot indexes could enhance the predictability of the model. Accordingly, a plane was fitted to the data to represent E_l as a function of E_{dyn} and knot indexes, and the outcomes are presented in Table 2 (Eqs. 6 and 7).

Integrating tKAR into the model raises R^2 from 0.53 to 0.60. Conversely, the introduction of mKAR to the model leads to a significant change, elevating R^2 to 0.66. This underscores that utilising both E_{dyn} and mKAR enables an accurate estimation of E_l.

Table 2. Revised model equations incorporating mKAR and tKAR

Equation	Equation Number
$E_g = 0.704E_{dyn} - 2465.8tKAR + 1031.7$	(4)
$E_g = 0.720E_{dyn} - 2074.9mKAR + 959.6$	(5)
$E_l = 0.767E_{dyn} - 4205.5tKAR + 808.0$	(6)
$E_l = 0.787E_{dyn} - 4539.6mKAR + 1246.2$	(7)
$fm = 0.0027E_{dyn} - 10.2tKAR + 7.3$	(8)
$fm = 0.0027E_{dyn} - 15.7mKAR + 11.0$	(9)

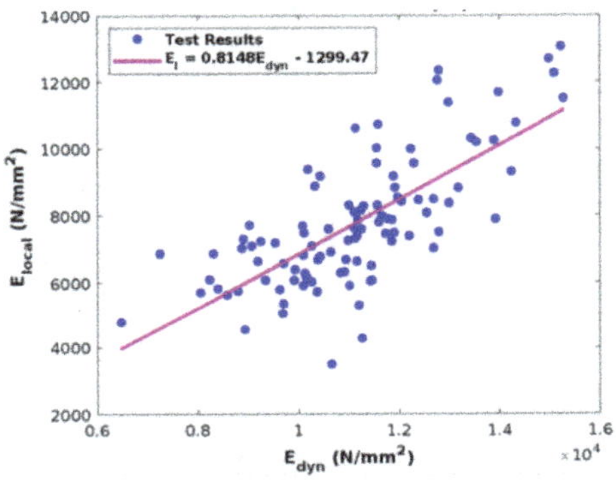

Fig. 9. E_{dyn} versus E_{l} (=E_{local}) ($R^2 = 0.53$)

Figure 10 depicts E_{l} against E_{dyn} and mKAR. In this figure, it can be inferred that as E_{dyn} increases and mKAR decreases, the E_{l} will increase.

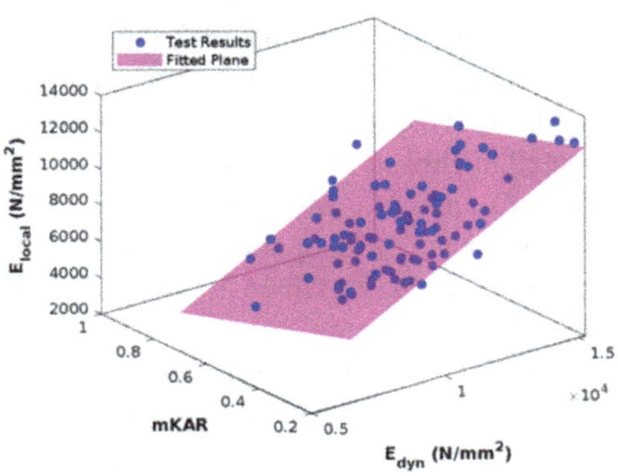

Fig. 10. E_{l} (=E_{local}) versus E_{dyn} and mKAR ($R^2 = 0.66$)

The same procedure was applied to investigate the relationship between f_m and E_{dyn}. As depicted in Fig. 11, f_m exhibits a moderate correlation with E_{dyn}, with an R^2 of 0.40. Utilising tKAR in the model yields only a slight increase in R^2 to 0.43. The inclusion of mKAR proves more effective, elevating the R^2 to 0.50. Thus, it can be inferred that a good estimate of f_m can be achieved using E_{dyn} and mKAR. Equations (7) and (8) (Table 2) represent the regression model of f_m with E_{dyn}, tKAR, and mKAR.

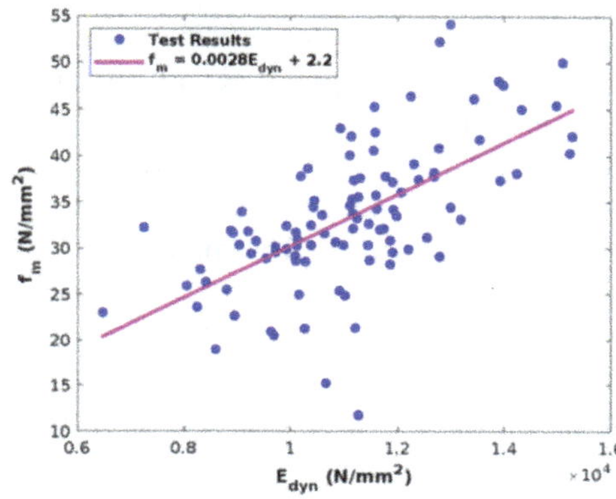

Fig. 11. E_{dyn} versus f_m ($R^2 = 0.4$)

3.3 Laboratory Results Discussion

As stated in earlier sections, the aim of this project is to examine if E_{dyn} evaluated using FAKOPP microsecond timer is a trustworthy parameter to estimate the mechanical properties of existing timber structures. Table 3 summarizes the R^2 values for different modeling scenarios to predict E_g, E_l and f_m.

Table 3. R^2 values for different models

	Independent Variables		
	E_{dyn}	E_{dyn}, tKAR	E_{dyn}, mKAR
E_g	0.62	0.65	0.66
E_l	0.53	0.60	0.66
f_m	0.40	0.43	0.50

E_g demonstrates the strongest correlation with E_{dyn} and can be estimated with high reliability. Incorporating tKAR and mKAR slightly improves the model, with the best results achieved by utilizing E_{dyn} and mKAR. E_l can also be predicted solely using E_{dyn}. However, significant enhancement in the model occurs when mKAR is integrated, increasing the R^2 from 0.53 to 0.66. Lastly, f_m exhibits only a moderate correlation with E_{dyn}. Although the inclusion of knot indexes could enhance the model, its correlation remains lower compared to E_g and E_l. By selecting E_{dyn} and mKAR as independent variables to estimate E_g and E_l, both dependent variables yield almost the same level of R^2, ensuring reliable predictions.

Arriaga et al. [4] examined 120 pieces of four distinct pine species grown in Spain. They reported higher R^2 values (0.76 to 0.85) compared to those in this study. This variation could be attributed to differences in species. In another study by Sequera et al. [28], E_{dyn} was determined using the longitudinal natural frequency of vibration for Salzmann pine timber pieces. They reported R^2 values ranging from 0.59 to 0.68, which align with those obtained in this study. Gil-Moreno et al. [8] reported R^2 values of 0.67 and 0.50 for the correlation between bending stiffness and bending strength of Douglas fir respectively with E_{dyn}, which align closely with the results of this paper. They employed frequency-based methods to determine E_{dyn} in their study.

In comparison to other methods such as resonance-based acoustic techniques, Llana et al. [5] have demonstrated that acoustic velocity determined using ToF is higher. They noted that with a piece length of 4 m, the error in computing acoustic velocity is only 3%. Given that the pieces' length in this study is 3.6 m, it is reasonable to assert that the values obtained for ToF, and acoustic velocity are reliable.

4 Results of the Case Study Assessment on an Existing Building

A descriptive summary of the measured properties of the timber elements in the historical building in Ireland, as well as the predicted values, is given below. The mean MC was 15.7%. The highest MC was 20.1% and 19.4%, both on the west side of the building which was more exposed to the weather conditions. The lowest MC was 12.9%. Essentially, all members had a MC below 20% at the measured locations, which is the limit above which timber is more prone to wet rot and insect attack.

Density values from 67 core samples extracted ranged from 441 to 781 kg/m^3, with a mean of 619 kg/m^3 and a characteristic value of 489 kg/m^3. The $E_{dyn,end}$ values ranged between 7.17 kN/mm^2 and 13.9 kN/mm^2 (average 12.0 kN/mm^2). These correspond to velocities of between 3877 m/s and 5403 m/s (mean 5005 m/s). These values are only slightly below the reported near-mean of 5300 m/s for Scots pine in Spain using the MST as given by [4] for new sawn timber.

Applying Eqs. (2) and (3), the modulus of elasticity values ranged between 5.7 and 9.9 N/mm^2 (average 8.7 N/mm^2), whereas strength ranged between 19 and 42 N/mm^2 (average 35.3 N/mm^2). It is worth noting that these are individual values for each assessed length and therefore they do not directly correspond to design values, which are typically based on sample population mean values for the modulus of elasticity ($E_{m,0,mean}$) and the fifth percentile of values for bending strength ($f_{m,k}$).

The density values measured from cores showed higher density values than that typically expected for Scots pine. Density values for Scots pine samples grown in Ireland had an approximate characteristic density value of $\rho_k = 440$ kg/m^3 and a mean value of $\rho_{mean} = 510$ kg/m^3 [4]. For comparison, the characteristic density value for timber of strength class C24 is $\rho_k = 350$ kg/m^3 and for C50 is $\rho_k = 430$ kg/m^3. Scots pine was the predominant conifer cultivated in Ireland until around 1950, although the forested area during the construction of the building was minimal. It can be speculated that the timber originated from Scandinavia or Scotland. Notably, the density of Scots pine in Scandinavia exhibits a wide range of mean values depending on tree age, ranging from 448 kg/m^3 for trees less than 50 years old [29] to 505 kg/m^3 for those aged 80 years [30].

4.1 Discussion on the Application of NDT for Existing Structures

The inspection of existing timber structures, whether historic or modern high rise, includes various phases, with the evaluation through non-destructive techniques being among the most crucial. It is essential, however, to acknowledge the limitations of these techniques, regardless of the specific method employed. Therefore, it is of utmost importance that contractors understand these limitations before commencing the NDT measurements, as their significance may vary depending on the intended use of the building.

For example, space limitations in the current study prevented the acoustic measurements from reaching the end of some members, affecting the consistency of readings. As well as that, and despite efforts to ensure reliable data collection, variations were observed in readings from certain members. This may also be a result of the variation of the moisture content within a member, but could also result from the presence of nails and other metallic elements, etc. To mitigate this, only readings deemed to be reliable and based on comparison with data from other members should be used.

While the primary purpose of NDT measurements often involves estimating mechanical properties, it is imperative to recognise that accurate determination can only be achieved through laboratory testing, particularly when assessing strength, which necessitates testing the piece to failure. In addition to the limitations of the techniques, and the understanding of the non-destructive values measured, the estimation of properties depends on statistical models that do not always fit the characteristics of the timber members, due to variations in geometries, moisture content, and other factors. In addition, wood properties can vary notably between species and even within different growing conditions, resulting also in changes in relationships with predictor variables. Furthermore, the setups in which measurements are conducted may vary across studies. Therefore, it is advisable to use models that are specifically developed for the timber species of interest and are established under as similar conditions as possible. However, this may pose challenges in existing buildings where prior studies were not feasible. This the motivation for the laboratory part of this study - to avoid such situations and ensure more accurate assessments in the future; the preliminary study was aimed at establishing the relationship between MST measurements and mechanical properties in Sitka spruce.

5 Conclusions and Recommendations

5.1 Laboratory Study of NDT

In the laboratory study presented here, the utilization of the FAKOPP microsecond timer as a tool for non-destructive evaluation of existing timber structures was examined. The ToF has been determined for 100 Irish-grown Sitka spruce timber of size 100 × 44 mm and subsequently E_{dyn} was calculated. The boards were then subjected to destructive testing in bending and E_g, E_l and f_m have been evaluated.

The laboratory results indicate that E_{dyn} can effectively estimate the mechanical properties of structural elements with a high degree of accuracy. The strongest correlation was observed between E_{dyn} and E_g. By incorporating mKAR in regression analysis, the model for predicting E_l can be improved to the level of E_g. Although f_m demonstrates a moderate correlation with E_{dyn}, introducing mKAR into the model can yield a reasonably accurate estimation of bending strength.

As stated earlier, this study was carried out in the laboratory where there is high control on important parameters. However, the results showed this method can be reliably employed for in-situ assessment of existing timber structures. Studies of this nature must be extended to real existing structures. In such cases, there is usually limited access to the structural elements and the weakest point of an element is not necessarily positioned at the mid-span. Such factors have an effect on final results and must be addressed in future studies.

Based on the results obtained in this study, ToF determined by the FAKOPP microsecond timer can be employed to accurately estimate the mechanical properties of structural elements as an indicator of structural safety, integrity, and planning for possible required maintenance operations.

Alternative devices employing different mechanisms can be employed to determine E_{dyn}. For instance, Gil-Moreno et al. [27] utilized the resonance technique and established strong correlations between E_{dyn} and the mechanical properties of spruce and Douglas fir. However, effective use of this device necessitates access to the full length of the piece and is primarily suitable for laboratory settings. On the other hand, the sensors of the FAKOPP microsecond timer can be easily implemented for visibly clear lengths of structural elements under consideration, providing an advantage over other ultrasonic devices that require access to the ends of the element.

It is worth noting that this study was carried out using Irish Sitka spruce, for which there is limited published knowledge on its mechanical properties. Consequently, exploring whether NDT methods can accurately estimate the mechanical properties of this species is of great importance as its use increases in construction.

Finally, the FAKOPP microsecond timer not only delivers precise measurements of E_{dyn} but also stands out for its ease of setup, making it an ideal option for non-destructive assessments. As depicted in Fig. 2, this device can be swiftly attached to structural elements within an existing timber building. This flexibility presents a significant advantage over other techniques and devices that may require application prior to incorporation into the structural frame. In this regard, a case study of an in-situ assessment in Ireland was also described where FAKOPP microsecond timer was employed for estimation of mechanical properties of structural elements of an existing timber building; associated conclusions are presented in the following section.

5.2 Case Study Assessment of the Existing Building

Timber structures traditionally played a minor role in Irish buildings. However, they were and still are widely used for roofing applications and roof trusses. Historically, timber has been imported mainly from Scandinavia, a region also with a long tradition of timber construction due to its significant timber production. There is now a growing interest in timber buildings, mainly constructed using light timber frame or mass timber. As a result, in-depth knowledge of local and imported timber properties and assessment tools will be increasingly required for in-situ timber and building inspections. Therefore, in addition to the laboratory investigation, this chapter also presented the application of a non-destructive technique in practice as part of the assessment of an existing timber building.

Overall, the timber members assessed in the historical building are in good condition. However, many of the timber members have had some of the original section reduced, either physically or functionally as a result of deterioration of the outer fibres on the surface as the result of previous uses of the building. Therefore, it is recommended that future structural calculations for such elements consider a smaller effective cross-section. Likewise, the acoustic assessment did not reach the full length of the members, and it is advisable to check these sections using alternative NDTs, such as resistance drill techniques to determine the condition of the timber.

To prevent measurement errors (time lag), it is advised to measure time-of-flight over distances of at least 2.5–3 m and always as long as is physically possible to improve the reliability of measurements. It must also be noted that the density value typically used for the calculation of E_{dyn} in the laboratory is that of the full timber board. In the absence of more accurate information for an in-situ assessment, it may be advisable to use the fifth percentile of the measured densities, which will provide conservative values of E_{dyn}.

Finally, it must be remembered that the mechanical properties reported are the result of the use of predictive models based on measurements with non-destructive techniques. Although they may be a good approximation, statistical models are subject to inaccuracies and use of an appropriate factor of safety by design engineers in structural calculations is to be recommended.

Acknowledgements. This research was conducted at the University of Galway as part of the BTIS project, funded under the 2022 Policy & Strategic Studies Research Call by the Forest Sector Development Division of the Department of Agriculture, Food, and the Marine in Ireland, Project Ref 2022PSS153. The authors extend their appreciation to Murray Timber Group for providing access to the Goldeneye 702, and Colm Walsh and Peter Fahy for assisting during the destructive testing phase. The authors also wish to express their gratitude to the Northern Research Station of Forest Research UK for lending the MST, and to Edinburgh Napier University for providing the resistograph, both of which were indispensable for the assessment presented.

References

1. Jiao, J., Xia, Q., Shi, F.: Nondestructive inspection of a brick–timber structure in a modern architectural heritage building: lecture hall of the Anyuan Miners' Club, China. Front. Archit. Res. **8**(3), 348–358 (2019). https://doi.org/10.1016/j.foar.2019.06.005
2. Mol, A., Cabaleiro, M., Sousa, H.S., Branco, J.M.: HBIM for storing life-cycle data regarding decay and damage in existing timber structures. Autom. Constr. **117**, 103262 (2020). https://doi.org/10.1016/j.autcon.2020.103262

3. Abraldes, P., Cabaleiro, M., Sousa, H.S., Branco, J.M.: Use of polar coordinates for improving the measurement of resistant cross-sections of existing timber elements combining laser scanner and drilling resistance tests. Measurement. **204**, 112027 (2022). https://doi.org/10.1016/j.measurement.2022.112027
4. Arriaga, F., Llana, D.F., Esteban, M., Íñiguez-González, G.: Influence of length and sensor positioning on acoustic time-of-flight (ToF) measurement in structural timber. Holzforschung. **71**(9), 713–723 (2017)
5. Llana, D.F., Iñiguez-Gonzalez, G., Arriaga, F., Wang, X.: Time-of-flight adjustment procedure for acoustic measurements in structural timber. Bioresources. **11**(2), 3303–3317 (2016)
6. Esteban, M., Bobadilla, I., Arriaga, F., Íñiguez-González, G., García, H.: NDT applied to estimate the mechanical properties of the timber of an ancient structure in Valsaín, Segovia (Spain). In: Proceedings of the 16th International Symposium on Nondestructive Testing of Wood. Beijing, China, pp. 152–157 (2009)
7. Gil-Moreno, D., Ridley-Ellis, D., O'Ceallaigh, C., Harte, A.M.: The relationship between bending and tension strength of Irish and UK spruce and pine. Eur. J. Wood Wood Prod. **80**(3), 585–596 (2022). https://doi.org/10.1007/s00107-022-01787-6
8. Gil-Moreno, D., Ridley-Ellis, D., Harte, A.M.: Timber grading potential of Douglas fir in the Republic of Ireland and the UK. Int. Wood Prod. J. **10**(2), 64–69 (2019). https://doi.org/10.1080/20426445.2019.1617984
9. Gil-Moreno, D., O'Ceallaigh, C., Ridley-Ellis, D., Harte, A.M.: Use of nondestructive techniques for determination of tension parallel-to-grain properties of spruce. In: 21st International Nondestructive Testing and Evaluation of Wood Symposium, ed, pp. 233–240. USDA Forest Service Forest Products Laboratory (FPL), Freiburg, Germany (2019)
10. Dackermann, U., Crews, K., Kasal, B., Li, J., Riggio, M., Rinn, F., Tannert, T.: In situ assessment of structural timber using stress-wave measurements. Mater. Struct. **47**, 787–803 (2014)
11. Arriaga, F., Osuna-Sequera, C., Esteban, M., Íñiguez-González, G., Bobadilla, I.: In situ assessment of the timber structure of an 18th century building in Madrid, Spain. Constr. Build. Mater. **304**, 124466 (2021). https://doi.org/10.1016/j.conbuildmat.2021.124466
12. Arriaga, F., Osuna-Sequera, C., Bobadilla, I., Esteban, M.: Prediction of the mechanical properties of timber members in existing structures using the dynamic modulus of elasticity and visual grading parameters. Constr. Build. Mater. **322**, 126512 (2022). https://doi.org/10.1016/j.conbuildmat.2022.126512
13. Ma, B., Kong, Q., Ding, Y., Chen, L., Gao, W.: A percussion method with attention mechanism and feature aggregation for detecting internal cavities in timber. NDT & E Int. **141**, 102986 (2024). https://doi.org/10.1016/j.ndteint.2023.102986
14. Wang, J., You, H., Qi, X., Yang, N.: BIM-based structural health monitoring and early warning for heritage timber structures. Autom. Constr. **144**, 104618 (2022). https://doi.org/10.1016/j.autcon.2022.104618
15. Riggio, M., Anthony, R.W., Augelli, F., Kasal, B., Lechner, T., Muller, W., Tannert, T.: In situ assessment of structural timber using non-destructive techniques. Mater. Struct. **47**(5), 749–766 (2014). https://doi.org/10.1617/S11527-013-0093-6/FIGURES/7
16. Íñiguez-González, G., Arriaga, F., Esteban, M., Llana, D.F.: Reference conditions and modification factors for the standardization of nondestructive variables used in the evaluation of existing timber structures. Constr. Build. Mater. **101**, 1166–1171 (2015). https://doi.org/10.1016/J.CONBUILDMAT.2015.05.128
17. European Committee for Standardization: EN 408 Timber structures — Structural timber and glued laminated timber — Determination of some physical and mechanical properties (2012)
18. National Standards Authority of Ireland (NSAI): IS 127:2015. Structural timber – Visual strength grading - Sawn softwoods with rectangular cross-section (2015)

19. Arriaga, F., Montón, J., Bobadilla, I., Llana, D.F.: Influence of length on acoustic time-of-flight (ToF) measurement in built-in structures of Norway spruce timber. Holzforschung. **73**(4), 339–352 (2019)
20. Rinn, F., Schweingruber, F.-H., Schär, E.: Resistograph and X-ray density charts of wood. Comparative evaluation of drill resistance profiles and X-ray density charts of different wood species. Holzforschung. **50**, 303–311 (1996)
21. Íñiguez, G., Esteban, M., Arriaga, F., Bobadilla, I., Gil, M.: Influence of specimen length on ultrasound wave velocity. In: 15th International Symposium on Nondestructive Testing of Wood, pp. 155–159 (2008)
22. Arriaga, F., Iñiguez, G., Esteban, M., Bobadilla, I.: Proposal of methodology for the assessment of existing timber structures in Spain. In: 16th International Symposium on Nondestructive Testing and Evaluation of Wood. Beijing, China 12–14 October 2009, pp. 152–157. USDA Forest Service Forest Products Laboratory (FPL) (2009)
23. Montero García-Andrade, M.J.: Clasificación de madera estructural de gran escuadría de Pinus sylvestris L. mediante métodos no destructivos - Grading of structural large cross-section timber of Pinus sylvestris L. by nondestructive methods. Doctoral Thesis. Universidad Politécnica de Madrid, ETS Ingenieros de Montes (2013)
24. Arriaga, F., Esteban, M., Iñiguez-Gonzalez, G., Bobadilla, I., Liana, D., Gonzalez-Sanz, M.: Structural assessment of the timber structure of the Casa Grande building in the Real Cortijo de San Isidro, Aranjuez, Madrid (Spain). In: Proceedings of the 18th International Nondestructive Testing and Evaluation of Wood Symposium, Madison, Wisconsin, WI, pp. 233–244 (2013)
25. CEN: Moisture content of a piece of sawn timber—Part 2: Estimation by electrical resistance method. EN 13183-2:2002. ed: European Committee for Standardisation, Brussels, p. 6. (2002)
26. CEN: European Committee for Standardization, Moisture content of a piece of sawn timber - Part 1: Determination by oven dry method. EN 13183-1 (2002)
27. Gil-Moreno, D., Ridley-Ellis, D., O'Ceallaigh, C., Harte, A.M.: The relationship between bending and tension strength of Irish and UK spruce and pine. Eur. J. Wood Wood Prod. **80**(3), 585–596 (2022). https://doi.org/10.1007/S00107-022-01787-6/FIGURES/6
28. Carlos Osuna-Sequera, D.F.L., Arriaga, F., Íñiguez-González, G.: Predicting the mechanical properties of timber from existing structures by the longitudinal vibration method, visual grading and definition of the nominal cross-section. Wood Mater. Sci. Eng. **19**(2), 366–374 (2024). https://doi.org/10.1080/17480272.2023.2243673
29. Fundova, I., Funda, T., Wu, H.X.: Non-destructive wood density assessment of Scots pine (Pinus sylvestris L.) using Resistograph and Pilodyn. PLoS One. **13**(9), e0204518 (2018)
30. Karlman, L., Mörling, T., Martinsson, O.: Wood density, annual ring width and latewood content in larch and Scots pine. Eurasian J. For. Res. **8**(2), 91–96 (2005)

How Long Can Timber Buildings Last? Strategies for Service Life and Sustainability

Shady Attia[1(✉)], Tomasso Verdier[1,2], and Joris Van Acker[3]

[1] Sustainable Building Design Lab, Department UEE, Faculty of Applied Sciences, Université de Liège, Liege, Belgium
shady.attia@uliege.be

[2] Energy Department, Politecnico di Torino, Polytechnic University of Turin, Technology Energy Building Environment Research Group, Turin, Italy

[3] UGent-Woodlab (Laboratory of Wood Technology), Department of Environment, Ghent University, Ghent, Belgium

Abstract. Lifecycle assessments (LCA) often estimate the service life of taller timber buildings at 50 years, but how long can they realistically last? More importantly, how can their lifespan be extended? This chapter addresses these questions by exploring strategies to extend service life and methods to predict durability during the planning phase. A survey of over 120 experts in timber construction was conducted to define the actual service life of medium-rise and tall timber buildings. The survey results highlight key measures for extending service life, including material selection, climate-adaptive design, proactive maintenance, and principles of circular construction. These findings serve as the foundation for refining service life prediction models and establishing sustainable design guidelines. This chapter provides actionable insights and expert-driven strategies to ensure the longevity and resilience of taller timber buildings, contributing to their role as a sustainable alternative in the global construction industry.

Keywords: environmental impact · GHG emission · carbon modelling · High-rise construction · end-of-life · engineered timber

1 Introduction

In Europe, over 75% of timber waste is either incinerated or sent to landfills, while most cascading recycling efforts result in material downgrading rather than high-value reuse [1, 2]. This inefficiency underscores the need for innovative strategies to maximize timber's lifecycle value. Simultaneously, there is an urgent demand for low-carbon construction solutions, particularly for buildings with a carbon intensity below 12 kg CO_2e/m^2/year across their entire lifecycle [3]. Engineered timber has emerged as a renewable material that not only meets these environmental goals but also facilitates the construction of taller buildings, such as HAUT in Amsterdam [4] and the Wood Hotel in Mjøstårnet [5], which serve as benchmarks for sustainable urban development.

G. Fink et al. (Eds.): *Holistic Design of Taller Timber Buildings*, SPRTRCIENG, pp. 205–212, 2025.
https://doi.org/10.1007/978-3-032-02098-7_16

While engineered timber offers significant environmental advantages, including lower greenhouse gas emissions and circular economy potential, its application in high-rise buildings faces several challenges. Existing research has largely focused on mid-rise or low-rise structures, leaving a critical gap in understanding the durability and service life of taller timber buildings. Additionally, current end-of-life practices for timber often fail to capitalize on their full potential, emphasizing the need for systematic approaches to extend material utility [6]. Understanding and improving the service life of tall timber buildings is vital for advancing their role as low-carbon alternatives to traditional materials like steel and concrete. Accurate predictions of service life and robust strategies for maintenance and reuse are essential to optimize the lifecycle carbon intensity of these structures. This study, conducted between 2023 and 2025 under COST Action HELEN, addresses two critical questions:

- How long is the lifespan of tall timber buildings?
- How can the service life of these buildings be accurately predicted during the planning process?

A survey of over 120 industry experts was undertaken to collect data on the lifecycle and durability of medium- and high-rise timber structures. The survey remains open until 2026 to ensure a comprehensive dataset and meaningful conclusions.

Combined with literature review and the creation of a dataset of building connections and case studies, the results will inform strategies to extend the service life of timber buildings by at least 150 years and reduce their lifecycle carbon intensity. This research aims to redefine timber's role in sustainable construction, offering practical guidelines for long-lasting, low-carbon timber buildings while addressing the environmental inefficiencies of current practices.

2 Challenges of Extending the Service Life

Tall timber buildings symbolize a paradigm shift in sustainable urban construction. Prominent examples include HAUT in Amsterdam, a 21-story hybrid timber structure, and Mjøstårnet in Norway, an 18-story glulam tower, both of which highlight the viability of engineered timber in achieving ambitious height while minimizing environmental impact. These buildings serve as benchmarks for sustainable design, leveraging engineered timber's lower carbon footprint compared to traditional materials like steel and concrete. These structures serve as benchmarks, yet ensuring their long service life and addressing their end-of-life (EoL) scenarios are vital to realizing timber's full environmental benefits.

2.1 Load-Bearing Capacity Over Time

Tall timber buildings endure higher vertical and lateral loads due to increased height and mass. Over time, these stresses can cause structural creep and deformation in timber components. Hybrid construction, such as the use of concrete cores in HAUT, mitigates these challenges by enhancing stability and load distribution. Also, tall structures are more exposed to strong winds, requiring advanced lateral bracing and connection systems.

In seismic regions, earthquake resistance must also be considered in the structural design of tall timber buildings. While engineered timber systems like CLT and glulam perform well in absorbing seismic energy due to their flexibility and low mass, seismic loads introduce complex lateral forces that require robust connection detailing, ductility, and energy dissipation capacity. Hybrid designs with concrete cores or steel reinforcement can improve seismic performance, but they may also affect circularity and end-of-life strategies.

2.2 Environmental Exposure and Fire Safety

Timber's susceptibility to moisture and water ingress can lead to swelling, shrinkage, and decay. Effective detailing, such as moisture barriers and well-sealed joints, is critical to preventing degradation. A moisture management plan is essential in timber buildings. A moisture management plan integrates design, material selection, construction, and maintenance to control moisture ingress and accumulation. Architectural strategies include overhangs, rain screens, and site drainage, while pre-dried, treated timber and vapor-permeable membranes enhance durability. Construction best practices, such as weather protection and sealed joints, minimize exposure. HVAC systems, dehumidifiers, and sensors regulate indoor humidity, preventing mold and swelling. Routine inspections, sealant reapplications, and moisture testing ensure long-term performance, mitigating decay and structural risks.

Prolonged exposure to sunlight and UV Radiation accelerates weathering and weakens timber surfaces. Protective coatings and regular maintenance are required to counteract this issue.

Fire safety is a critical concern for tall timber buildings. Products like cross-laminated timber (CLT) exhibit predictable charring behavior, but achieving compliance with stringent fire codes necessitates additional fire-resistant layers or encapsulation.

2.3 End-of-Life Scenarios

In Europe, over 75% of timber waste is incinerated or landfilled, and most recycling efforts result in material downgrading rather than high-value reuse [2]. A timber building designed with poor disassembly potential will have higher embodied GHG emissions than a design optimized for recovery and reuse [7]. This highlights the need for improved end-of-life strategies to maximize the utility and carbon sequestration potential of timber. Designing for deconstruction (DfD) with reversible connections and modular components enables the reuse of timber compnents in new construction, reducing environmental impact and promoting circularity [8]. Advanced lifecycle assessments (LCA) that include deconstruction and recycling pathways can better inform planning and design phases.

2.4 Importance of Addressing these Challenges

Extending the service life of tall timber buildings is crucial to their sustainability and lifecycle performance. Current lifecycle assessments often limit predictions to 50 years, yet evidence suggests that with proper interventions, timber structures can surpass 150 years of service life. Key strategies include:

- **Monitoring and Maintenance**: IoT-enabled sensors, such as the WM1 moisture monitoring system, provide real-time data to address structural risks preemptively [9]. See Sect. 2.2.
- **Circular Design**: Incorporating end-of-life scenarios into the initial design phase, such as modular construction and material recovery strategies, ensures that timber components retain value beyond their initial use.
- **Policy and Guidelines**: Clear standards for durability, fire safety, and deconstruction will drive adoption and ensure the long-term viability of tall timber buildings.

By addressing these technical and environmental challenges, tall timber buildings can achieve the durability and low carbon intensity needed to meet the demands of modern urban development. This research under COST Action HELEN contributes to identifying practical, scalable solutions for sustainable timber construction.

3 Survey Methodology and Results

The research employed a mixed-methods survey approach, integrating closed-ended and open-ended questions to comprehensively collect data on the service life of components used in medium-rise (3–8 stories) and tall (above 8 stories) timber buildings. The survey aimed to define the actual service life of key building components, including structural components, envelope components, and connection systems.

3.1 Survey Design and Objectives

A pilot study involving 12 timber construction experts was conducted in September 2023 to ensure clarity and relevance. The pilot validated the survey's design, ensuring that the questions were appropriately formulated and aligned with the study's objectives. Based on the feedback, minor adjustments were made before the full-scale launch.

The finalized survey, designed to be concise and accessible, was distributed in English and targeted architects, engineers, and industry professionals globally. Recruitment leveraged the COST Action HELEN network, employing snowball sampling techniques and LinkedIn announcements to reach potential participants. Specific outreach included professionals attending the Holzbau conference and experts familiar with timber construction across Europe, North America, and Australia. The survey featured a detailed questionnaire asking participants to estimate the lifespan range (minimum and maximum years) for various timber components without differentiating between specific timber materials (https://form.jotform.com/232473794563365). Assessed components included structural components such as timber columns, beams, and panels; envelope components like wall claddings, waterproofing layers, and insulation materials; and interior finishes such as plasterboards, floor finishings, and glazed products. Participants could leave fields blank for components they lacked experience or knowledge about, ensuring the data collected was both accurate and focused.

3.2 Key Findings

The survey, completed by over 120 participants from Europe, North America, and Asia, revealed significant insights into the service life of tall timber building components. The aggregated responses suggest that taller timber buildings can achieve a lifespan exceeding 150 years, with some components potentially lasting up to 300 years under optimal conditions, as shown in Fig. 1. Timber columns, beams, and solid panels demonstrated high durability, with life expectancy often surpassing 150 years when properly maintained. External timber claddings and coatings showed shorter lifespans, typically ranging from 50 to 75 years due to environmental exposure, while the performance of waterproofing layers and insulation materials varied depending on material type and maintenance practices. Interior finishes, such as plasterboards and floor finishes, were identified as requiring periodic replacement, with average lifespans between 50 and 70 years. The survey also highlighted region-specific challenges, including climate-related impacts in humid or arid regions, underscoring the critical importance of regular maintenance to extend the service life of these components.

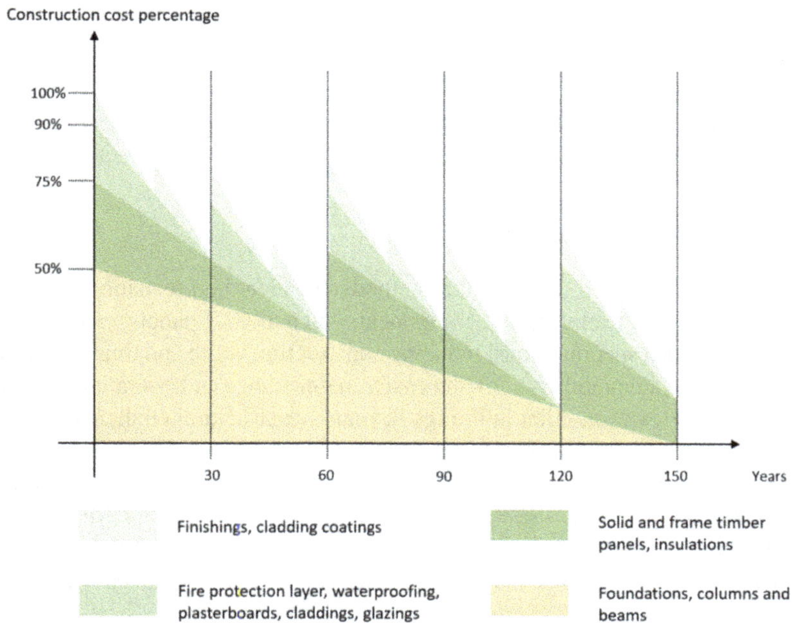

Fig. 1. Tall timber building components expected life span

3.3 Validation and Dissemination

Eight timber construction experts were interviewed in-depth to validate the findings, focusing on their interpretations of the survey data and practical implications for design and maintenance. Additionally, the initial results were shared during an international webinar in October 2023 [10], where industry professionals provided further feedback and discussed the potential applications of the findings.

4 Strategies for Achieving a 150-Year Lifespan in Tall Timber Buildings

4.1 Material Innovations

Engineered timber products such as Cross-Laminated Timber (CLT) and Glued Laminated Timber (glulam) are revolutionizing the construction of tall timber buildings. These materials are designed to provide exceptional structural performance, enabling timber to replace carbon-intensive materials like concrete and steel in high-rise applications. CLT, for instance, combines layers of wood arranged perpendicularly and bonded with adhesives, offering high strength and stability, even for tall structures.

Despite these advancements, challenges remain in achieving long lifespans for timber in demanding environments. Timber's susceptibility to fire, moisture, and decay necessitates innovative treatments. Protective coatings enhance fire resistance by slowing charring rates, while moisture barriers and UV-blocking finishes mitigate environmental degradation. However, as building heights increase, hybrid systems integrating concrete or steel become essential. While these materials improve load-bearing capacity, fire resistance, and stability under lateral forces, they also offset some of timber's environmental advantages, posing a trade-off in lifecycle carbon intensity.

4.2 Structural Design Strategies

Modular and prefabricated construction methods enhance the precision and efficiency of tall timber projects. Prefabricated components, such as CLT panels, can be manufactured off-site to exact specifications, reducing construction waste and time. Modular systems also facilitate adaptability, allowing easier maintenance and potential reuse of timber components. However, taller buildings face unique structural challenges. Increased exposure to wind and seismic forces requires robust design solutions. Hybrid systems combining timber with concrete or steel cores are critical for ensuring lateral stability and mitigating vibrations. While these approaches enhance structural integrity, they introduce higher embodied carbon compared to all-timber designs. Projects like HAUT in Amsterdam demonstrate how balancing these trade-offs can achieve both performance and sustainability goals.

In conclusion, the integration of material innovations and hybrid systems addresses the limitations of timber in tall structures, ensuring longer lifespans and enhanced safety. However, the environmental trade-offs of incorporating concrete and steel highlight the need for ongoing research and innovation to optimize sustainability in tall timber construction.

5 Circularity and End-of-Life Strategies for Tall Timber Buildings

The design of tall timber buildings must prioritize circular construction principles to maximize material reuse and minimize waste. Reversible connections and modular components facilitate disassembly, preserving timber's value for repurposing in future projects [11]. However, not all timber buildings and connections can be dismantled in practice. Demolition contractors report that screw, nut, and bolt connections frequently deform, making disassembly difficult and limiting material recovery. Our database on demountable connections supports this finding, indicating that most timber connections are not easily reversible, emphasizing the need for improved fastening systems and alternative connection methods to enhance reuse potential [12]. While residual materials can support energy recovery, optimizing connection design remains crucial to reducing reliance on virgin materials and lowering lifecycle emissions. Lessons from projects like HAUT demonstrate the economic and environmental advantages of integrating circularity in tall timber construction [13] reinforcing timber's role in sustainable building practices aligned with global carbon reduction goals [14].

6 Conclusion

Tall timber buildings represent a promising solution to the environmental challenges of urban construction. Achieving a 150-year lifespan requires integrating durable materials, advanced structural designs, proactive maintenance, and predictive planning. Interdisciplinary collaboration and the adoption of circular principles are essential for maximizing the environmental and economic benefits of these structures. The findings from this research emphasize the potential of tall timber buildings to serve as long-lasting, low-carbon alternatives to traditional construction methods. By prioritizing sustainability at every stage, the construction industry can redefine its approach to urban development and pave the way for a greener future.

References

1. Ahn, N., Bjarvin, C., Riggio, M., Muszynski, L., Pestana, C., Dodoo, A., Puettmann, M.: Envi sioning mass timber buildings for circularity: Life cycle assessment of a mass timber building with different end-of-life (eol) and post-eol options. In: Proceedings of the WCTE (2023)
2. Baldassarri, C., Allacker, K., Reale, F., Castellani, V., Sala, S.: Consumer Footprint. Basket of Products Indicator on Housing. European Commission, Joint Research Centre, Publication Of fice of the European Union, Luxembourg. ISBN 978-92-79-73195-2 (2017)
3. Attia, S.: Net Zero Energy Buildings (NZEB): Concepts, Frameworks and Roadmap for Project Analysis and Implementation, 2nd edn. Elsevier, Cambridge, MA (2025)
4. Verhaegh, R., Vola, M., de Jong, J.: Haut: a 21-storey tall timber residential building. Int. J. High-Rise Build. **9**(3), 213–220 (2020)
5. Abrahamsen, R.: Mjøstårnet-construction of an 81 m tall timber building. In: Internationales Holzbau-Forum IHF, vol. 2017. Garmisch-Partenkirchen, Germany (2017)
6. Bahr, C., Lennerts, K.: Lebens-und nutzungsdauer von bauteilen. Bundesinstituts Für Bau-, Stadt-Und Raumforschung (BBSR) Und Bundesamtes Für Bauwesen Und Raumordnung (BBR), Berlin (2010)

7. Abu-Ghaida, H., Ritzen, M., Hollberg, A., Theissen, S., Attia, S., Lizin, S.: Accounting for product recovery potential in building life cycle assessments: a disassembly net work-based approach. Int. J. Life Cycle Assess. **29**, 1151–1176 (2024). https://doi.org/10.1007/s11367-024-02324-8
8. Al-Obaidy, M., Carens, H., Campain, C., Giannasi, E., Mori, M., van Vliet, M., Attia, S.: Expert system for building disassembly potential evaluation and inspection. J. Build. Eng. **103**, 112148 (2025). https://doi.org/10.1016/j.jobe.2025.112148
9. Oy, W.: WM1-WAN: Wireless Moisture Meter for Wood Structures, Tampere, Finland (2023). Available: https://www.wiiste.com/en/products
10. Verdier, T., Attia, S.: Lifespan of high-rise timber building components (Webinar 1), 2024, [Video recording]. https://www.youtube.com/watch?v=zFED83Z5w-A
11. Cabrero, J.M., Ottenhaus, L., Palma, P., Attia, S.: Disassembly and reuse in tall TimberBuildings: advancing circular construction practices. In: Fink, G., Jocker, R., et al. (eds.) Design of Taller Timber Buildings. Springer, Berlin (2025)
12. Sandin, Y., Shotton, E., Cramer, M., Sandberg, K., Walsh, S.J., Östling, J., Zabala Mejia, A.: Design of Timber Buildings for Deconstruction and Reuse—Three Methods and Five Case Studies. Rise Report, RISE Research Institutes of Sweden, Ghotenburg (2022)
13. Attia, S.: Lifecycle impact assessment for tall timber building: learning from HAUT. In: Fink, G., Jocker, R., et al. (eds.) Design of Taller Timber Buildings. Springer, Berlin (2025)
14. Martien van den Oever, M., Telleman, Y., van Kampen, A., van der Voort, M., van der Weide, R., van Baren, S., Jacobs, S.: Biobased grondstoffen voor hoogbouw: Geïndustrialiseerde modulaire en lage emissie hoogbouw in de G4. Wageningen University & Research, Wageningen (2024)

The New Model Building - A Research Initiative Piloting Post-Grenfell Holistic Design of Medium-Rise Residential Buildings in Timber

Matt Caldwell(✉)

Department of Architecture and Civil Engineering, University of Bath, Buro Happold, Bath, UK
matthew.caldwell@burohappold.com

Abstract. In September 2023, Waugh Thistleton Architects published the New Model Building Guide. It demonstrates a pathway to realising safe medium-rise residential buildings with a timber superstructure which comply with the current UK building regulations. It represented the culmination of more than 2 years' collaboration between a multi-disciplinary team of designers and stakeholders. The following chapter highlights the key points for design teams, and in particular structural engineers, to learn and implement from the initiative.

Keywords: Mass Timber · Wood · Glulam · GLT · Cross laminated Timber · CLT · Holistic design · Fire · Moisture damage · Hackitt · Grenfell

1 Background

In the second decade of the twenty-first century, the UK was at the global forefront in the development, expertise and construction of medium-rise engineered-timber buildings. Confidence was high and bold predictions were made on the potential future trajectory of the trend [1].

The landscape surrounding design of timber buildings in the UK changed significantly between 2017 and 2018 as a result of the fire at the Grenfell Tower, the subsequent Hackitt Report [2] and the UK Government response [3]. For an excellent narrative on the events during release of the Hackitt Report, as well as an insight into its recommendations and how the UK Government response runs contrary to many of them, the reader is referred to *"Prescription in English fire regulation: treatment, cure or placebo?"* [3]. In the following years, the UK government implemented amendments to the UK Building Regulations, specifically the definition and requirements of 'Relevant Buildings'. Subsequent to the publication of the New Model Building documentation [4], the UK Building Safety Act 2022 [5] also came into force.

"A cultural and behavioural change is now required across the whole [construction] sector to deliver an effective system that ensures complex buildings are built and maintained so that they are safe for people to live in for many years after the original construction. The mindset of doing things as cheaply as possible and passing on responsibility for problems and shortcomings to others must stop."—Dame Judith Hackitt, December 2017.

G. Fink et al. (Eds.): *Holistic Design of Taller Timber Buildings*, SPRTRCIENG, pp. 213–222, 2025.
https://doi.org/10.1007/978-3-032-02098-7_17

For timber-structured buildings, one of the most significant results of the UK regulatory response goes under the colloquial name of 'the ban on combustibles in the façade'. In more technical terms this means the façade meeting Class A2-s1 in accordance with EN 13501-1:2018 [6] and the Building Regulations Part B Regulation 7 [7]. The definition of 'façade' is adopted from Regulation 2 [8] and extends inward as far as the inside face of the final finishes on the inside of the perimeter wall. This is a key point to note for timber-structured buildings.

The principal mode of fire propagation in the Grenfell Tower fire was through retro-fixed external cladding and the changes to the regulations were ostensibly related to the façade. However, a significant number of building forms have at least some of the primary structure of the building within the 'façade', as defined in the previous paragraph. It follows that any of these structural elements formed in timber are now outlawed within buildings meeting the amended definition of a 'Relevant Building', principally those over 11 m in height. A whole series of exemptions to 'the ban' were subsequently brought into force (regulation 7, paragraph 3) [7], though timber structural elements were notably absent.

The result of the amendments was that from around the middle of 2018, the UK construction of engineered timber residential buildings over 11 m in height very drastically reduced. The UK's once world-leading drive for low-carbon renewable residential building structures in timber was at a virtual standstill.

2 Scope and Purpose

Prompted by frustrations in navigating the new regulations, in 2021 Waugh Thistleton Architects made a successful application to build by Nature for funding to undertake a research project to examine and address the issues. They assembled a team of collaborators including Buro Happold, providing structural and embodied carbon consultancy, University College London providing fire engineering consultancy, Gardiner & Theobald providing cost and project management consultancy and the National House Building Council (NHBC) providing building control assessment and insurance and warranty advice.

The intent of the initiative was to investigate the challenges, constraints, and opportunities for delivering timber-structured multi-storey residential buildings within the new UK regulatory environment. Understandably, clients on live commissioned projects were nervous about investing their time and money into schemes with an as-yet untested path to compliance. The New Model Building (NMB) initiative would provide the opportunity to navigate this route so that others could follow. The following 2 years saw the team explore this new territory.

The NMB initiative provides the outline design of a theoretical building, which would meet all the technical, regulatory and viability requirements of the current UK context. The principles and philosophy of the design could then be applied to buildings of varying forms. The form of the theoretical building would be chosen to be representative of the most common typologies seen in the UK multi-stored residential market. It also would encompass enough geometric and architectural features to sufficiently test the design response regarding both performance-based and prescriptive regulation.

The building would be 5 to 7 storeys, no more than 18 m to the uppermost finished floor level and to be of multi-residential occupancy. This would place it within the scope of a 'Relevant Building', though outside that of a 'High-Risk Building'. It would need to comply with the Building Regulations of England and Wales, via the Approved Documents, and thus achieve an External Wall System (EWS)1 Certificate and meet the contemporary Greater London Authority (GLA) funding requirements. The design would also be reviewed by NHBC and achieve a 'letter of comfort' stating that it meets the NHBC technical requirements.

The design was also to meet or better an up-front embodied carbon target of 326 $kgCO_2e/m^2$ (Modules A1-A5) and a Whole Life (Modules A-C) embodied carbon target of 271 $kgCO_2e/m^2$. The latter requirement would allow it to meet the Royal Institution of British Architects (RIBA) 2030 target. The design was also to be as product-agnostic as possible, such that future users of the information were not limited in terms of procurement.

3 The Building Design

The final building design comprised 6 storeys with a central circulation and services core. Each storey comprised 5 separate residential units. Four of these units typically had an external balcony. The superstructure of the building, from ground floor upwards, was tested as a Glulam (GLT) frame with Cross Laminated Timber (CLT) slabs and with an alternative system of CLT walls in lieu of Glulam columns. Lateral stability was via the walls of the central core, which were justified variously as balloon-frame CLT, reinforced concrete or braced steel. Foundations were assumed to be concrete augered piles with reinforced concrete piles caps and ground floor slab.

4 Timber Structure and Fire

Once the overall geometry of the building was decided, a significant part of the challenge concerned its fire performance. There were several points most directly affecting the structural design, as described below.

It should be noted that the intent of the design was to demonstrably meet the requirements of the new regulations, including the prescriptive amendments. This does arguably result in a conservative 'belt-and-braces' solution to the fire engineering design in some cases. However, it establishes a baseline beyond which it is anticipated that project-specific designs will justify refinements.

As previously mentioned, the definition of the 'façade' in this context meant that all of the timber structure needed to stay inboard of the finished inner face of the external walls. In order that design unambiguously met this requirement, fire encapsulation was specified to wrap right around glulam columns (Fig. 1), and the end of CLT walls (Fig. 2), prior to abutting the non-combustible façade. The mounting system for the façade therefore needs to cater for a slightly increased eccentricity of the support reaction from the façade, though it is relatively small and within the bounds of many conventional designs.

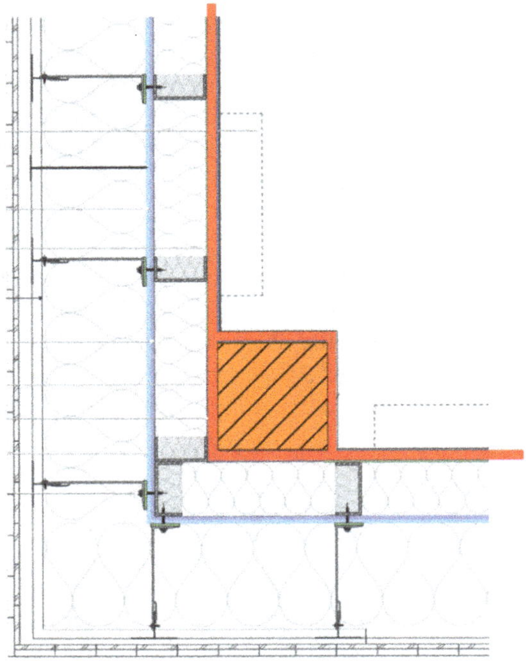

Fig. 1. Plan detail view at building corner, showing full encapsulation (red) of a glulam column (orange & hatched)

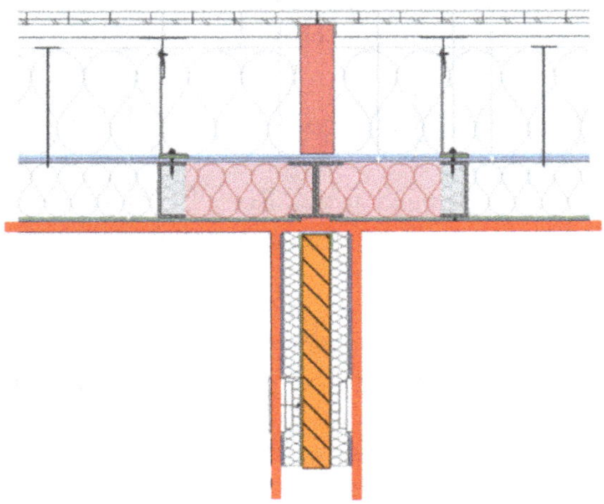

Fig. 2. Plan detail view where a CLT wall meets the zone of external façade, showing encapsulation (red) wrapping the end of the CLT wall (orange & hatched)

Although not explicitly mentioned by the regulations, consideration must be given to encapsulation of metallic elements and fasteners which meet or are embedded in the structural timber. Where the metallic element is exposed to elevated temperatures, it readily transmits this to the adjacent timber. Timber begins to suffer irreversible damage above 100 °C and pyrolysis can be instigated above 200 °C. It should be noted that intumescent paints typically trigger at around 100 °C and may not provide their full insulating effect until 200–300 °C. Therefore, by themselves, they are not typically an effective solution to the problem.

An over-riding principle of the fire engineering design was that the structural timber should be fully encapsulated such that it makes no contribution to the fuel load and thus dynamics of the fire. This approach removed the need for complex parametric and performance-based fire engineering analysis. Further, it therefore simplified regulatory approval allowing the prescriptive UK Approved Documents route to approval to be applied, in accordance with the recommendations of the UK Structural Timber Association [9]. However, this approach was taken in order to illustrate the simplest route to approval. With more detailed fire engineering modelling it is possible justify safe design of a building of this type with certain amounts of exposed timber.

The degree of encapsulation to elements performing a fire separating function can be defined using the K_2 class of clause 7.6.4.2 of EN 13501-2. The requirement for designation K_2 states that the temperature rise at face of the substrate must be less than 250° for a defined period and there should be no burned or charred material behind the encapsulation. The designation is also suffixed with the relevant period of protection. In the case of the New Model Building, all encapsulation to the structural timber is to meet $K_2$60 and an additional requirement was imposed in line with Structural Timber Association (STA) guidance such that the substrate temperature was further limited to 200 °C.

Applying $K_2$60 encapsulation to all timber surfaces does mean that none of the structural timber is visible. This is an important point to note in the early-stage design of a building following this model. In some instances, other stakeholders may question the value of using timber when you can't experience it as a building user. However, if the timber is being used for sustainability and constructional reasons, then timber is still a justifiably beneficial material. The embodied carbon of the encapsulation accounted for an uplift of approximately 8% on the timber A1-A3 total. It should also be noted that in most cases, buildings of this typology constructed in other structural materials also have a full lining of plasterboard, for a combination of visual, acoustic and fire performance reasons.

The design of the vertical circulation core was left flexible as many contemporary designs prefer a hybrid construction with the core being reinforced concrete or occasionally braced steel. The core is however possible to implement in Cross Laminated Timber (CLT), following the same fire-encapsulation principles as the rest of the building. A balloon-frame style of construction, with the CLT panels orientated vertically, allows the number of tension-compression connections to be minimised and overall lateral stiffness to be improved.

A notable challenge at the time regarding encapsulation of the timber structure was obtaining the required documentation for the specific geometry and substrates in any one project. Basic tested and approved details are available from parties such as the Structural Timber Association [10]. However, any small departure from these requires additional analytical justification or project-specific fire testing. Similar principles apply to fire-rated penetrations through timber walls. Certain manufacturers do have timber-specific fire test certificates available. In all instances the approach for the NMB was to design-out the issues wherever possible and to keep the detailing simple where it was unavoidable.

5 Timber Structure and Moisture

The challenge for regulatory compliance was almost exclusively focussed on fire design. However, the challenge for insurance and warranty purposes was at least equally concerned with design for moisture control. An accepted, though likely conservative, threshold for avoiding onset of fungal decay in engineered softwood timber is 20% relative moisture content [11, 12]. The effects of undetected moisture on CLT are well documented [13, 14] and litigation for large sums of money related to moisture damage to mass timber buildings in the UK have already occurred.

There are several mechanisms that could lead to exceedance of the safe moisture content threshold, therefore different mitigations were proposed for each in the design.

Construction-stage wetting is a significant risk in the UK and there are a variety guidance documents on managing it [15–20]. The design of the NMB was developed to avoid or minimise moisture traps during construction and to facilitate an efficient process of construction and follow-on with the building envelope. This includes simple, repeated geometry and keeping the façade zone in one plane and wholly out-with the primary structure.

Options are given for the roof planes. The thickness of CLT generally means that passive detection of moisture incident on its upper surface is much more difficult. Any CLT roof is therefore designed with a minimum 10° fall to it (Fig. 3). Lightweight roofs such as joisted and boarded constructions are assessed as being able to reduce in pitch down to 1:40 (Fig. 4) as passive moisture detection is more likely to be successful. It is important to note however, that the fall must be formed by the structural plane, as opposed to there being a horizontal structural plane with the falls created by tapered firings or insulation above. Any roof that had less than a 1:40 pitch to it was deemed to be too high a risk in timber and materials with a longer resistance to moisture ingress were required to be used. For example, concrete or metallic decks.

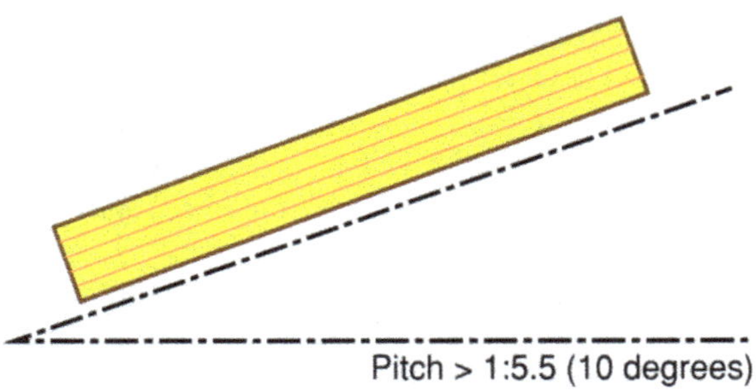

Fig. 3. A minimum pitch of 10° was deemed necessary for roof to be constructed from CLT

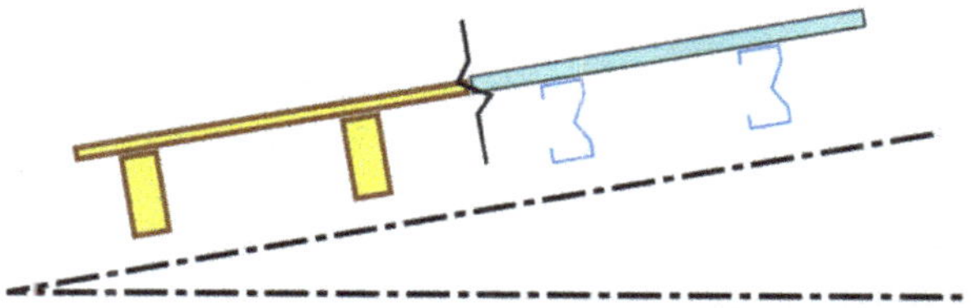

Fig. 4. Where the structural roof deck pitch was less than 10° but greater than 1.5 a sheathed timber joist deck was deemed acceptable

Adopting a fall to the structural deck offers benefits to construction-stage moisture control as well as in the permanent case. In both conditions the principle is to avoid standing water. During the construction stage rainwater is passively drained to an outlet without wholly relying on active removal such as squeegees and wet vacs. This is also further facilitated using self-adhesive breather membranes or sprayed hydrophobic coatings to minimise moisture uptake by panel faces and more importantly edges.

In the permanent condition it was assumed that some part of the roof's waterproofing membrane system will fail at some point within the potential serviceable life of the building structure. The adoption of a fall to the structure means that any accidental water ingress is encouraged to move to a location where it can be detected, either passively through seepage in thin sheathing, weepholes in CLT, or actively through electronic leak or moisture detection.

Wet services within the building present another potential source of moisture. The design principles included overhead horizontal distribution, so that maintenance was more readily undertaken, and leaks were more detectable and repairable. In a similar way vertical distribution is clustered within dedicated risers for the same reasons. Flow detection is, in principle, a method of detecting leaks, though the system would need to be zoned to the level of the individual apartment and the flow rate sensitivity high enough to detect slow leaks.

Areas with active water fixtures, principally bathrooms and kitchens in this instance, present additional challenges as there is likely to be daily user-spillages as well as any systemic leaks. Multiple options were explored, and each project would be required to adopt at least two of the following mitigations: Passive measures include adopting a joist-and-board floor so that leaks are visually obvious sooner than with CLT, and repair is also potentially more straightforward. This could incorporate preservative-treated timber to buy additional time. Adopting a false floor in the bathroom with a ventilated void below opens the opportunity for leak detection tapes as well as a route of escape for small amounts of moisture in the short term.

A potential risk area during construction and in use is where the timber structure meets the concrete foundation. In this instance standard guidelines were followed in placing all timber onto a concrete plinth raised above FFL and providing end-grain sealant [19]. Detailing of the perimeter wall also followed standard guidance for tracking of moisture down and away from the timber.

As previously noted, the primary superstructure was located wholly within the demise of the façade zone, principally for adherence with the fire regulations. However, a second benefit is that the timber is also therefore wholly within the thermal and moisture control lines of the envelope. This ensures that all of the timber is within Use Class 1 as defined in EN335:2013 cl. 4.2 with consequent benefits for structural design and durability. None of the internal spaces were deemed to generate humidity in excess of Use Class 1 from within. Design and analysis therefore adopted Service Class 1 as defined in EN 1995-1-1:2004 + A2:2014 cl. 2.3.1.3.

As the structure is not integrated with layers of the façade, it also facilitates simpler inspection of the timber frame and carrying out of any potential repairs. The serviceable life of façade systems is also usually much shorter than that of the primary frame, typically 25 years. By the façade having its own dedicated zone, outside the structure, the efficiency of façade replacement and the variety of options for its replacement type are also improved.

6 Documents and Output

The published output of the NMB initiative is freely available as three volumes of information. These can be accessed via TDUK [4].

The NMB Guide Book provides a general introduction to the initiative, its objectives and the underlying principles of the design. It also summarises the embodied carbon metrics for the building, including comparison with an equivalent flat slab reinforced concrete building. An outline cost comparison was also produced by Gardiner & Theobald for two variations of the timber structure and for the concrete flat-slab.

The NMB Details Book provides more than 70 pages of construction details developed for the building. These are illustrated as isometric cut-aways and annotated detail sections. They each have a checklist of the essential properties of the detail and cross reference to performance criteria for each.

The NMB Evidence Book explains the philosophy behind the Architectural, Fire and Structural design. It outlines the assumptions and requirements to stay within the remit of the project, as well as reference to other relevant documents.

The design and Evidence Book documentation of for the NMB initiative was reviewed by the National House Building Council (NHBC) to ensure it aligned with their requirements. A letter of comfort is provided by NHBC stating that projects delivered to the NMB design philosophy align with the NHBC technical standards. As a national insurer and warranty provider for residential properties, NHBC were ideally placed to help represent the perspective of these two key stakeholders in the process of getting schemes built and insured.

7 Summary

The NMB initiative provides a comprehensive blueprint for medium-rise residential timber buildings in the contemporary UK regulatory and industry context. The process and the output very much underscore the need for, and value of, holistic and multidisciplinary thinking.

This is particularly the case for timber structures, where the knock-on effects and inter-relationship between disciplines is especially important. The timber principally forms the building structure, but its effective functioning is undermined if other aspects of the design are not cognisant of its requirements and potential vulnerabilities.

The building geometry and construction should facilitate a construction methodology which avoids the timber getting excessively wet or trapping moisture. Any trapped moisture can lead to loss of strength and stiffness, mould, fungal decay, corrosion and ultimately to structural failure.

The design of the building façade and how it relates and is fixed to the structure can have a significant influence on the overall durability and longevity of the building. A system that can enclose the structure quickly, ideally as the structure above is still being erected, will greatly help management of construction-stage moisture. The same principles apply to future re-cladding operations and the moisture risks to the timber structure at that stage.

By employing a simple, collated, and coherent wet services distribution strategy, the maintenance, leak-detection, and repair of systems can be simple and efficient. This mitigates a significant moisture-damage risk to the timber structure.

The same principles applied to the rest of the building services distribution result in fewer and more easily managed penetrations in the timber structure. This is particularly significant where fire collars are required, and enclosed voids are to be managed with regard to fire spread.

If the fire engineering strategy is decided and communicated at an early stage, relatively minor adjustments to the architectural and structural design can facilitate a much more consistent and coherent approach to structural fire safety.

In some areas of the building, mass timber may not be the optimum solution holistically. Bathroom floors, low-pitch roofs and external balconies or terraces are exposed to constant water risk. In these areas, combinations of different materials or form of construction along with facilitating efficient detection and remediation are essential and have a direct effect on the reliability of the structure itself.

These principles of multi-disciplinary thinking are very much at the core of the Hackitt Report and its recommendations. They are most effectively achieved by holistic performance-based design strategies.

Acknowledgments. The author would like to acknowledge the significant contributions to the New Model Building initiative made by a number of individuals from several organisations including Alistair Ogle at Waugh Thistleton Architects and Professor Jose Terero Cullen at University College London.

References

1. Ramage, M., Foster, R., Smith, S., Flanagan, K., Bakker, R.: Super tall timber: design research for the next generation of natural structure. J. Archit. **22**(1), 104–122 (2017)
2. Hackitt, J.: Building a safer future. In: *Independent Review of Building Regulations and Fire Safety: Final Report* (2018)
3. Law, A., Butterworth, N.: Prescription in English fire regulation: treatment, cure or placebo? Proc. Inst. Civ. Eng. Foren. Eng. **172**(2), 61–68 (2019)
4. Timber Development UK: New Model Building Guide. https://timberdevelopment.uk/resources/new-model-building-guide/

5. UK Government: The Building Safety Act.. The Law Society, 2022
6. B. EN: BS EN 13501-1: 2018 Fire Classification of Construction Products and Building Elements-Classification Using Data from Reaction to Fire Tests. British Standards Institution, London (2018)
7. UK Government: The Building Regulations 2010, Part 2, Regulation 7 (2023)
8. UK Government, "The Building Regulations 2010, Part 2, Regulation 2," 2023
9. Structural Timber Association: Structural Timber Buildings Fire Safety in Use Guidance. Volume 6 - Mass Timber Structures; Building Regulation compliance B3(1). Structural Timber Association (2023)
10. Stuctural Timber Association: Structural Timber Buildings Fire Safety in Use Guidance. Structural Timber Association (2022)
11. BM TRADA: Wood Information Sheet 4-14. Moisture in Timber. BM TRADA (2019)
12. BM TRADA: Wood Information Sheet 4-28. Durability by Design. BM TRADA (2019)
13. CROSS: Safety Report 852 - Rotting of Cross-Laminated Timber (CLT) Roof Panels. CROSS (2019)
14. CROSS: Safety Report 1124 - Water Ingress to Cross-Laminated Timber Structural Frame. CROSS (2022)
15. Structural Timber Association: Moisture Management Strategy. Process Guidance for Structural Timber Buildings. Structural Timber Association (2022)
16. RDH Building Science: Mass Timber Moisture Management Guide. RDH Building Science (2021)
17. FP Innovations: Construction Moisture Management - Cross Laminated Timber. FP Innovations (2020)
18. Karacabeyli, E., Gagnon, S.: CLT Handbook. FP Innovations (2019)
19. Structural Timber Association: Advice Note 14 - Robustness of CLT Structures. Structural Timber Association (2017)
20. Caldwell, M.: Timber Knowledge Sheet – Moisture Management During Construction. Timber Development UK (2024)

Hazard Criteria

Structural Robustness: Design Framework for Enhanced Robustness

Maria Felicita[1,3](✉), Luka Vojnovic[2], Katharina Sroka[1,3], and Pedro Palma[3]

[1] Empa - Swiss Federal Laboratories for Materials Science and Technology, ETH Zurich, Zurich, Switzerland
{maria.felicita,katharina.sroka}@empa.ch
[2] University of Bristol, Bristol, UK
nf21539@bristol.ac.uk
[3] Empa - Swiss Federal Laboratories for Materials Science and Technology, Dubendorf, Switzerland
pedro.palma@empa.ch

Abstract. Structural robustness refers to the ability of a structure to withstand unforeseen adverse events without being damaged to an extent disproportionate to the original cause. Robustness design strategies include redundancy and segmentation, which are assessed based on threat-independent scenarios. Redundancy provides alternative load paths (ALP) to bypass damaged components, while segmentation isolates damaged areas to prevent failure progression. Current design guidelines categorize buildings by importance and risk levels to determine robustness requirements, and propose appropriate design strategies. These strategies must take into account the brittle failure modes and high variability of mechanical properties of timber members, and the often limited deformation capacity of timber connections, amongst other aspects. Case studies show various implementations of such strategies.

1 Introduction and Terminology

A structure and its members must be able to withstand unforeseen adverse events without being damaged to an extent disproportionate to the original cause. Exposure refers to the occurrence of a hazardous event, while vulnerability refers to the occurrence of damage given a hazardous event. Design strategies related to reducing exposure and vulnerability focus on preventing an initial damage from occurring, and thus are assessed in a threat dependent framework, where hazardous events are identified, quantified, and accounted for in the design. Robustness refers to the ability of a structure to limit damage propagation given an initial damage. The nature of the triggering event or initial damage can be unforeseen or even unforeseeable, and thus the corresponding design strategies are based a threat- independent approaches. A disproportionate damage process and the corresponding prevention strategies are presented in Fig. 1.

G. Fink et al. (Eds.): *Holistic Design of Taller Timber Buildings*, SPRTRCIENG, pp. 225–237, 2025.
https://doi.org/10.1007/978-3-032-02098-7_18

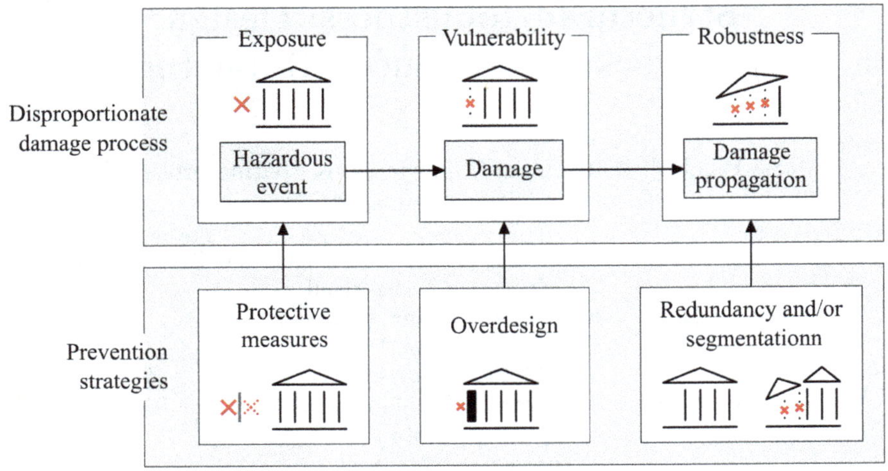

Fig. 1. Disproportionate damage process and corresponding prevention strategies, based on [10,12,14]

Design strategies against disproportionate consequences commonly fall into the following categories (Fig. 1):

1. Preventing local damage:
   a. Protection measures aimed at reducing the probability of occurrence of the hazardous event or of its intensity, such as barriers against impacts, or active fire protection systems;
   b. Overdesign measures aimed at reducing the vulnerability of key elements against specific hazardous events, such as overdesigning load-carrying columns to withstand vehicle impacts, the use materials with higher durability and the use of effective firestops.
2. Assume initial local damage and limit damage propagation:
   a. Robustness measures aimed at limiting damage propagation through:
      i. Redundancy (e.g. design beams to carry vertical loads through catenary action in case a supporting load-carrying column is damaged, therefore creating an ALP); and/or
      ii. Segmentation (e.g. design a fire compartment for full burnout, thus isolating it from its surroundings).
3. Prescriptive rules: usually simple and general guidance related to structural detailing, such as enabling continuity and providing ductility. The application of prescriptive rules should be limited to buildings of minor importance, since their effectiveness is often unclear.

In the context of structural robustness, a simplified design framework commonly used to simulate a threat-independent initial damage is the sudden removal of a notional vertical element, such as a column [8]. Given that the loss of the column is not related to a specific hazardous event, the disproportionality of the consequences is often assessed in relation to predefined requirements or expectations (e.g. area that is allowed to collapse given a notional initial damage scenario). However, specifying these requirements is not only an engineering problem [14], since it must reflect the will of the owner, the concerns of the stakeholders that might be affected by the potential consequences of a disproportionate collapse (e.g. civil authorities, insurance companies, users), and even public opinion. Therefore, agreeing on acceptable direct and indirect consequences might require administrative or even political decisions [14].

Robustness-related strategies focus on limiting damage propagation through either redundancy or segmentation. Redundancy strategies are based on providing alternative ways to fulfill the performance requirements bypassing the damaged component by either developing sufficient ductility, resistance or deformation capacity. Segmentation strategies are based on isolating the damaged areas, either through weak fuse elements or strong isolating elements. Most redundancy strategies are better suitable to prevent vertically propagating failures, while segmentation strategies are better suitable to prevent horizontally propagating failures. However, appropriate robustness strategies will depend not only on the geometry of the structure and whether vertical or horizontal collapse propagation is to be expected, but also on the mechanical characteristics of the system. In the context of timber structures the key mechanical characteristics influencing the performance of such strategies are brittle failure modes, low connection stiffness and deformation capacity.

Redundancy strategies must bypass the damaged element in order to redistribute the loads to the surrounding structure. This can be done by flexural mechanisms such as bending and cantilevering, or tensile force development mechanisms such as catenary and membrane action. Flexural mechanisms occur at low deformation levels and rely on the bending resistance. Due to the low rotational capacity of timber connections, these mechanisms would rely on the element continuity, which might not always be the case in timber structures, mainly due to acoustic requirements. Mechanisms that rely on the development of tensile force in the system rely on large deformations, which imposes high deformation requirements on the system. Timber members fail mostly in a brittle manner, which makes timber connections a critical component in ductile load redistribution. However, connections in traditional timber construction are characterized by low rotational stiffness and deformation capacity, which could therefore lead to significant impact on the design of such connectors to accommodate robustness strategies. Experiments on novel connections have shown that timber assemblies could potentially achieve the same mechanisms as reinforced concrete and steel buildings when sufficient ductility is provided (Mpidi Bita et al. 2022).

Segmentation strategies aim to compartmentalize the structure in such a way that failure is prevented from progressing at predefined locations. This can be done by designing segment borders by using fuse-type elements or control joints which physically separate the compartments. Vertical segmentation often relies on shock-absorbing zones with sufficient load-carrying and energy dissipation capacity, such as "power-storeys" to carry debris from the collapsed storeys above. Horizontal segmentation often relies on fuse elements to interrupt structural continuity. Segmentation can be achieved by limiting the upper value of the load-carrying capacity of fuse elements. Given the large variability of mechanical properties in timber elements and their brittle behavior, fuses are often introduced at the connections and based on the concept of capacity design. Their load carrying capacity $R_{d,upp}$ should be smaller that the design effect of the action which will act on the fuse in case of damage scenarios $E_{d,fuse}$: $R_{d,upp} \leq E_{d,fuse}$ [11].

2 Design Framework

The level of design requirements should be determined based on the risk assessment of the structure. This involves classifying the exposure and the importance of the building. For buildings with low importance and exposure, it should be possible to achieve an adequate level of resistance to disproportionate collapse without the need for explicit design verifications. However, for buildings with higher levels of importance or exposure, increasingly complex verifications are often required. The specification of hazardous events should involve defining both threat-specific scenarios, such as the impact of a car on a ground-floor column, and non-threat-specific scenarios, such as notional damage like the sudden removal of a structural component. Performance objectives, which define the acceptable level of damage or consequences, may need to be set in consultation with stakeholders beyond the owner and design team. These stakeholders may include relevant civil and building authorities as well as the insurance company. For large projects, specifying hazardous scenarios might require experience, as establishing general rules is difficult due to the wide range of possible scenarios and the project-specific nature of many of them. Once the hazard scenarios are defined and the performance objectives are set, the structural design team can then select appropriate design strategies. These strategies may include measures to prevent local damage, as overdesigning structural members and connections, but should focus on measures to enhance robustness by limiting damage propagation. The design verification procedures are then based on structural modelling and checking design code requirements, physical testing, or even past experience. A schematic overview of this design framework for achieving resistance to disproportionate collapse is provided in Fig. 2.

3 Robustness Stakeholders

Most aspects of the presented design framework are engineering problems that can be tackled by the structural design team, but some aspects, namely the specification of performance objectives (e.g. acceptable levels of damage), must involve other stakeholders. Stakeholders are the *promoters and owners* of the structure and the *structural design team*, but also those directly or indirectly

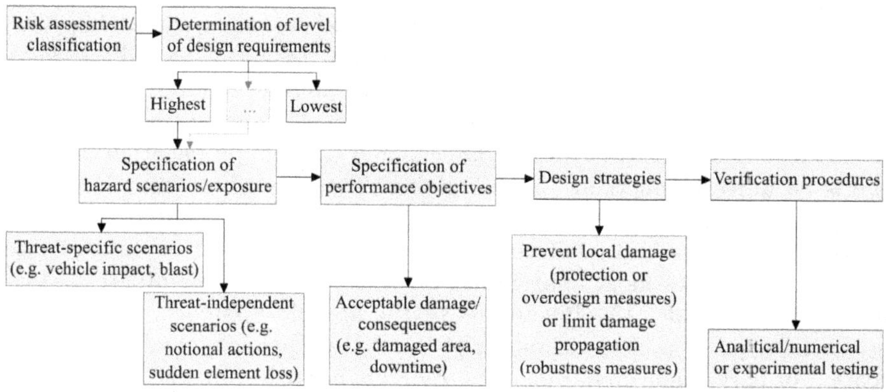

Fig. 2. Design framework for resistance to disproportionate collapse, based on [13].

affected by such project, such as *civil and building authorities, users, insurance companies, contractors*, and even *neighbors.*

The *promoters and owners* are naturally interested in optimising monetary costs of safety measures and in reducing downtime in case of damages, namely downtime needed for repairs and the corresponding economic and reputational losses. Promoters and owners often have no incentive to avoid externalising consequences (e.g. impairment of public infrastructure, or spill of hazardous materials into the environment after a collapse) and it is often up to the civil and building authorities to mitigate any consequences of such incidents as a result of insufficient robustness.

Civil and building authorities take the wider public interest into account and may set additional performance objectives, in particular acceptable levels of damage, and define what the acceptable verification procedures are.

The *structural design team* is interested in achieving a safe, economical and sustainable design, preferably by following rational and well-established design frameworks, in which their responsibilities are clearly defined. If relevant, the structural designer must account for resistance to disproportionate collapse from the early stages of conceptual design and develop a design strategy and the corresponding verification procedures. This should take into account the owner's preferences and the requirements set by the authorities regarding acceptable levels of damage. There can be a tendency to focus on simply overdesigning key structural elements, which might indirectly lead to some increase in resistance to disproportionate collapse in some scenarios, but does not ensure robustness (Fig. 1).

4 Robustness Quantification

Robustness is best achieved when considered from the early stages of conceptual design. While robustness is often approached qualitatively, quantitative approaches can be useful in specific scenarios, such as when comparing different strategies for the same structure or optimizing design for cost-effectiveness. These measures typically involve comparing the extent of damage within a system or analyzing its response to initial damage [3]. Different types of buildings are susceptible to different types of damage and damage propagation, meaning there is no universal method for quantifying robustness. For example, energy-based measures are particularly suited for assessing impact-related progressive collapses, such as pancake failures, whereas measures based on reserve load-carrying capacity are more appropriate for redistribution-type progressive collapses involving ALP [13]. Regardless of the specific approach, a robustness assessment requires a clear definition of the system being evaluated, the identification of performance requirements, relevant hazards, and the consequences of potential damages [9]. Various methods for quantifying structural robustness have been proposed, including deterministic, reliability-based, and risk-based indexes [15]; [6]. Among these, deterministic-based indexes are the simplest to calculate. However, as single-value indexes that aim to summarize complex systems, they provide only limited insight into the susceptibility of the structure to disproportionate collapse. Moreover, these indexes are generally not practical for ordinary design processes. Another limitation is the lack of standardized target values for robustness indexes. This limits their use to comparisons between similar scenarios, such as assessing the same structure under different initial damage conditions or with varying connectivity between elements [12]. These measures of robustness should have a simple and clear definition and be based on quantifiable properties or behavior of the structure [13].

5 Design Codes

5.1 Overview

Robustness design frameworks are incorporated in several current design standards and guidelines. In Europe, the most widely used framework is provided by EN 1991-1-7:2006, which addresses accidental actions on structures. In the upcoming versions of the Eurocodes, EN 1991-1-7 will focus on identified and quantified accidental actions, for which the structure shall be designed, whereas robustness aspects following unforeseen events should be addressed by EN 1990 *Basis of Design* and EN 1992 - 1999. In North America, multiple standards and guidelines provide recommendations for accidental actions on federal and military structures, including UFC 4-023-03 and GSA 2016. Additionally, a newer standard, ASCE 76-23, focuses on mitigation of disproportionate collapse potential of civilian buildings and structures. In this section, a brief overview of the robustness design frameworks provided in EN 1991-1-7:2006, UFC 4-023-03, and ASCE 76-23 will be provided.

5.2 EN 1991-1-7:2006

EN 1991-1-7:2006 [8] provides a robustness assessment method as one of the several methods to mitigate the risk of accidental actions (framework summarized in Table 1). The framework is based on the importance of the structure which is reflected into consequence classes CC1, CC2, and CC3. CC1 buildings are those linked to low consequences, such as small buildings with low occupancy. CC2 buildings refer to medium consequences, such as mid-rise office and residential buildings. An additional distinction is made within CC2, namely CC2a and CC2b, which is based on the use type of the structure. Lastly, CC3 buildings refer to high consequences, such as high occupancy buildings with public access.

Table 1. Robustness design framework provided by EN 1991-1-7:2006.

CC	Hazard scenarios definition	Design strategies	Acceptable level of damage
CC1	Design according to EN 1990 to EN 1999.		
CC2	*CC2a:* • Prescriptive rules.	*CC2a:* • Minimum horizontal ties.	*CC2a:* • No damage acceptance defined.
	CC2b: • Prescriptive rules. • Notional removal of each supporting beam and column.	*CC2b:* • Minimum horizontal ties. • Vertical tying of columns and walls from foundations to roof level. • ALP analysis.	*CC2b:* • Damage threshold $S_{\lim}$ is the smaller of 15% of floor area or 100 m^2 in each adjacent storey to notional damage.
CC3	Systematic risk assessment for foreseeable and unforeseeable events.		

5.3 ASCE 76-23

ASCE 76-23 [4] is a performance based design standard for mitigating disproportionate collapse (framework summarized in Table 2). It is based on collapse-resistant design categories (CRDC), defined based on the building risk category. The risk category is determined using ordinal ranks of the hazard, vulnerability, and consequences. Four different hazard independent damage scenarios (HIDS) are defined, with the descriptions defined below, to be applied at locations defined by designers to have the most detrimental effect. Additionally, four different areas of influence (AI) are defined to quantify damage tolerance, and can be related to the HIDS as described below.

- HIDS H1 (AI-1): Component level notional damage $D_V = 4 \times 4 \times 4\ in$,
- HIDS H2 (AI-2): Damage to individual secondary members or notional damage $D_V = 2 \times 2 \times 2\ ft$,

Table 2. Robustness design framework provided by ASCE 76-23.

CRDC	Hazard scenarios definition	Design strategies	Acceptable level of damage
A	Design according to ASCE 7-16 1.4.		
B	HIDS H-1		• No failure of primary components allowed. • Failure of secondary components adjacent to H1 allowed.
	HIDS H-2		• Failure ≤ 2× AI-2
C	HIDS H-1	HIDS H-1: • There are no deemed to comply solutions.	Same as HIDS H1 for risk CRDC B.
	HIDS H-2	HIDS H-2: • ALP analysis.	Same as HIDS H2 for risk CRDC B.
	HIDS H-3	HIDS H-3: • ALP analysis.	• Failure ≤ 2× AI-3
D	HIDS H-1	HIDS H-4: • There are no deemed to comply solutions.	Same as HIDS H1 for risk CRDC B.
	HIDS H-2		• No failure of components from H2 allowed. • Failure ≤ AI-3
	HIDS H-3		• Failure ≤ 2× AI-3
	HIDS H-4		• Failure ≤ 2× AI-4

- HIDS H3 (AI-3): Damage to individual primary member or notional damage D_V. Wall-type structures minimum damage 144 ft^2 with no floor-wall connections. $D_V = 4 \times 4 \times 4\ ft$,
- HIDS H4 (AI-4): Simultaneous damage to two primary members or notional damage D_V. Wall-type structures minimum damage 288 ft^2 with no floor-wall connections. $D_V = 4 \times 8 \times 20\ ft$.

5.4 UFC 4-023-03

UFC 4-023-03 [16] is a standard for the design of military facilities to resist progressive collapse (framework summarized in Table 3). It is based on risk categories (RC), which are defined based on the occupancy and risk of the structure.

Table 3. Robustness design framework provided by UFC 4-023-03.

<table>
<tr><th>RC</th><th>Hazard scenarios definition</th><th>Design strategies</th><th>Acceptable level of damage</th></tr>
<tr><td>I</td><td colspan="3">No specific progressive collapse design requirements.</td></tr>
<tr><td>II</td><td>• Prescriptive rules.</td><td>• Minimum tying force requirements.
• Local strengthening of ground-floor corner columns.
• ALP analysis if unable to fulfill minimum ties.</td><td rowspan="4">• No damage allowed.</td></tr>
<tr><td>III</td><td rowspan="3">• Column or wall loss scenarios at specified locations.</td><td>• ALP analysis.
• Local strengthening of all ground-floor columns or walls.</td></tr>
<tr><td>IV</td><td>• Minimum tying force requirements.
• Local strengthening of all ground-floor columns or walls.</td></tr>
<tr><td>V</td><td>• ALP analysis.</td></tr>
</table>

6 Case Studies

A selection of case studies of robustness strategies in taller timber buildings from Europe and North America are presented, including a brief description of the structural design, and the design strategies adopted to mitigate disproportionate damage propagation.

Mjøstårnet (Brumunddal, Norway, 18 storeys)

Project description:
The structure comprises large-scale GLT trusses along the façades as well as internal columns and beams. Floors 211 were made with prefabricated wooden decks, while floors 12-18 were made with concrete [1].

Design strategies:

- *Vulnerability:*
  - GLT columns were designed to resist a pressure of 34 kPa, based on EN 1991-1-7:2006.
- *Robustness:*
  - Connections between GLT elements were designed to exhibit a ductile failure mode (redundancy).
  - The structure was designed to withstand the impact of a falling concrete floor (segmentation).

HoHo Wien (Vienna, Austria, 24 storeys)

Project description:
The primary load-bearing structure is a combination of a stiffening concrete core and a simple GLT edge-beam and column structure. The floor system used prefabricated timber-concrete composite elements, connected to internal concrete beams [18].

Design strategies:

- *Robustness:*
  - Column-to-column connections were designed to carry tension forces and support the floor below in case of column loss.
  - Horizontal and vertical ties were used to provide alternative load paths:
    · Vertical ties: glued-in steel rods connected to concrete beams.
    · Horizontal ties: on-site casted reinforcement bars within concrete beam connections (redundancy).

HAUT (Amsterdam, Netherlands, 21 storeys)

Project description:
Lateral stability is provided by a concrete core and two CLT shear walls. Vertical loads are transferred through load-bearing CLT walls, supporting timber-concrete composite floors. Where floor edges are not supported by load-bearing walls, glulam downstand beams are introduced [17].

Design strategies:

- *Robustness:*
  - Tension ring around the floor perimeter acting as a horizontal structural tie (redundancy).

Treet (Bergen, Norway, 14 storeys)

Project description:
The structure consists of vertical GLT trusses and intermediate concrete storeys, on which prefabricated apartment modules are stacked [2].

Design strategies:

- *Robustness:*
  - Concrete power storeys were designed to withstand failures in prefabricated modules and secondary members, also accounting for debris loading (segmentation).

H1-Zwhatt (Regensdorf, Switzerland, 24 storeys)

Project description:
The load-bearing structure consists of a reinforced concrete core and three lower post-beam storeys. The 21 residential storeys above are built with GLT and LVL columns and beams, and biaxial load-bearing prefabricated timber-concrete floor elements [5].

Design strategies:

- *Robustness:*
  - Vertical tension anchoring of timber columns using glued-in threaded rods at the top and dowel connections at the base.
  - Flexural rigid connections between secondary beams via steel tension/compression zones, ensuring continuity in case of column loss.

Brock-Commons (Vancouver, Canada, 18 storeys)

Project description:
The structural system comprises two concrete cores and CLT floor panels which are point-supported on timber columns [7].

Design strategies:

- *Robustness:*
  - Multi-span CLT floor panels were designed for two-way action and cantilevering in case of loss of a column. Column- to- column connections were designed to carry tension forces to hold the floor below (redundancy).

References

1. Abrahamsen, R.: Mjøstårnet – construction of an 81m tall timber building. In: International Holzbau-Forum IHF 2017 (2017)
2. Abrahamsen, R., Malo, K.A.: Stuctural design and assembly of 'Treet' – a 14-storey timber residential building in Norway. In: Proceedings World Conference on Timber Engineering. WCTE 2014, 8, Quebec City, Canada (2014)
3. Andre, J., Faber, M.H.: Proposal of guidelines for the evolution of robustness framework in the future generation of Eurocodes. Struct. Eng. Int. **29**(3) (2019)
4. American Society of Civil Engineers ASCE/SEI 76-23: standard for mitigation of disproportionate collapse potential in buildings and other structures (2023)
5. Brühwiler, I.: Holz-hochhaus in zürich-regensdorf – tragwerksentwurf, variantenstudium, robustheid (timber high-rise building in zürich-regensdorf – structural design, study of variants, robustness). In: International Holzbau-Forum IHF 2022 (2022)
6. Chen, Y.L., Huang, L., Lu, Z.Q., Deng, L., Tao H.Z.: Assessment of structural robustness under different events according to vulnerability. J. Perform. Constructed Facil. **30**(5) (2016)
7. Fast, P., Jackson R.: Brock Commons – A case study in tall timber. Struct. Mag. June 2017(2017)
8. European committee for standardization . Eurocode 1: Actions on structures – Part 1-7: General actions – Accidental actions (2006)
9. Maes, M.A., Fritzsons, K.E., Glowienka S.: Structural robustness in the light of risk and consequences analysis. Struct. Eng. Int. 12 (2): 101-107 (2006)
10. Mpidi Bita, H., Huber, J., Palma, P., and Tannert T. Prevention of disproportionate collapse for multi-story mass timber buildings: review of current practices and recent research. J. Struct. Eng.**148** (7)(2022)
11. Palma, P., Munch-Andersen, J., Dietsch, P.: Updating Eurocode 5 – Design guidance for increasing the robustness of timber structures. In: World conference on Timber Engineering 2023. Oslo, Norway (2023)
12. Palma, P., Steiger, R., Jockwer, R.: Addressing design for robustness in the second generation EN 1995 Eurocode 5. In: International Network Timber Engineering Research – Meeting fifty-two. Tacoma, United States (2019)
13. Starossek, U.: Progressive collapse of structures. In: ICE Publishing (2018)
14. Starossek, U.,Haberland, M.: Disproportionate collapse: terminology and procedures. J. Perform. Constr. Facil. **24**(6) (2010)

15. Sørensen, J.D.: Framework for robustness assessment of timber structures. J. Eng. Struct. **33**(11)(2011)
16. United states department of defense . UFC 4-023-03: Design of buildings to resist progressive collapse (2013)
17. Verhaegh, R., Vola, M., De Jong, J.: HAUT – A 21-storey tall timber residential building. J. High-Rise Buildings. **9**(3) (2020)
18. Woschitz., Zotter, J.: Holzhochhaus HoHo Wien – Das Tragwerkskoncept (High-rise timber building HoHo Wien – The structural concept). Österr. Ing. – Archit – Z., **162** (1-12) (2017)

Lateral-Load-Resisting Systems (LLRSs) in Taller Timber Buildings

Michele Mirra[1(✉)], Giacomo Iovane[2], Beatrice Faggiano[2], Saeid Javidi[3], Victor Belafonte[4], Gabriele Tamagnone[5], Jorge Branco[4], Tuba Tatar[6], Wenchen Dong[7], and Giuseppe D'Arenzo[8]

[1] Delft University of Technology, Delft, The Netherlands
m.mirra@tudelft.nl
[2] University of Naples Federico II, Naples, Italy
[3] Birmingham City University, Birmingham, UK
[4] University of Minho, Guimarães, Portugal
[5] University of Montpellier, Montpellier, France
[6] Sakarya University, Sakarya, Turkey
[7] StructureCraft, Abbotsford, BC, Canada
[8] University of Enna Kore, Cittadella Universitaria, Enna, Italy

Abstract. The present chapter examines different types of lateral-load-resisting systems (LLRSs) for taller timber buildings (TTBs). The LLRSs are presented with specific reference to their role in the seismic response of TTBs, and have been classified according to the main structural elements composing them. Therefore, timber-framed LLRSs, consisting of one-dimensional members, are first examined, followed by panelised LLRSs, featuring two-dimensional elements; mixed timber-framed and panelised LLRSs are then addressed. Next, hybrid timber-steel and -concrete, as well as volumetric modular LLRSs, are presented. Finally, a separate section is dedicated to the interaction of timber diaphragms with LLRSs in TTBs. Besides a comprehensive literature review, for each LLRS type, relevant examples of buildings realized in practice have been reported. In the final section, based on the analysed research studies from literature, the main future challenges and research directions for an effective, holistic design of LLRSs in TTBs, are addressed.

Keywords: Timber-framed structures · panelised structures · modular construction · hybrid timber buildings · timber diaphragms · seismic loading

1 Introduction

The structural behaviour of Taller Timber Buildings (TTBs) subjected to earthquake actions is governed to a large extent by the Lateral-Load-Resisting Systems (LLRSs), whose performance is mainly dependent on the structural type, the geometry, the Engineered Wood Products (EWPs) adopted for the different structural elements, and the

G. Fink et al. (Eds.): *Holistic Design of Taller Timber Buildings*, SPRTRCIENG, pp. 238–252, 2025.
https://doi.org/10.1007/978-3-032-02098-7_19

type of connections. While the latter aspect is discussed in the dedicated chapter *High-performance connections for seismic resistance*, in this chapter an overview of the typical LLRSs is presented, some of which are summarized in Fig. 1, and corresponding examples of constructed buildings are introduced.

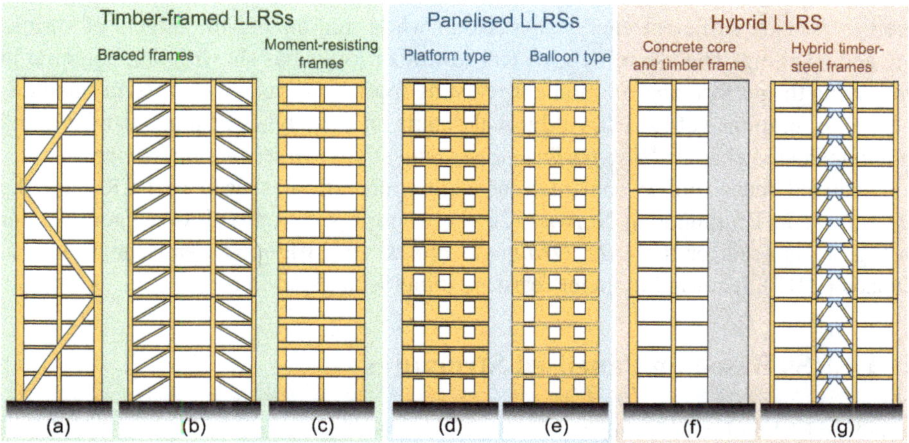

Fig. 1. Examples of LLRS for taller timber buildings based on: 1D elements, such as braced frames (a, b) or moment-resisting frames (c); 2D elements, such as platform-type (d) or balloon-type (e) CLT structures; hybrid systems involving concrete (f) or steel (g) elements.

2 LLRSs Based on Timber-Framed Structures

Lightweight timber frames are widely used and highly adaptable construction types around the world. All vertical and horizontal loads are carried and transferred through vertical (columns), horizontal (beams) and diagonal small-size timber members. The lightweight nature of this LLRS makes it efficient and practical for various building types and structural requirements, but not suited for TTBs, where this system has evolved into "heavy" timber-framed LLRSs.

In the context of timber-framed structures, it is possible to classify two types of LLRSs [1]: braced frames (e.g. Fig. 1a, b), and moment-resisting frames (e.g. Fig. 1c). In braced frames, resistance to lateral loads is developed through axial forces in the structural elements; ductile mechanisms in the connections at the ends of the diagonal braces dissipate energy, while other members and connections remain elastic [2–4]. The connections used in braces may be considered relatively ductile if they are properly conceptualized following capacity design, without risk of brittle failures in the wooden members. Within the LLRS, the braces can be arranged in different configurations: the braces can be concentric (e.g. in Fig. 1b, [3]) or eccentric (e.g. in [1, 5]), placed at every storey or as outriggers for more efficient lateral stability (e.g. in Fig. 1a, [6]). Recent developments in braced timber frames consist of the use of buckling-restrained braces (discussed in Sect. 4), or the shape optimization of the braced frames [7].

Moment-resisting frames (Fig. 1c) are composed of the assemblage of beams and columns through connections designed to transfer axial, shear and bending moment. They resist the lateral forces mainly through bending of members. Some examples of such connections are bolts with tight-fit pins, or glued connections. However, moment-resisting timber-to-timber connections feature difficult joint detailing and some extent of rotations and associated drifts under seismic loads [8]. In addition, these connections need to provide sufficient energy dissipation while minimizing the effects of shrinkage and swelling. The use of hybrid steel-to-timber joints has shown great potential in improving the performance of moment-resisting frames but has not resulted in standardized configurations yet [9–11]. Additional details on these types of joints can be found in the dedicated chapter *High-performance connections for seismic resistance.*

Realized mass timber constructions with timber frames used as LLRS are the *Mjøstårnet* in Brumunddal, Norway (2019), where a solution with outriggers similar to Fig. 1a is used, or the *UBC Okanagan Fitness and Wellness Centr*e in Kelowna, Canada (2013), featuring moment-resisting joints with glued-in rods.

3 LLRSs Based on Panelised Structures

Panelised LLRSs for TTBs are usually made of Cross-Laminated Timber (CLT) [12–16] or, in some cases, Laminated Veneer Lumber (LVL) [17]. In mid-rise buildings up to six storeys, also light timber-frame with oriented strand board (OSB) or plywood panels exhibit excellent seismic performance [18–20]. Depending on the design approach, CLT LLRSs could be platform- or balloon- type, meaning the shear walls are interrupted at each storey (Fig. 1d) or go bottom-up (Fig. 1e), respectively.

Platform-type shear walls can be monolithic or segmented, namely, made of one single panel or two or more panels, respectively [2, 14, 21–25]. Structural systems made with single-panel CLT shear walls provide low-to-medium energy dissipation, due to the limited number of connections. Segmented walls ensure a higher energy dissipation thanks to the contribution of panel-to-panel vertical joints. Moreover, the more slender the panels are, the more rocking-governed the drift is, leading potentially to no or minimal residual drift after the seismic action.

Balloon-type structures take advantage of the continuity of the CLT panels along the height of the shear walls in resisting lateral in-plane loads [26–29]. In these cases, the LLRS and the vertical load resisting system are separated, with shear walls only withstanding lateral actions. The slender aspect ratio renders these shear walls more prone to rocking than sliding, so most solutions used focus on this mechanism to resist actions and dissipate energy. The specific resisting mechanism of these walls depends on the connections and devices used. The wall can be connected to the foundation with connections similar to those used in platform-type shear walls (i.e., hold-down and angle brackets [30]), but, understandably, in larger dimensions.

The optimal performance of CLT LLRS requires the application of capacity design in the joints [31], whose effect could yet result in larger and more expensive use of material, potentially removing the advantage of using higher behaviour factors for dissipative CLT structures [32].

Advanced connections, such as slip friction or other dissipative [33] or high-capacity joints [34], can be also used in CLT LLRSs for TTBs. Another implemented technique consists of unbonded post-tensioned cables [9, 35, 36], anchored to the foundation and fixed at the top of the wall, placed in the centre of it, at least one per side.

Regardless of the system, shear walls can also be coupled along the length to benefit from additional dissipation given by the connections between the walls [33, 35], which is usually performed through purposely designed and shaped steel plates that are able to deform plastically without hindering the rocking mechanism, although, in some cases, fuse steel elements have also been used, such as coupling beams or purposely perforated flat plates [37]. Finally, more sophisticated solutions, combining two or more of the above approaches, can be used [33, 36, 38].

Many examples of low-to-mid-rise buildings composed of CLT shear walls are built worldwide, such as *Bridport House* in London, United Kingdom (2010), or *Origine* in Quebec City, Canada (2017).

4 Mixed Timber-Framed and Panelised LLRSs

The two former types of LLRSs based on timber-framed and panelised structures can be combined in core- or shear-wall-framed constructions [1]. These structures are composed of two coupled structural systems: the vertical load-resisting system, generally a beam-column structure and the LLRS, such as, for example, CLT or mass timber walls or cores.

One of the most common solutions involves the use of CLT for the LLRS, consisting of cores (including elevator shafts) and perimeter walls, in combination with timber beams and columns, as well as CLT floors [39, 40]. A building realised by employing these mixed LLRSs can benefit from the above mentioned advantages of both timber-frame and panelised structural systems. Yet, it is important to note that the position and layout in plan and elevation of the cores and perimeter walls can strongly influence the response of the whole structural system. These TTBs might become flexible and exhibit inter-storey drifts greater than the limits prescribed in the standards; when irregular configurations of the LLRS are considered, a lower lateral deformability is observed, but greater uplift loads are transferred to the foundations [39]. Hence, in seismic regions, the feasibility of designing multi-storey framed timber buildings braced by CLT shear-walls, depends on the possibility to control lateral drifts and provide anchoring connections capable of handling large uplift forces between the superstructure and the foundations.

Other investigated mixed LLRSs consist of shear walls realised by incorporating CLT [41] or light-frame [42] infills, within glulam beams and columns. These systems show enhanced ductility and energy dissipation, but their efficiency is highly dependent on the aspect ratio and the presence and dimensions of openings of the infills, all aspects influencing the interaction and effective collaboration with the glulam frame.

A recent, relevant example of the combination of timber-framed and panelised LLRSs consists of the *Sara Kulturhus Center* in Skellefteå, Sweden (2021), where this solution allowed to build a 20-storey structure.

5 Hybrid LLRSs

Hybrid systems represent those LLRSs, in which timber elements are combined with steel and/or concrete elements (e.g. Fig. 1f, g). These LLRSs show smart and more efficient use of the materials to improve the global and local behaviour with respect to timber-only structures. The advantages of the hybridization lie in the possibility to design optimized structures, where the lightness of timber is combined with the stiffness of steel and concrete and the ductility of steel [43, 44]. A large variety of hybrid systems can be adopted as LLRS, for instance:

- Steel-timber composite shear walls featuring mass timber panels [38, 45–49] or lightweight timber walls [50–54], as well as cold-formed steel diaphragms with bracing timber-based panels [55]. In particular, the combination of steel moment-resisting frames and light timber frame structures [50] can also be integrated with more advanced seismic protection technologies, such as friction dampers [54] or self-centring capacity [56]. Another variation falling under this category is a mass timber and podium integration, consisting of mass timber panels supported by a stiff podium constructed from concrete or steel. The podium functions as a base that provides essential lateral stiffness and strength while acting as a critical interface between the foundation and the lightweight timber superstructure. The effectiveness of this hybrid system in mitigating lateral deformation under extreme loading conditions has been demonstrated by [19, 26];
- Steel/concrete LLRSs with mass timber [57–63] or lightweight timber (-composite) diaphragms and timber walls/columns [64–67]. One of the most adopted solutions consists of a reinforced concrete core LLRS with timber members used to carry vertical loads (e.g. Fig. 1f). The concrete core fulfils the structural performance and simultaneously is the stair and lift shaft, as it is the case with conventional reinforced concrete high-rise buildings. Such a system presents several advantages, such as: (i) it eases the limitation related to fire safety of TTBs as stairs and lift shafts are constructed of non-combustible materials, (ii) its performance under seismic actions is significantly less complex to be designed for the majority of practitioners, (iii) it represents a cost-optimized solution compared to other timber alternatives [40, 61, 68];
- Hybrid timber-steel framed systems consisting of steel/concrete columns and timber beams, slabs and/or walls [69, 70], or timber frames with steel members and/or high-performance steel joints (e.g. Fig. 1g [10, 71–76]);
- Timber buckling-restrained braced frames [77–81], consisting of a steel core and a timber casing acting as restraint for the steel core buckling. This system is popular because it provides enhanced strength and stiffness [82–84] and it can be combined with more advanced joints or seismic protection technologies [73, 85, 86].

Examples of realized TTBs with hybrid systems are the *Haut* project in Amsterdam, The Netherlands (2022), which is one of the tallest timber hybrid buildings with a 73 m height; the *Brock Commons Tallwood House* (2016) and the *MEC head office* (2014), both in Vancouver, Canada.

6 Volumetric Modular LLRSs

The revision of conventional models replicated by the construction sector is contributing to the expansion of volumetric modular systems. Several advantages come from this industrial approach, starting with manufacturing control, quality in a shortened time, accuracy and an increase in the speed of construction [87]. The use of timber, as the main structural material, in a modular system, gives a highlight to the environmental concern, based on the introduction of Engineered Wood Products (EWPs) to achieve the carbon neutrality strategy and the demand for buildings into opportunities for sustainable development [88].

In volumetric modular construction, each module is designed as an independent unit, incorporating load-bearing and non-load-bearing walls [89]. When assembled and stacked, these modules must function cohesively as an integrated structural system. This requires seamless interaction between the vertical and horizontal load-resisting systems to ensure overall stability and structural performance.

The schematic design and development stages on the erection procedure of modular constructions demand specific attention to ensure the proper connections of the LLRS [90]. The floor plan needs to align with the building's functional use, considering various configurations of the modular units and the coordination of development phases [91].

The connections between modular units play a fundamental role in the structural performance of modular constructions, ensuring stability and resistance under load demands [92]. In tall buildings, the mechanism of force transfer of horizontal loads to the stabilizing system presents additional challenges regarding the connection between the module and the core, normally made by concrete or steel [93].

In practice, timber modular buildings predominantly utilize CLT panels as both the LLRS and the gravity load-bearing system [94]. Moreover, the evaluation of the LLRS definitions on a volumetric timber modular system is made regarding the easy handling and assembly of those connections combined with the perspective of structural effectiveness.

In volumetric modular systems, the LLRS is typically formed by interconnecting the walls of individual modules, as illustrated in Fig. 2. In tall buildings, this configuration often follows a particular case of balloon-type approach [95], where walls are directly joined using plates (shear connectors) and tie-downs (tension and compression connectors) until they are connected to the foundation typically with hold-downs and angle brackets.

Within this design, the shear diaphragm plays a critical role in transferring forces to the walls and down to the foundation. This behaviour is effectively achieved by connecting ceiling slabs to form diaphragms, offering a practical and efficient solution due to their ease of installation and assembly. Nonetheless, achieving a reliable floor-to-floor

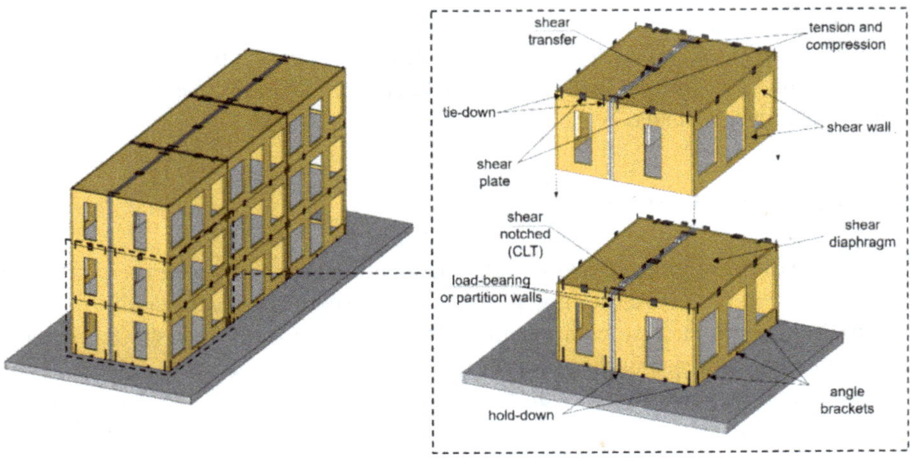

Fig. 2. Typical Volumetric Modular Timber Configuration.

connection presents significant challenges. Since volumetric modules are fully assembled in the factory, including interior finishes and services, introducing robust connections between floor diaphragms must not compromise the integrity of these prefinished components. The need for rapid and straightforward on-site assembly further limits the use of complex connection systems. Additionally, tolerances between modules can lead to misalignments, making it difficult to ensure continuous load transfer and diaphragm action, especially when combined with limited space for proper connections. Moreover, not all walls contribute to the LLRS since some cannot be directly anchored to the foundation (for example, partition or load-bearing walls). For instance, when a new module is placed beside an existing one, the adjacent wall of the new module may lack sufficient space for a direct ground connection. This scenario highlights the importance of following the correct erection sequence to ensure that proper wall-to-foundation connections are made.

These challenges underline the need for innovative solutions and refined construction strategies to fully harness the potential of modular systems. The modular construction approach offers numerous advantages, making it well-suited for high-rise buildings and potentially shaping the industry's future. Nevertheless, while volumetric modular systems are widely used in low-rise structures, their application in high-rise construction remains relatively limited, where several challenges arise. One key issue is inter-modular connectivity, which becomes more complex in tall buildings. Additionally, geometric imperfections, common due to construction tolerances, are often amplified in hybrid systems, such as those incorporating concrete core structures. The assembly process and detachable connections also pose challenges, particularly in relation to Design for Manufacturing and Assembly (DFMA) principles. Furthermore, there is a lack of comprehensive structural guidelines for volumetric modular systems, especially concerning their behaviour under lateral forces, robustness, and the effective integration of the LLRS, all of which are crucial for ensuring the overall stability and performance of the structure.

7 Interaction of Floor Diaphragms with LLRSs

Floor diaphragms are crucial components in LLRSs, as they allow the transfer of horizontal loads (wind, earthquake) to the vertical structural elements, ensuring that the entire structural system responds under seismic actions uniformly. A key role in this regard is played by the in-plane stiffness of the diaphragms, since it influences the distribution of inertia forces to the vertical elements of the LLRS and the development of energy dissipation mechanisms. When a floor diaphragm has significantly greater in-plane stiffness than the vertical elements, seismic forces are distributed to the vertical elements in proportion to their stiffness and a controlled building response under earthquake can be achieved. The critical role of diaphragms in LLRSs has been widely recognized, hence particular attention in their design has been recommended [96]. The main factors influencing the effective action of the diaphragms under seismic loading are the stiffness of timber members and that of their connections [97], which are also responsible for providing energy dissipation under cyclic in-plane loading [98–100]. Too flexible diaphragms enhance higher mode effects and alter the load redistribution in the vertical elements of the LLRS [97]. Thus, to ensure sufficient in-plane stiffness, the quantification of overstrength factors between vertical elements of the LLRS and diaphragms has been recommended [96], to enable ductile behaviour at building level for higher-than-expected forces or deformation capacity under extreme loading conditions.

These design considerations are accounted for in Eurocode 8 [101] for light-frame floors, but for CLT floors no prescriptive rigid diaphragm conditions are currently included, with appropriate capacity design rules being absent as well [24]. In this context, previous studies have emphasized that the in-plane stiffness of CLT diaphragms is primarily governed by floor panel-to-panel joints and by floor-to-wall connections [102]. CLT diaphragms and their connections to the wall LLRSs have also been found to influence their rocking mechanism, although further investigations are required to explore the influence of the panel-to-panel joints in floor diaphragms [103]. From a preliminary study on CLT building archetypes, an increase in the stiffness of the floor panel-to-panel joints, and a decrease in the stiffness of the shear walls, result in a rigid in-plane response of the floor [104].

8 Challenges and Outlook

The previous description of the typical LLRS in TTBs has already highlighted some of the future challenges that need to be addressed in future research, with specific reference to holistic design procedures.

LLRSs composed of timber-framed structures constitute an effective option, yet showing as main challenge the need for adopting high-performance joints with sufficient strength and hysteretic energy dissipation, supported by appropriate capacity design rules, and in combination with, in some cases, the need to meet comfort requirements, for instance when sound insulation is needed.

Two-dimensional LLRSs, mostly involving CLT structures, are very efficient and have already found broad practical application in TTBs. However, especially in regions prone to seismic events, the use of traditional connections (hold downs and angle brackets) results in a limitation in the number of storeys, because of the large lateral loads involved, in combination with capacity design requirements. However, possible advancements in this regard may involve the use of high-performance connections. The same limitations apply to mixed timber-framed and panelised LLRSs.

Hybrid systems are also a viable option to realize TTBs, as highlighted in the related state of the art. However, similarly to the previous case, the efficiency of these systems often relies on high-performance connections, which needs to be further developed and standardized for an easier and more accurate design, modelling and application, including compliance to capacity design and clear guidelines on how this can be achieved.

Effective joints constitute the main challenge also when volumetric systems are used in TTBs, since inter-modular connectivity needs to be realized for efficient lateral load transfer. Additionally, geometric imperfections, common due to construction tolerances, are often amplified in hybrid volumetric systems, such as those incorporating concrete core structures. From the point of view of holistic design, the assembly process and the conceptualization of detachable connections also pose challenges, particularly in relation to DFMA principles. Furthermore, comprehensive structural guidelines are lacking for volumetric modular systems, especially concerning their behaviour under lateral forces, robustness, and the effective integration of the LLRS, all of which are crucial for ensuring the overall stability and performance of TTBs realized with these systems.

Finally, lack of prescriptions is observable also in the case of one of the main components of LLRSs, the floor diaphragms. In fact, guidelines are only available in case of light-frame timber floor systems, but conditions for in-plane diaphragm rigidity in case of CLT floors are not yet specified. Additional investigations on how panel-to-panel floor joints and wall-to-floor connections in CLT diaphragms influence the response of the walls' LLRS are also needed.

References

1. Faggiano, B., Iovane, G.: First considerations on the design approach and criteria for seismic resistant moment resisting and bracings timber frames. In: World Conference on Timber Engineering, Vienna, Austria (2016)
2. Popovski, M.: Lateral load resisting systems for engineered wood construction. In: Final Report 2008/09, Canadian Forest Service No. 27, Project No. 4625 (2009)
3. Popovski, M., Karacabeyli, E., Prion, H.G.L.: Dynamic response of braced timber frames. In: 8th Canadian Conference on Earthquake Engineering, Vancouver, Canada (1999)
4. Chen, Z., Popovski, M.: Connection and system ductility relationship for braced timber frames. J. Struct. Eng. **146**(12) (2020). https://doi.org/10.1061/(ASCE)ST.1943-541X.0002839
5. Cao, J., Xiong, H., Ghahari, S.F., Taciroglu, E.: A validated lateral response model for mass timber frames with knee-braces. Eng. Struct. **239**, 112278 (2021). https://doi.org/10.1016/j.engstruct.2021.112278
6. Liven, H., Abrahamsen, R.: Mjøstårnet: the world's tallest timber building. In: World Conference on Timber Engineering, Oslo, Norway (2023)
7. Mam, K., Douthe, C., Le Roy, R., Consigny, F.: Shape optimization of braced frames for tall timber buildings: influence of semi-rigid connections on design and optimization process. Eng. Struct. **216**, 110692 (2020). https://doi.org/10.1016/j.engstruct.2020.110692
8. Heiduschke, A., Kasal, B., Haller, P.: Shake table tests of small- and full-scale laminated timber frames with moment connections. Bull. Earthq. Eng. **7**, 323–329 (2009). https://doi.org/10.1007/s10518-008-9075-4

9. Palermo, A., Pampanin, S., Buchanan, A., Newcombe, M.: Seismic design of multi-storey buildings using laminated veneer lumber (LVL). In: 2005 NZSEE Conference, Taupo, New Zealand (2005)
10. Andreolli, M., Piazza, M., Tomasi, R., Zandonini, R.: Ductile moment-resistant steel–timber connections. Proc. Inst. Civ. Eng. Struct. Build. **164**, 65–78 (2015). https://doi.org/10.1680/stbu.9.00098
11. Faggiano, B., Iovane, G., Tartaglia, R., Ciccone, G., Landolfo, R., Mazzolani, F.M., Andreolli, M., Tomasi, R., Piazza, M.: Numerical simulation of monotonic tests on beam-column timber joints equipped with steel links for heavy timber seismic resistant MRF. In: International Conference on Numerical Analysis and Applied Mathematics 2018, ICNAAM 2018, Rhodes, Greece (2019)
12. Ceccotti, A., Sandhaas, C., Yasumura, M.: Seismic behaviour of multistory cross-laminated timber buildings. In: Proceedings of the International Convention of Society of Wood Science and Technology and United Nations Economic Commission for Europe – Timber Committee, Geneva, Switzerland, Oct. 11-14 (2010)
13. Gavric, I., Fragiacomo, M., Ceccotti, A.: Cyclic behavior of CLT wall systems: experimental tests and analytical prediction models. J. Struct. Eng. **141** (2015). https://doi.org/10.1061/(ASCE)ST.1943-541X.0001246
14. van de Lindt, J.W., Furley, J., Amini, M.O., Pei, S., Tamagnone, G., Barbosa, A.R., Rammer, D., Line, P., Fragiacomo, M., Popovski, M.: Experimental seismic behavior of a two-story CLT platform building. Eng. Struct. **183**, 408–422 (2019). https://doi.org/10.1016/j.engstruct.2018.12.079
15. Sandoli, A., D'Ambra, C., Ceraldi, C., Calderoni, B., Prota, A.: Sustainable cross-laminated timber structures in a seismic area: overview and future trends. Appl. Sci. **11**(5), 2078 (2021). https://doi.org/10.3390/app11052078
16. D'Arenzo, G., Ruggeri, E.M., Fossetti, M.: Lateral performance of cross-laminated timber shear walls connected to perpendicular walls: experimental tests and analytical modeling. J. Struct. Eng. **150**(8) (2024). https://doi.org/10.1061/JSENDH.STENG-13151
17. D'Arenzo, G., Seim, W.: Cyclic behaviour of Glulam, LVL and GLVL shear walls and their base connections. Eng. Struct. **315**, 118493 (2024). https://doi.org/10.1016/j.engstruct.2024.118493
18. van de Lindt, J.W., Pei, S., Pryor, S.E., Shimizu, H., Isoda, H.: Experimental seismic response of a full-scale six-story light-frame wood building. J. Struct. Eng. **136**(10) (2010). https://doi.org/10.1061/(ASCE)ST.1943-541X.0000222
19. van de Lindt, J.W., Pryor, S.E., Pei, S.: Shake table testing of a full-scale seven-story steel–wood apartment building. Eng. Struct. **33**(3), 757–766 (2011). https://doi.org/10.1016/j.engstruct.2010.11.031
20. Tomasi, R., Sartori, T., Casagrande, D., Piazza, M.: Shaking table testing of a full-scale prefabricated three-story timber-frame building. J. Earthq. Eng. **19**, 505–534 (2015). https://doi.org/10.1080/13632469.2014.974291
21. Reynolds, T., Foster, R., Bregulla, J., Chang, W.S., Harris, R., Ramage, M.: Lateral-load resistance of cross-laminated timber shear walls. J. Struct. Eng. **143**(12) (2017). https://doi.org/10.1061/(ASCE)ST.1943-541X.0001912
22. D'Arenzo, G.: An elastic model for the prediction of the lateral response of cross-laminated timber shear walls with openings. Eng. Struct. **274**, 115055 (2023). https://doi.org/10.1016/j.engstruct.2022.115055
23. Rinaldi, V., Casagrande, D., Cimini, C., Follesa, M., Fragiacomo, M.: An upgrade of existing practice-oriented FE design models for the seismic analysis of CLT buildings. Soil Dyn. Earthq. Eng. **149**, 106802 (2021). https://doi.org/10.1016/j.soildyn.2021.106802

24. Izzi, M., Casagrande, D., Bezzi, S., Pasca, D., Follesa, M., Tomasi, R.: Seismic behaviour of Cross-laminated timber structures: a state-of-the-art review. Eng. Struct. **170**, 42–52 (2018). https://doi.org/10.1016/j.engstruct.2018.05.060
25. Casagrande, D., D'Arenzo, G., Masroor, M., Gavrić, I., Doudak, G.: An analytical matrix approach for the prediction of the elastic lateral displacements of CLT platform-type lateral load resisting systems. Structure. **64**, 106490 (2024). https://doi.org/10.1016/j.istruc.2024.106490
26. Chen, Z., Chui, Y.: Lateral load-resisting system using mass timber panel for high-rise buildings. Front. Built Environ. **3**, 40 (2017). https://doi.org/10.3389/fbuil.2017.00040
27. Dickof, C., Shahnewaz, M., Bevilacqua, N., Tannert, T.: Experimental investigations on balloon frame CLT shearwalls. In: World Conference on Timber Engineering, Santiago, Chile (2021)
28. Yang, T.Y., Lepine-Lacroix, S., Ramos Guerrero, J.A., McFadden, J.B.W., Al-Janabi, M.A.Q.: Seismic performance evaluation of innovative balloon type CLT rocking shear walls. Resil. Cities Struct. **1**(1), 44–52 (2022). https://doi.org/10.1016/j.rcns.2022.03.004
29. Xing, D., Casagrande, D., Doudak, G.: Investigating the deformation characteristics of balloon-type CLT shearwall. Can. J. Civ. Eng. **51**(7), 704–722 (2024). https://doi.org/10.1139/cjce-2023-0233
30. Gavric, I., Fragiacomo, M., Ceccotti, A.: Cyclic behaviour of typical metal connectors for cross-laminated (CLT) structures. Mater. Struct. **48**, 1841–1857 (2015). https://doi.org/10.1617/s11527-014-0278-7
31. Casagrande, D., Doudak, G., Polastri, A.: A proposal for the capacity-design at wall- and building-level in light-frame and cross-laminated timber buildings. Bull. Earthq. Eng. **17**, 3139–3167 (2019). https://doi.org/10.1007/s10518-019-00578-4
32. Cenci, G., Follesa, M., Rinaldi, V., Casagrande, D., Fragiacomo, M.: Seismic design of a CLT multi-storey building in different ductility classes. In: World Conference on Timber Engineering, Oslo, Norway (2023)
33. Hashemi, A., Zarnani, P., Masoudnia, R., Quenneville, P.: Seismic resilient lateral load resisting system for timber structures. Construct. Build Mater. **149**, 432–443 (2017). https://doi.org/10.1016/j.conbuildmat.2017.05.112
34. Loss, C., Pacchioli, S., Polastri, A., Casagrande, D., Pozza, L., Smith, J.: Numerical study of alternative seismic-resisting systems for CLT buildings. Buildings. **8**(11), 162 (2018). https://doi.org/10.3390/buildings8110162
35. Brown, J.R., Li, M., Palermo, A., Pampanin, S., Sarti, F.: Investigating CLT lateral load resisting systems for taller timber buildings. In: 2021 NZSEE Conference, Christchurch, New Zealand (2021)
36. Brown, J.R., Li, M., Palermo, A., Pampanin, S., Sarti, F.: Experimental testing of a low-damage post-tensioned C-shaped CLT core-wall. J. Struct. Eng. **147**(3), 1–16 (2021). https://doi.org/10.1061/(ASCE)ST.1943-541X.0002926
37. Daneshvar, H., Niederwestberg, J., Dickof, C., Jackson, R., Chui, Y.H.: Perforated steel structural fuses in mass timber lateral load resisting systems. Eng. Struct. **257**, 114097 (2022). https://doi.org/10.1016/j.engstruct.2022.114097
38. Hashemi, A., Masoudnia, R., Quenneville, P.: Seismic performance of hybrid self-centring steel-timber rocking core walls with slip friction connections. J. Constr. Steel Res. **126**, 201–213 (2016). https://doi.org/10.1016/j.jcsr.2016.07.022
39. Polastri, A., Izzi, M., Pozza, L., Loss, C., Smith, I.: Seismic analysis of multi-storey timber buildings braced with a CLT core and perimeter shear-walls. Bull. Earthq. Eng. **17**, 1009–1028 (2019). https://doi.org/10.1007/s10518-018-0467-9
40. Connolly, T., Loss, C., Iqbal, A., Tannert, T.: Feasibility study of mass-timber cores for the UBC tall wood building. Buildings. **8**(8), 98 (2018). https://doi.org/10.3390/buildings8080098

41. Xue, J., Ren, G., Qi, L., Wu, C., Yuan, Z.: Experimental study on lateral performance of glued-laminated timber frame infilled with cross-laminated timber shear walls. Eng. Struct. **239**, 112354 (2021). https://doi.org/10.1016/j.engstruct.2021.112354
42. Cao, J., Xiong, H., Liu, Y., Yu, D., Chen, J.: Seismic performance of glulam timber post and beam structures with and without light frame timber shear wall infill. J. Build. Eng. **57**, 104965 (2022). https://doi.org/10.1016/j.jobe.2022.104965
43. Premrov, M., Žegarac Leskovar, V.: Innovative structural systems for timber buildings: a comprehensive review of contemporary solutions. Buildings. **13**, 1820 (2023). https://doi.org/10.3390/buildings13071820
44. Pastori, S., Mazzucchelli, E.S., Wallhagen, M.: Hybrid timber-based structures: a state of the art review. Construct. Build Mater. **359**, 129505 (2022). https://doi.org/10.1016/j.conbuildmat.2022.129505
45. Dickof, C.: CLT infill panels in steel moment resisting frames as a hybrid seismic force resisting system. Ph.D. dissertation, University of British Columbia, Vancouver, Canada (2013)
46. Tesfamariam, S., Stiemer, S.F., Dickof, C., Bezabeh, M.A.: Seismic vulnerability assessment of hybrid steel-timber structure: steel moment-resisting frames with CLT infill. J. Earthq. Eng. **18**(6), 929–944 (2014). https://doi.org/10.1080/13632469.2014.916240
47. Bezabeh, M.A., Tesfamariam, S., Stiemer, S.F., Popovski, M., Karacabeyli, E.: Direct displacement-based design of a novel hybrid structure: steel moment-resisting frames with cross-laminated timber infill walls. Earthq. Spectra. **32**(3), 1565–1585 (2016)
48. Loss, C., Piazza, M., Zandonini, R.: Connections for steel-timber hybrid prefabricated buildings. Part I: Experimental tests. Construct. Build Mater. **122**, 781–795 (2016). https://doi.org/10.1016/j.conbuildmat.2015.12.002
49. Loss, C., Piazza, M., Zandonini, R.: Connections for steel-timber hybrid prefabricated buildings. Part II: Innovative modular structures. Construct. Build Mater. **122**, 796–808 (2016). https://doi.org/10.1016/j.conbuildmat.2015.12.001
50. He, M., Li, Z., Lam, F., Ma, R., Ma, Z.: Experimental investigation on lateral performance of timber-steel hybrid shear wall systems. J. Struct. Eng. **140**(6), 04014029 (2014). https://doi.org/10.1061/(ASCE)ST.1943-541X.0000855
51. Kho, D., Dong, W., Li, M., Lee, C.-L.: Cyclic behaviour of timber-steel hybrid shear walls. In: Proceedings of the 2018 NZSEE Conference, Auckland, New Zealand (2018)
52. Cui, Y., Chen, F., Li, Z., Qian, X.: Self-centering steel–timber hybrid shear wall: experimental test and parametric analysis. Materials. **13**(11), 2518 (2020). https://doi.org/10.3390/ma13112518
53. Li, Z., Wang, X., He, M., Ou, J., Li, M., Luo, Q., Dong, H.: Structural damage evaluation of multistory timber–steel hybrid structures through shake table tests. J. Perform. Constr. Facil. **35**(1) (2020). https://doi.org/10.1061/(ASCE)CF.1943-5509.0001555
54. Dong, H., He, M., Christopoulos, C., Li, Z.: Quasi-static tests and parametric simulations of hybrid steel frame and light wood shear walls with frictional dampers. Eng. Struct. **228**, 111485 (2021)
55. Landolfo, R., Fiorino, L., Della Corte, G.: Seismic behavior of sheathed cold-formed structures: physical tests. J. Struct. Eng. **132**(4) (2006). https://doi.org/10.1061/(ASCE)0733-9445(2006)132:4(570)
56. Chen, F., Li, Z., He, M., Wang, Y., Shu, Z., He, G.: Seismic performance of self-centering steel-timber hybrid shear wall structures. J. Build. Eng. **43**, 102530 (2021)
57. Moore, M.: Scotia place-12 storey apartment buildings. A case study of high-rise construction using wood and steel. N. Z. Timber Des. J. **10**(1), 5–12 (2000)
58. Asiz, A., Smith, I.: Connection system of massive timber elements used in horizontal slabs of hybrid tall buildings. J. Struct. Eng. **137**(11) (2011). https://doi.org/10.1061/(ASCE)ST.1943-541X.0000363

59. Hassanieh, A., Valipour, H., Bradford, M.: Experimental and numerical investigation of short-term behaviour of CLT-steel composite beams. Eng. Struct. **144**, 43–57 (2017). https://doi.org/10.1016/j.engstruct.2017.04.052
60. Loss, C., Rossi, S., Tannert, T.: In-plane stiffness of hybrid steel–cross-laminated timber floor diaphragms. J. Struct. Eng. **144**(8) (2018). https://doi.org/10.1061/(ASCE)ST.1943-541X.0002105
61. Pan, Y., Tannert, T., Kaushik, K., Xiong, H., Ventura, C.E.: Seismic performance of a proposed wood-concrete hybrid system for high-rise buildings. Eng. Struct. **238**, 112194 (2021). https://doi.org/10.1016/j.engstruct.2021.112194
62. Roncari, A., Loss, C.: Cross-laminated timber buildings equipped with ductile multiple shear slotted-in steel plates connections: proof of concept. In: World Conference on Timber Engineering, Santiago, Chile (2021)
63. Dellepiani, M.G., Roa Munoz, G., Yanez, S.J., Guzman, C.F., Saavedra Flores, E.I., Pina, J.C.: Numerical study of the thermo-mechanical behavior of steel–timber structures exposed to fire. J. Build. Eng. **65**, 105758 (2023). https://doi.org/10.1016/j.jobe.2022.105758
64. Xu, L., Tangorra, F.M.: Experimental investigation of lightweight residential floors supported by cold-formed steel C-shape joists. J. Constr. Steel Res. **63**(3), 422–435 (2007). https://doi.org/10.1016/j.jcsr.2006.05.010
65. Kyvelou, P., Gardner, L., Nethercot, D.A.: Composite action between cold-formed steel beams and wood-based floorboards. Int. J. Struct. Stab. Dyn. **15**(8) (2015). https://doi.org/10.1142/S0219455415400295
66. Vella, N., Gardner, L., Buhagiar, S.: Experimental analysis of cold-formed steel-to-timber connections with inclined screws. Structure. **24**, 890–904 (2020). https://doi.org/10.1016/j.istruc.2020.02.009
67. Karki, D., Far, H., Saleh, A.: Numerical studies into factors affecting structural behaviour of composite cold-formed steel and timber flooring systems. J. Build. Eng. **44**, 102692 (2021). https://doi.org/10.1016/j.jobe.2021.102692
68. Wu, K., Iqbal, A.: Braced frame system for tall timber building. SESOC J. **33**(1), 58–62 (2020)
69. Zhang, X., Fairhurst, M., Tannert, T.: Ductility estimation for a novel timber–steel hybrid system. J. Struct. Eng. **142**(4) (2015). https://doi.org/10.1061/(ASCE)ST.1943-541X.0001296
70. Jamil, M., Quenneville, P., Clifton, G.C.: Scaled shake table tests of a novel gravity rocking moment frame system. In: Proceedings of the New Zealand Society for Earthquake Engineering Annual Conference, Rotorua, New Zealand (2015)
71. Ascione, F., Esposito, F., Iovane, G., Faiella, D., Faggiano, B., Mele, E.: Sustainable and efficient structural systems for tall buildings: exploring timber and steel–timber hybrids through a case Study. Buildings. **14**(2), 524 (2024). https://doi.org/10.3390/buildings14020524
72. Miller, S., Woods, J.E., Erochko, J., Lau, D.T., Gilbert, C.F.: Experimental and analytical fragility assessment of a combined heavy timber–steel-braced frame through hybrid simulation. Earthq. Eng. Struct. Dyn. **50**(2), 270–289 (2020). https://doi.org/10.1002/eqe.3329
73. Gilbert, C.F., Erochko, J.: Development and testing of hybrid timber-steel braced frames. Eng. Struct. **198**, 109495 (2019)
74. Chen, F., Li, Z., Li, M., Shu, Z.: Experimental testing of post-tensioned steel-timber hybrid frames equipped with energy-dissipating braces. Eng. Struct. **296**, 116960 (2023). https://doi.org/10.1016/j.engstruct.2023.116960
75. Iovane, G., Oliva, V., Faggiano, B.: Design and analysis of dissipative seismic resistant heavy timber frame structures equipped with steel links. Procedia Struct. Integr. **44**, 1864–1869 (2023). https://doi.org/10.1016/j.prostr.2023.01.238

76. Li, Z., Chen, F., He, M., Long, W., Ou, J., Li, M.: Experimental investigation on self-centering steel-timber hybrid beam-column connections. J. Struct. Eng. **149**(3) (2023). https://doi.org/10.1061/JSENDH.STENG-11570
77. Blomgren, H.E., Koppitz, J.-P., Diaz Valdes, A.: The heavy timber buckling-restrained braced frame as a solution for commercial construction in regions of high seismicity. In: World Conference on Timber Engineering, Vienna, Austria (2016)
78. Murphy, C., Pantelides, C.P., Blomgren, H.E., Rammer, D.: Development of timber buckling restrained brace for mass timber-braced frames. J. Struct. Eng. **147**(5) (2021). https://doi.org/10.1061/(ASCE)ST.1943-541X.0002996
79. Kia, L., Valipour, H.R.: Composite timber-steel encased columns subjected to concentric loading. Eng. Struct. **232**, 111825 (2021). https://doi.org/10.1016/j.engstruct.2020.111825
80. Williamson, E., Pantelides, C.P., Blomgren, H.E., Rammer, D.: Mass timber braced frames with mass timber buckling restrained braces. In: World Conference on Timber Engineering, Oslo, Norway (2023)
81. Luo, L., Sun, X., Song, X., Ou, Z., Zhang, L.: Experimental and numerical study on the hysteretic behavior of a hybrid timber buckling-restrained brace with a cross-shaped steel core. Soil Dyn. Earthq. Eng. **162**, 107461 (2022). https://doi.org/10.1016/j.soildyn.2022.107461
82. Dong, W., Li, M., Lee, C.L., MacRae, G., Abu, A.: Experimental testing of full-scale glulam frames with buckling restrained braces. Eng. Struct. **222**, 111081 (2020)
83. Dong, W., Li, M., Lee, C.L., MacRae, G.: Numerical modelling of glulam frames with buckling restrained braces. Eng. Struct. **239**, 112338 (2021)
84. Badal, P.S., Tesfamariam, S.: Seismic collapse performance of high-rise mass timber building with buckling-restrained braced frames. Soil Dyn. Earthq. Eng. **177**, 108369 (2024)
85. Yousef-beik, S.M.M., Veismoradi, S., Zarnani, P., Hashemi, A., Quenneville, P.: Experimental study on cyclic performance of a damage-free brace with self-centering connection. J. Struct. Eng. **147**(1), 1–14 (2021)
86. Yousef-beik, S.M.M., Bagheri, H., Veismoradi, S., Zarnani, P., Hashemi, A., Quenneville, P.: Seismic performance improvement of conventional timber brace using re-centring friction connection. Structure. **26**, 958–968 (2020)
87. Svatoš-Ražnjević, H., Orozco, L., Menges, A.: Advanced timber construction industry: a review of 350 multi-storey timber projects from 2000-2021. Buildings. **12**(4) (2022). https://doi.org/10.3390/buildings12040404
88. Tenório, M., Ferreira, R., Belafonte, V., Sousa, F., Meireis, C., Fontes, M., Vale, I., Gomes, A., Alves, R., Silva, S.M., Leitão, D., Fontes, A., Maia, C., Camões, A., Branco, J.M.: Contemporary strategies for the structural design of multi-story modular timber buildings: a comprehensive review. Appl. Sci. **14**(8) (2024). https://doi.org/10.3390/app14083194
89. Thai, H.T., Ngo, T., Uy, B.: A review on modular construction for high-rise buildings. Structure. **28**, 1265–1290 (2020). https://doi.org/10.1016/j.istruc.2020.09.070
90. Li, Z.: Innovative Interlocking Connection System for Medium-Rise Timber Modular Structures. Ph.D. dissertation, University of Leeds (2024)
91. Jaramillo, J.S.Z., Fischer, E.C.: The role of intermodular connections in the global behavior of high-rise mass timber buildings. In: 13th World Conference on Timber Engineering, WCTE 2023, vol. 5, pp. 2650–2659 (2023). https://doi.org/10.52202/069179-0348
92. Lacey, A.W., Chen, W., Hao, H., Bi, K.: Structural response of modular buildings—an overview. J. Build. Eng. **16**, 45–56 (2018). https://doi.org/10.1016/j.jobe.2017.12.008
93. Lawson, R.M., Ogden, R.G., Bergin, R.: Application of modular construction in high-rise buildings. J. Archit. Eng. **18**(2), 148–154 (2012). https://doi.org/10.1061/(asce)ae.1943-5568.0000057

94. Bhandari, S., Fischer, E.C., Riggio, M., Muszynski, L.: Numerical assessment of In-plane behavior of multi-panel CLT shear walls for modular structures. Eng. Struct. **295** (2023). https://doi.org/10.1016/j.engstruct.2023.116846
95. Chao, T., Zhang, G.L., Lam, F.: Connections for Stackable Heavy Timber Modules in Midrise to Tall Wood Buildings. Timber Engineering and Applied Mechanics (TEAM) Laboratory (2018)
96. Moroder, D., Smith, T., Pampanin, S., Palermo, A., Buchanan, A.H.: Design of floor diaphragms in multi-storey timber buildings. In: Proceedings of the 2015 NZSEE Conference, pp. 291, 2015–301
97. Moroder, D.: Floor Diaphragms in Multi-Storey Timber Buildings. Ph.D. Dissertation, University of Canterbury, New Zealand (2016). https://doi.org/10.26021/2006
98. Jayamon, J., Line, P., Charney, F.A.: State-of-the-art review on damping in wood-frame shear wall structures. J. Struct. Eng. **144**(12) (2018). https://doi.org/10.1061/(ASCE)ST.1943-541X.0002212
99. Mirra, M.: A set of calculation tools supporting the design, modelling and application of plywood-based seismic retrofitting interventions on timber floors in existing buildings. Structure. **63**, 106378 (2024). https://doi.org/10.1016/j.istruc.2024.106378
100. Mirra, M., Ravenshorst, G., van de Kuilen, J.-W.: An analytical model describing the in-plane behaviour of timber diaphragms strengthened with plywood panels. Eng. Struct. **235**, 112128 (2021). https://doi.org/10.1016/j.engstruct.2021.112128
101. EN1998–1, Eurocode 8: Design of Structures for Earthquake Resistance - Part 1: General Rules, Seismic Actions and Rules for Buildings (2013)
102. D'Arenzo, G., Casagrande, D., Reynolds, T., Fossetti, M.: In-plane elastic flexibility of cross-laminated timber floor diaphragms. Construct. Build Mater. **209**, 709–724 (2019). https://doi.org/10.1016/j.conbuildmat.2019.03.060
103. Tamagnone, G., Rinaldin, G., Fragiacomo, M.: Influence of the floor diaphragm on the rocking behavior of CLT walls. J. Struct. Eng. **146**(3) (2020). https://doi.org/10.1061/(ASCE)ST.1943-541X.0002546
104. D'Arenzo, G., Nicolussi, V., Rigo, P., Pozza, L., Polastri, A., Casagrande, D.: Investigating the in-plane rigid diaphragm conditions of cross-laminated timber floors for seismic design. In: Proceedings of the 11th International Network on Timber Engineering Research (INTER) Meeting, Padua, Italy (2024)

Seismic Protection Technologies

Luka Naumovski[1] (✉), Onur Kaplan[2], Vojko Kilar[3], Erkan Çelebi[4], Giuseppe D'Arenzo[5], Beatrice Faggiano[6], Giacomo Iovane[6], Saeid Javidi[7], and Daniele Casagrande[8]

[1] Slovenian National Building and Civil Engineering Institute, Ljubljana, Slovenia
luka.naumovski@zag.si
[2] Institute of Earth and Space Sciences, Eskisehir Technical University, Eskişehir, Turkey
[3] Faculty of Architecture, University of Ljubljana, Ljubljana, Slovenia
[4] Department of Civil Engineering, Sakarya University, Sakarya, Turkey
[5] Department of Civil and Architectural Engineering, Aarhus University, Aarhus, Denmark
[6] Department of Structures for Engineering and Architecture, University of Naples Federico II, Naples, Italy
[7] Faculty of Computing, Engineering and the Built Environment, Birmingham City University, Birmingham, UK
[8] Department of Civil, Environmental and Mechanical Engineering, University of Trento, Trento, Italy

Abstract. Taller timber buildings (TTBs) offer sustainability benefits but pose unique seismic challenges. The following chapter reviews state-of-the-art seismic protection technologies (SPTs) for TTBs, including low-damage self-centering systems, post-tensioned systems, supplemental damping systems, passive and active control systems and base isolation. It discusses the principles, applications, and future challenges of each technology. While significant progress and innovative solutions have been achieved, outstanding challenges include scaling the technology, optimizing cost-effectiveness, and managing interactions between structural and non-structural elements to enhance functional recovery, damage limitation, and acceleration reduction. By examining current practices and future directions, this review facilitates a broader understanding and implementation of SPTs, promoting the sustainable growth of TTBs in seismic-prone regions.

Keywords: Seismic protection technologies · Low-damage and self-centring · Post-tensioned systems · Supplemental damping systems · Passive and active control · Base isolation

1 Introduction

Timber has long been a material of choice in construction due to its renewability, excellent strength-to-weight ratio, and aesthetic appeal. In recent years, timber has grown in popularity, driven by the growing emphasis on sustainable construction practices and innovative mass timber products. Among these developments, the construction of taller

G. Fink et al. (Eds.): *Holistic Design of Taller Timber Buildings*, SPRTRCIENG, pp. 253–271, 2025.
https://doi.org/10.1007/978-3-032-02098-7_20

timber buildings (TTBs) has emerged as a significant innovation, showcasing advancements in engineering and design. While the adoption of TTBs offers numerous benefits, including reduced carbon footprints and enhanced architectural versatility, their performance under seismic conditions presents unique challenges that necessitate further exploration.

Historically, timber buildings have demonstrated varying levels of seismic resilience, influenced by factors such as construction techniques, material properties, and design standards. Traditional timber structures often relied on flexibility and low mass to withstand seismic forces, yet they frequently suffered irreversible damage that was difficult to repair. Modern engineering approaches have introduced new opportunities for designing timber buildings to meet stringent seismic performance requirements, particularly in Serviceability Limit State (SLS) and Ultimate Limit State (ULS) scenarios. However, as urbanization accelerates and the demand for mid to high-rise timber structures increases, addressing the seismic demands and challenges associated with TTBs becomes paramount.

The inherent characteristics of timber budlings, while beneficial in many aspects, also introduce specific complexities for seismic design. Their lightweight nature, while beneficial in reducing inertial forces, necessitates careful consideration of dynamic properties to avoid potential amplification of accelerations during seismic events. Furthermore, integrating innovative materials and hybrid systems has added layers of complexity, requiring interdisciplinary collaboration and development of new devices and design methodologies.

Following the afore-discussed challenges, Seismic Protection Technologies (SPTs), such as low-damage self-centering systems, post-tensioned systems, supplemental damping systems, and base isolation technologies, have been adapted and developed to enhance the resilience of timber buildings. These technologies focus on minimizing structural and non-structural damage, improving energy dissipation, and enabling self-centering behaviour after seismic events. Despite significant progress, challenges remain in scaling these solutions for TTBs, optimising cost-effectiveness, and addressing the interactions between structural and non-structural elements. Therefore, this chapter explores the state-of-the-art on seismic protection technologies for TTBs, highlighting both established and emerging solutions. It also identifies critical difficulties and research gaps, including the need for comprehensive empirical data, refined design parameters, and the development of performance-based frameworks tailored to contemporary timber structures. By examining current practices and future directions, this study aims to contribute to the broader understanding and implementation of SPTs, facilitating the growth of TTBs in seismic regions. In addition to discussing the dynamic behaviour and energy dissipation principles of TTBs equipped with SPTs, this paper presents an overview of various seismic protection technologies and their practical applications. The discussion is further contextualized through a review of current trends, advancements, and outlook, setting the stage for future innovations in this fast-evolving field.

2 Concept and Methodology of Seismic Protection Technologies

The recent shift in contemporary seismic design—from focusing solely on structural safety to prioritizing damage limitation and full functional recovery with an emphasis on performance-based loss design - was both highly anticipated and necessary. Although significant research in recent years has focused on the structural safety of timber buildings - particularly on connections, lateral load-resisting systems, design codes, and construction techniques - SPTs are an emerging area in timber engineering design aimed at further enhancing their resilience.

Generally, connections play a crucial role in timber structures by transferring forces between members and dissipating energy. When designed to yield, these connections are susceptible to significant non-linear deformations, and non-recoverable damage.

Consequently, SPTs should prioritize reducing the demand on connections and prevent non-linear deformation and unrecoverable damage, ensuring all structural elements and connections to behave elastically under strong ground motions. Additionally, to guarantee financial justification for their implementation in TTBs, these technologies should also minimize large accelerations in upper stories, which are recognised as a primary source of damage to non-structural elements.

In recent years various seismic protection technologies have been developed and are being studied numerically and experimentally. In this sense, seismic protection technologies and devices for TTBs can be broadly categorised into five major groups: *Low-damage and self-centring systems*; *Post-tensioned systems*; *Supplemental damping systems*; *Passive and active control systems*, and *Base isolation systems*. An illustrative overview of the types of SPTs and the number of original publications, which were found (per type and year), is presented in Fig. 1. Although a clear boundary between low-damage self-centering solutions and post-tensioned systems, as well as motion dependant and some passive control systems, could not be easily established, it is evident that the former constitute the majority of publications. Nonetheless, in recent years, there has been an overall increase in research, particularly in the areas of supplemental damping and isolation technologies. However, hybrid or combined approaches are still lacking.

As fundamental principles in SPTs, three key approaches can be identified: i) increasing damping ii) increasing flexibility (elongating the fundamental vibration period) and iii) hybrid approach which combines the two afore mentioned principles.

Although shortening the period usually means greater acceleration, it leads to smaller displacements, and therefore lower drifts, which is definitely an advantage for timber buildings, as structural damage correlates more strongly with displacement and interstorey drifts than with forces [1]. Following, shifting the fundamental vibration period to a large value to limit the amount of transferred force to the building, results in higher displacements and much smaller accelerations of the superstructure, which protects the acceleration-sensitive content in the building. In spite of the increased displacement demand during an earthquake, most of the demand should be met by the SPT system, while the superstructure typically shows much smaller peak floor accelerations and so less non-structural damage, which in contemporary construction constitutes up to 90% of the building cost. However, at this point it is important to note that accelerations, even though highly related to non-structural damages, are not generally been studied and picked up to such an extent in the scientific literature.

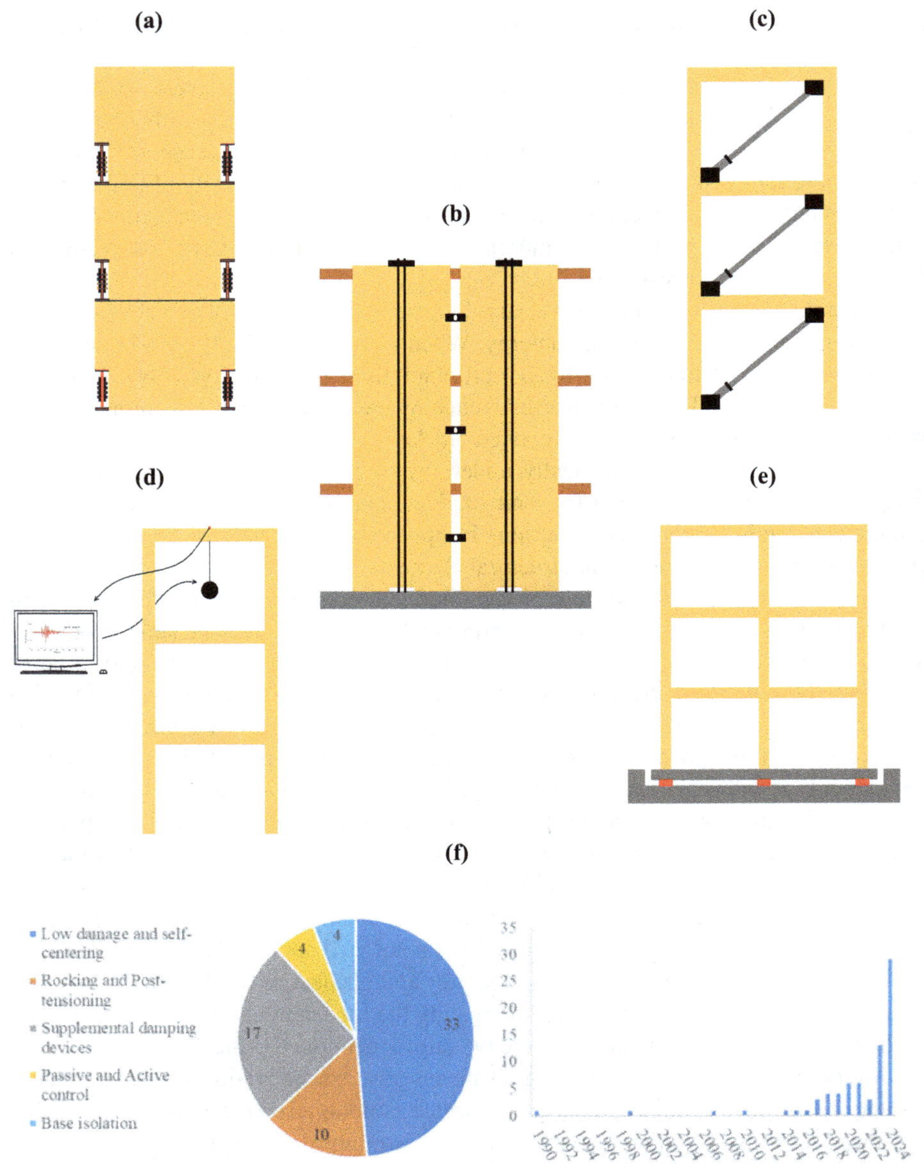

Fig. 1. Types of SPTs for TTBs: (a) Low-damage and self-centring e.g. RSFJ walls; (b) Post-tensioned systems e.g. PT walls; (c) Supplemental damping e.g. Viscous dampers; (d) Passive and Active control e.g. Active mass damper; (e) Base isolation e.g. Elastomeric isolation; (f) Number of recent research publications per SPT type and total number of publications per year.

Furthermore, it is argued that timber buildings would not benefit from additional damping due to their high effective damping ratio, which correlates with their inelastic behaviour associated with structural damage, primarily connections. Therefore, the main principle that should be followed for the implementation of SPTs into the design process is to reduce or exclude the energy dissipated by the inelastic behaviour of the structure (high damage and failure of connections). Finally, in such a way their economic investment would be easily negotiable.

3 Types of Seismic Protection Technologies

3.1 Low Damage and Self-Centering Systems

The systems concepts are based on automatic mechanisms for returning the main irreplaceable parts of the structural systems to their initial positions after earthquakes. In this regard, to achieve the self-centring capacity of TTBs, several innovative systems and models have been proposed and studied, experimentally and numerically. Table 1 briefly summarises the state-of-the-art for low-damage and self-centring systems and solutions that are included and discussed in this chapter.

Steel and abrasive-resistant Slip-Friction Devices (SFDs) for mass timber structures, with prestressed bolts, were initially proposed by Loo et al. [2]. The innovative hold-down devices consist of steel plate encased in the structural element and an envelope of two abrasive-resistant steel plates, which form two shear planes held together by prestressed bolt-springs. The results of cyclic tests showed a high energy dissipation capability without pinching and a characteristic flag-shaped hysteresis. A further step ahead towards an improved low-damage connection solution was the adaptation of the system into a damage-free self-centring solution for CLT shear walls – the resilient slip friction joint (RSFJ) introduced by Zarnani and Quenneville [3], which is still being investigated and highly tested for various structural systems and materials. The RSFJ has two middle and two cap serrated plates compressed by a bolt and disc spring system, enabling a high self-centring capacity of the system. The device was furthermore verified by an implementation in a Self-Centring Rocking timber Wall system by Hashemi et al. [4]. In addition, Hashemi et al. [5] used the RSFJ in a hybrid damage-avoidant steel-timber wall systems to prevent residual displacement and minimize damage. Finally, Hashemi et al. [6] utilized RSFJ as hold-downs in rocking CLT walls to reduce the residual displacement and improve the self-centring capacity of the rocking CLT walls.

More recently, Assadi et al. [7] tested the performance of the RSFJ as a low-damage floor connection. The implemented solution significantly improved the seismic performance. It increased the damping capacity, reduced displacement and force demands, and enhanced the overall economical design by reducing the size of wall hold-downs. In addition, Assadi et al. [8] conducted a comprehensive full-scale shake-table test on a 3-story steel structure equipped with different configurations of RSFJs. The results confirmed the reliability and predictability of the RSFJ system in reducing structural damage and ensuring quick reoccupation of buildings post-earthquake.

Solutions based on a similar operating principle to the RSFJ include: the slip-friction moment-resisting connection using screwed-in threaded rods introduced by Hegeir et al. [9] and Self-Centring Timber Brace (SC-TB) by Yousef-Beik et al. [10]. In the former solution, the moment-resisting connection was implemented in CLT beams and columns. Four full-scale tests were conducted, under service loads and destructive cyclic loads. The connection showed high stiffness and moment capacity, with ductile behaviour and

Table 1. Low-damage and self-centring systems literature review.

System type	Relevant research
Shear walls	1. Slip-Friction Connectors (SFC) [2]
	2. Resilient Slip Friction Joint (RSFJ) [3]
	3. Self-Centring Rocking Wall system (SC-RW) [4]
	4. RSFJ as hold-downs in rocking CLT walls [6]
	5. Elliptically profiled CLT walls [11]
	6. Rocking CLT walls with Uplift Friction Dampers (UFD) [12]
	7. High performance rocking timber wall with innovative low-damage floor connections [7]
Post & Beam	1. Hysteretic Performance of Self-Centering Glulam Beam-to-Column Connections [13]
	2. Self-centering timber moment resisting frames [14]
	3. Self-Centring Timber Brace (SC-TB) [10]
	4. Slip-friction moment-resisting connection using screwed-in threaded rods [9]
	5. Timber Beam–Column Connections with SMA Bars [15]
	6. Mass timber frames with timber buckling restrained braces [16]
	7. Glulam frames with dual-tube self-centering buckling-restrained braces [17]
Hybrid and Buildings	1. Hybrid steel-timber wall system using the RSFJs [18]
	2. Self-Centring Steel-Timber Hybrid Shear Wall (SC-STHSW) [19]
	3. Performance of self-centering hybrid damping systems under far-field and near-field ground motions [20]
	Seismic Fragility Assessment of a Balloon-framed CLT Building with Self-centering Hold-down [21]

failure through plastic hinging in the rods. The Self-Centring Timber Brace (SC-TB) was primarily aimed to improve the conventional timber braces. Experimental and numerical studies validate the SC-TB's effectiveness. A comparative study on a four-story building showed that SC-TB outperforms conventional braces in terms of base shear, pinching, and damping. In addition, the SC-TB system demonstrated superior performance with zero residual drift, lower base shear, and no pinching or timber crushing.

To address the limitations of conventional steel–timber hybrid systems Cui et al. [19] introduced a self-centering steel–timber hybrid shear wall (SC-STHSW) system. The proposed system integrates post-tensioned (PT) technology and slip friction dampers to improve both energy dissipation and self-centering capacity. A full-scale experimental test demonstrated its effectiveness by producing a flag-shaped hysteresis curve that indicates strong energy dissipation and minimal residual deformation.

An Uplift Friction Damper (UFD) composed of tension bolts and disc springs combined with an angled abrasive friction interface has been used as a low-damage and energy dissipater connector in rocking shear walls [12]. In the proposed UFD, post-tensioned (PT) rods provide the self-centring response while two angled steel wedges slide on each other to dissipate energy. Ricco et al. [11] proposed an elliptically profiled CLT walls to form a rocking story in multi-story CLT buildings. The system is a rocking soft story based on the elliptical rolling rod isolation, which also reminds of base isolation system. The use of a soft storey as a seismic isolator buffers the upper stiff portion of the building from ground motions during earthquakes. Two types of connectors with different restriction mechanisms, No-Slip Traction Rolling (NSTR) and Slip Friction Rocking (SFR) connection, were tested to assess the lateral behaviour of the elliptical walls, both types exhibited similar rotational capacities. Finally, an equivalent lateral force procedure, was proposed, for effectively estimating the seismic demands for CLT buildings with elliptical rocking walls, and simplifying the design by using a static force procedure to approximate the complex dynamic effects, providing a practical tool for design.

3.2 Post-Tensioned Systems

Post-tensioned timber technology can provide increased strength and stiffness for mass timber seismic load resisting systems while also providing energy dissipation and re-centring capabilities for large lateral deformation without significant residual damage. The state-of-the-art research and initial implementation of the post-tensioned timber systems to real world construction projects, which gained in popularity after the 2011 Christchurch Earthquake, was reported by Granello et al. [22]. Other key research projects and publications are listed in Table 2.

The response has been extensively studied on full-scale structures as part of the University of Canterbury's (Department of Civil Engineering) project in New Zealand reported in Newcombe et al. [23], Natural Hazard Engineering Research Infrastructure Tall Wood Project (NHERI-TWP) described in Pei et al. [24], and the Seismology and Earthquake Engineering Infrastructure Alliance of Europe Project (SERA) described by Pampanin et al. [25].

The research findings show that the controlled rocking mechanisms of elastic post-tensioning provide self-centring action to eliminate residual drifts, while additional dissipative devices, such as replaceable steel fuses, increase the damping and reduce lateral displacements of multi-storey buildings. Although no severe structural damage was observed, non-structural damage occurred even at low levels, partially caused by floor accelerations. The previous findings were in line with recently conducted shake-table tests on a 10-story building, which demonstrated that such a building can withstand design based and maximum considered earthquake level events repeatedly without notable residual drift, structural member damage, or connection damage. The building exhibited only moderate non-structural damage that would be repairable, meeting the intended design goals [26]. Consequently, improving the detailing of the connections and the sacrificial ductile elements between structural and non-structural elements was highlighted as a key opportunity to enhance the resistance [25, 26]. Furthermore, the interaction between structural and non-structural elements under seismic loading requires

Table 2. Post-tensioned systems literature review.

System type	Relevant research
Shear walls	1. Laminated veneer lumber (LVL) walls [30]
	2. CLT walls [27]–[31]
	3. Single and coupled Post-Tensioned CLT (PT-CLT) walls [32]
	4. Post-Tensioned CLT Shear Walls with Energy Dissipators [33]
	5. Higher-mode effects of rocking mass timber walls with controlled overturning moments [28]
Post & Beam	1. Beam-column connections [13–34]
	2. Frames [29–35]
	3. Post-tensioned glulam frame [36]
	4. Fragility functions for low-damage post-tensioned timber frames [37]
Hybrid and Buildings	1. Two storey Pres-Lam timber building [29]
	2. SERA Project - Integrated Low-Damage Building System [25]
	3. Vertical steel ties for stiffening earthquake-resistant CLT buildings [38]
	4. 10 - story mass timber building [24, 26, 39]

further considerations to achieve a holistic design [27]. This should include refined low-damage design methodologies and technologies, ensuring their cost-effectiveness and widespread adoption [25]. In addition, higher mode shapes influence and modifications of current design principles were emphasised as a further research area [26–28]. Furthermore, research on the stiffness of connections and long-term performance data is still lacking and limited, health monitoring, environmental effects and maintenance of the post-tensioned steel tendons also require further investigations [22]. In this sense, research is needed on the effects of anchorages in highly stressed beam-column joints to prevent issues like column fractures or failures [29].

Recent numerical research by Matteoni et al. [37] developed a framework for fragility functions evaluation of low-damage post-tensioned timber frames, using nonlinear static and time history dynamic analyses and a Python-based parametric workflow.

In the area of mass timber shear wall platform type buildings stiffened with vertical steel ties or tension rods, as reported in Keskisalo [40], Gräfe et al. [41] and Pacchioli et al. [38], further research and development is needed for a successful implementation of the technology in seismic prone areas. An emphasis is in investigating the coupling of these systems with ductile devices to dissipate energy and reduce inertial forces and minimise possible brittle failures.

Finally, an extensive review of recent progress of post-tensioned and self-centering systems for mass timber structures can be found in Ugalde et al. [42] and Chen et al. [43].

3.3 Supplemental Damping Systems

Seismic protection through supplemental damping aims at decreasing the structural demands and consequently inter-story drifts while potentially adding stiffness in the structural systems by increasing inherent damping dissipation through the addition of supplemental devices called dampers. These devices can increase the equivalent viscous damping and are activated by: (i) displacement (e.g. metallic dampers, hysteretic devices, friction dampers); (ii) velocity (e.g. viscoelastic dampers (VEDs) or viscous fluid (VF)), or (iii) motion (e.g. tuned mass dampers).

The first reported study on the subject of supplemental damping in timber structures was carried by Filiatrault [44], which analytically predicted the seismic response of friction damped shear walls. Nonetheless, the study presented only analytical results and did not include experimental validation of the proposed solution. Since then, significant progress and research have been conducted. A summary of recent state-of-the-art research can be found in Table 3.

Table 3. Supplemental damping systems literature review.

System type	Relevant research
Displacement dependant dampers	1. Perforated steel structural fuses in mass timber lateral load resisting systems [57, 58] 2. Highly ductile hold-down with adaptive stiffness for timber seismic bracing walls [59] 3. Timber frame with novel energy-dissipation joints [60] 4. Light wood shear wall structures with slotted-bolted dampers [61] 5. Superelastic SMA Dowel Connections for Braced Timber Frame Applications [62] 6. Timber Beam-to-Steel Column Connection with Replaceable U-Shaped Fuses [63] 7. Friction-based connectors with wooden dowels for timber shear walls [45]
Velocity dependant dampers	1. CLT shear walls with novel dissipative angle brackets and hold-downs [51] 2. Dissipative angle brackets and hold-downs with soft-steel and rubber for CLT buildings [52] 3. Sound-insulated joints and dynamic behaviour of CLT structures [64] 4. Pre-damaged Two-Story Traditional Timber Frame reinforced using Viscoelastic dampers [53]
Motion dependant dampers	1. High-Rise Hybrid-Timber Building: A Comparative Study [55]
	2. Tuned Liquid Column Damper (TLCD) for Taller Timber Buildings [56]
Hybrid approach	1. Low-cost and sustainable timber-based energy dissipation system with recentering ability [54]

The studies showcase advancements in hybrid steel-to-timber dissipative connections and displacement dependant energy dissipation systems. The presented innovative devices, on one hand enhanced seismic performance by introducing higher ductility, whereas on the other reduced total costs. The hybrid steel-to-timber approaches demonstrated stable performance and excellent ductility, while SMA fasteners show additional self-centering potential but need further research. Finally, a recent study on friction-based supplemental damping connectors with wooden dowels for timber shear walls, tested on a shake table by Wakashima et al. [45], reports improved seismic performance by providing high damping and minimal stiffness/strength degradation - connectors maintained their performance over time. Furthermore, the results highlight the potential of these dissipative connectors, especially in regions prone to repeated seismic events.

Supplemental velocity-dependent damping systems, such as viscoelastic-viscous, or tuned mass dampers, have been for mass timber structures, mostly studied through numerical research. Although, numerical analyses have demonstrated the effectiveness of velocity-based visco-elastic dampers in controlling the floor accelerations of such structures (e.g., [46–48]), experimental research on applying passive viscous or viscoelastic damping to mass timber structures is currently lacking. It has only been reported on light wood-frame shear walls by Dinehart et al. [49, 50], demonstrating their efficiency. In addition, Chen et al. [51] introduced dissipative angle brackets and hold-downs with encased rubber layer, which showed superior lateral performance, including higher ductility and energy dissipation compared to conventional brackets. Damage, depending on the load intensity, was localized at the rubber layer or brackets, with CLT walls exhibiting excellent deformation capacity and an increased inter-story drift ratio. Furthermore, full-scale shaking table tests of a complete building confirmed the effectiveness of these dissipative connections under various ground motions, when compared to traditional solutions [52]. Finally, Yi et al. [53] investigated the seismic performance of traditional timber frames, focusing on the effectiveness of VEDs in recovering the seismic performance of earthquake-damaged structures. The full-scale shake-table tests showed that VEDs significantly improved the seismic performance, reducing deformation and enhancing energy dissipation.

Low-cost and sustainable hybrid system approaches, such as the timber-based energy dissipation system with recentering ability - Dovetail with Springs (Dove-SP), which uses two timber slabs that slide against each other, restrained by a dovetail joint and recentred by low-cost steel springs, as presented by Tsiavos et al. [54], offer promising solutions for mitigating both residual sliding displacement and impact accelerations, deserving further research.

Motion-dependent supplemental damping systems for TTBs have been under-investigated, both numerically and experimentally. A recent study by Chapain et al. [55] presents a dynamic analysis of a 42-story hybrid-timber building and investigates the performance of three motion-dependent damping devices: (i) pendulum pounding tuned mass damper (PTMD), (ii) tuned mass damper inerter (TMDI), and (iii) tuned mass damper (TMD). The study evaluates the vibration reduction capability of these devices under filtered white noise and variable frequency sinusoidal excitations. The results show that the pendulum PTMD has higher performance in reducing peak accelerations, base shear, base moment, and inter-story drift ratio compared to TMD and TMDI. Meanwhile,

Çelebi [56] examined the use of Tuned Liquid Column Dampers (TLCDs) in multi-story timber buildings, highlighting their ability to dissipate energy and improve structural stability. TLCDs offer adaptability, low maintenance, and potential fire suppression benefits Finally, it should be mentioned that the former motion-dependent systems could also be partially classified as passive-active control systems, as a clear boundary between such supplemental damping devices and passive-active control systems is not distinct.

3.4 Passive and Active Control Systems

With the new challenge for designers to accommodate higher demand requirements, especially in SLS, innovative dissipating devices and self-centering systems have been combined in mass timber structures as active or passive control systems, as listed in Table 4. Control system can minimize damages to structural and non-structural elements when providing enough structural performance. The systems are designed to absorb and dissipate energy during seismic events, of various magnitudes, thereby reducing the impact on the primary or secondary structure. However, one significant challenge with traditional systems, if they are not combined with other technologies, is residual deformation, which complicates the reuse or reoccupation of structures after major earthquakes. In addition, unlike steel structures, which have established some guidelines for integrating control systems, mass timber structures require more tailored and material specific solutions. The absence of standards or implementation methods for these systems in TTBs highlights a critical gap in the field.

Table 4. Passive and Active control systems literature review.

System type	Relevant research
Inerter approach	1. Rocking timber buildings equipped with inerters [65] 2. Inerter-equipped rocking structures [66] 3. Fluid inerter-based vibration control [67]
Semi-Active approach	1. Timber Frame equipped with Semi-Active Resettable Devices [68]
SMA approach	1. Multi-outrigger tall-timber building: Using SMA-based damper and Lagrangian model [69] 2. CLT structure with shape memory alloy-based semi-active tuned mass damper (SMA-STMD) [70] 3. Pre-strained shape memory alloy-tuned mass dampers for CLT floor vibration mitigation [71]

Seismic performance of rocking post-tensioned timber buildings with rocking walls, combined with inerters to control rotation amplitude and suppress higher-mode effects were investigated by Thiers-Moggia and Málaga-Chuquitaype [65, 66]. This combination of approaches significantly improved the seismic performance by reducing floor accelerations. Nonetheless, further experimental research is needed and recommended to optimize inerter design and validate their effectiveness across various seismic conditions and structural configurations, as part of ERIES-TRUST-Truly resilient timber buildings (2024). In addition, data drive approaches to evaluate intensity measures and develop simplified regression models or parametric models for predicting peak floor accelerations, drift demands and displacements of TTBs, as employed by Junda and Málaga-Chuquitaype [72] or Demirci et al. [73], can serve as entry points for future development and implementation of more robust active or semi-active control systems or technologies such as Magnetorheological Fluid Dampers [74], active mass dampers [75] or Fluid inerter-based vibration control systems [67].

Other studies by Das and Tesfamariam [69], Yan et al. [70] and Jiang et al. [71], investigated shape memory alloy-based (SMA) passive and semi-active dampers for CLT structures, demonstrating their effectiveness in reducing seismic vibrations by using the temperature-dependent properties of SMAs. However, further research, for these technologies is needed to optimize the design process and potentially integrate them with other SPTs.

Finally, a timber frame equipped with a semi-active experimentally validated resettable tendon device to reduce seismic response was implemented by Franco-Anaya and Iqbal [68]. The research involved analytical studies of a four-story timber frame subjected to seismic excitations and controlled by the semi-active device, which greatly improved the structural performance under eight different ground motions.

3.5 Base Isolation Systems

Although light-frame timber buildings have been deemed to perform well during strong earthquakes, due to their low mass and ability to deform inelastically without collapsing [76], the Northridge Earthquake's consequences revealed that light-frame timber buildings are prone to significant damage when subjected to strong ground motions. The inelastic response of such buildings is generally linked with significant structural and non-structural damage that may be costly to repair. Filiatrault et al. [77] stated that large inter-story drifts are the primary cause of significant damage. In order to limit the excessive inter-story drifts in the horizontal direction, base isolation systems, elastomeric bearings with low lateral stiffness or sliding isolation bearings with relatively low level of friction, can be implemented between the foundation system and superstructure to shift the fundamental vibration period of the building and reduce inter-story drift [76–78]. In spite of being common for concrete and steel structures, installing base isolation systems in wood frame buildings can be challenging, as noted by Symans et al. [76]. This is because the floor diaphragms often lack sufficient in-plane stiffness to effectively transfer forces to the base isolation system, while ensuring uniform motion across all bearings. Additionally, in the case of low-friction sliding isolation bearings, the lightweight nature of wood frame buildings may lead to undesirable sliding during strong windstorms. The low mass can also require the use of slender and potentially unstable elastomeric bearings. Despite these practical challenges, recent limited studies, listed in Table 5, suggest that base isolation has the potential to improve the seismic performance of wood frame and contemporary mass timber buildings.

From a historical perspective, the first study by Delfosse [79], demonstrated, through an example design, that it is feasible to utilise an elastomeric base isolation system for a single-story wood-framed house. Following, Reed and Kircher [80] discussed a seismic retrofit study on a five-story wood-frame building using two different isolation system

Table 5. Base isolation systems literature review.

System type	Relevant research
Light timber frame	1. Half-Scale Base-Isolated Wood Frame Building [78] 2. Base-isolated and base-fixed Ancient timber buildings in hanging-wall/footwall Earthquakes [84] 3. Cost-efficient isolation system for light frame timber buildings [85] 4. Impact-resilient seismic isolation system [86]
Cross-laminated timber	1. Predictive models for mid-rise base-isolated CLT buildings in Chile [87] 2. Performance of CLT Structures with base isolation [88]

configurations: one with elastomeric bearings and the other with flat sliding bearings. Sakamoto et al. [81] presented an experimental and analytical study of a two-story light-framed wood building supported on a base isolation system. The building was constructed at the University of Tokyo for in-field (not laboratory) experimental testing. Furthermore, in 1988, a base isolation system was implemented within a two-story light-frame wood house in Montreal, Canada [82]. Finally, Zayas and Low [83] discussed retrofitting a four-story wood-framed residential building, implementing sliding friction pendulum system (FPS) bearings.

In more recent studies, J. W. Van de Lindt et al. [78] presented the issues related to the application of base isolation in light-frame wood buildings by conducting shake table tests on a half-scale, two-storey base-isolated (with FPS) residential building. The study results demonstrated that FPS bearings offer a technically viable passive seismic protection system for light-frame wood buildings in high seismic zones. Seismic performance of base-isolated and base-fixed Ancient timber buildings in hanging-wall/footwall earthquakes was studied by Ou and Wang [84]. The study found that base isolation greatly reduced the response of the four models, whereas the relative peak displacement decreased by more than half compared with the base-fixed model, indicating that isolation improved the overall seismic performance. In addition, Quizangha et al. [85, 86] investigated a 3-story Light Frame Timber Building (LFTB) with Impact-Resilient Double Concave Frictional Pendulum (IR-DCFP) isolators on a shake table. The IR-DCFP isolators provided effective seismic protection for the constructed LFTBs. Furthermore, the experimental and numerical results confirm the system's performance and highlight the importance of accurate damping modelling. Finally, despite the extreme excitation, the superstructure remained in the elastic range, with peak acceleration ratios (i.e., peak floor acceleration to peak ground acceleration) not exceeding 0.75 and story drift ratios smaller than 0.52% in most cases.

Despite only being limited to numerical studies, the implementation of base isolation in mass timber multi-storey structures promises to be a viable type of SPT for improving the seismic performance and structural safety. With that in mind, two types of friction pendulum isolators and an elastomeric isolator's performance were compared by modelling 3-, 6- and 12-storey CLT buildings in Poshtiban [88]. The effect of the aspect ratio of the CLT panels on the seismic performance was also studied. The study concluded that all three isolation systems significantly reduce base shear, acceleration, and interstory drift, with friction pendulum systems generally outperforming rubber isolators. The effectiveness of these systems varied with the number of stories and ground motion types. The panel aspect ratio plays a crucial role in the seismic performance of non-isolated CLT buildings, but it is less influential in base-isolated buildings. Furthermore, Medel-Vera et al. [87] presented predictive models for the seismic behaviour of mid-rise, base-isolated CLT buildings by performing nonlinear dynamic analyses on four building models. The study focuses on the fundamental period of vibration, a parameter that must be selected in the early stages of a seismically isolated building's design process. It concludes that selecting the fundamental period of vibration significantly impacts potential economic and environmental losses in seismically isolated CLT buildings. Therefore, this selection should consider resilience and sustainability parameters in addition to the technical aspects.

4 Challenges and Outlook

Considering the afore-discussed earthquake-related challenges of TTBs, holistic multi-hazard high-performance seismic protection solutions, which address both earthquake functional recovery with damage limitation and vibration and acceleration reduction, under various load amplitudes, should be explored as promising potential avenues to improve the structural resilience and viability of such structures. The primary role of TTBs SPTs should be transferring forces and accommodation of displacements between structural and non-structural elements, as well as ensuring appropriate dynamic behaviour, ductility with additional energy dissipation under different load amplitudes and additional self-centering without significant residual displacement. Moreover, the exposure of such seismic protection technologies to high amplitude cyclic loads should not be a dominant issue related with stiffness loss and irreversible residual damage. Following, hybrid systems, which combine more than one protection mechanism, offer a promising alternative. These systems provide greater flexibility and responsiveness to various intensity earthquake events, primarily by using the strengths of different approaches. As such, hybrid systems represent a forward-looking solution for enhancing the seismic performance of TTBs. Finally, considering construction techniques and cost competitiveness, SPTs assembly and installation procedures should adapt to contemporary mass timber building techniques and practices.

References

1. Priestley, M.J.N., Calvi, G.M., Kowalsky, M.J.: Direct displacement-based seismic design of structures. In: NZSEE Conference, vol. 30, pp. 1–23 (2007)
2. Loo, W.Y., Kun, C., Quenneville, P., Chouw, N.: Experimental testing of a rocking timber shear wall with slip-friction connectors. Earthq. Eng. Struct. Dynam. **43**(11), 1621–1639 (2014)
3. Zarnani, P., Quenneville, P.: A resilient slip friction joint. Patent No. WO2016185432A1, NZ IP Office (2015)

4. Hashemi, A., Masoudnia, R., Quenneville, P.: Seismic performance of hybrid self-centring steel-timber rocking core walls with slip friction connections. J. Constr. Steel Res. **126**, 201–213 (2016)
5. Hashemi, A., Zarnani, P., Masoudnia, R., Quenneville, P.: Seismic resilient lateral load resisting system for timber structures. Construct. Build Mater. **149**, 432–443 (2017)
6. Hashemi, A., Zarnani, P., Masoudnia, R., Quenneville, P.: Experimental testing of rocking cross-laminated timber walls with resilient slip friction joints. J. Struct. Eng. **144**(1), 04017180 (2018)
7. Assadi, S., Hashemi, A., Quenneville, P.: High performance rocking timber wall with innovative low-damage floor connections. In: Structures, vol. 57, p. 105075. Elsevier, Amsterdam (2023)
8. Assadi, S., Bagheri, H., Yan, Z., Zarnani, P., Hashemi, A., MacCrae, G., Clifton, C., Jia, L.J., Quenneville, P.: Shake-table testing of the resilient slip friction joints. In: NZSEE 2024 Annual Conference (2024)
9. Hegeir, O.A., Malo, K.A., Stamatopoulos, H.: An innovative slip-friction moment-resisting connection using screwed-in threaded rods in cross laminated timber and steel coupling parts: an experimental study. Eng. Struct. **318**, 118654 (2024)
10. Yousef-beik, S.M.M., Bagheri, H., Veismoradi, S., Zarnani, P., Hashemi, A., Quenneville, P.: Seismic performance improvement of conventional timber brace using re-centring friction connection. In: Structures, vol. 26, pp. 958–968. Elsevier, Amsterdam (2020)
11. Ricco, M.L., Rammer, D., Amini, M.O., Ghorbanpoor, A., Pei, S., Zimmerman, R.B.: Equivalent lateral force procedure for a building with a self-centering rocking story of cross-laminated timber (CLT) walls. In: Proceedings Paper, pp. 1–10 (2021)
12. Tatar, A., Dowden, D.M.: Analytical and numerical investigation of a low-damage uplift friction damper for self-centering cross-laminated timber rocking walls. Eng. Struct. **254**, 113836 (2022)
13. Li, Z., He, M., Wang, K.: Hysteretic performance of self-centering glulam beam-to-column connections. J. Struct. Eng. **144**(5), 04018031 (2018)
14. Shu, Z., Li, Z., He, M., Zheng, X., Wu, T.: Seismic design and performance evaluation of self-centering timber moment resisting frames. Soil Dyn. Earthq. Eng. **119**, 346–357 (2019)
15. Huang, J., Wang, B., Chen, Z.P., Zhu, S.: Development of novel self-centering timber beam–column connections with SMA bars. J. Struct. Eng. **150**(8), 04024078 (2024)
16. Williamson, E., Pantelides, C.P., Blomgren, H.E., Rammer, D.: Seismic performance of timber frames with timber buckling-restrained braces. J. Struct. Eng. **150**(6), 04024048 (2024)
17. Wang, X., Li, Z., Xie, P., He, M.: Seismic assessment of glulam frames with dual-tube self-centering buckling-restrained braces. J. Build. Eng. **97**, 110753 (2024)
18. Hashemi, A., Quenneville, P., Zarnani, P.: Seismic resilient structures with Cross Laminated Timber (CLT) walls coupled with innovative Resilient Slip Friction (RSF) joints. In: Proc. New Zeal. Soc. Earthq. Eng. Conf.(NZSEE), Wellington, New Zealand (2017)
19. Cui, Y., Shu, Z., Zhou, R., Li, Z., Chen, F., Ma, Z.: Self-centering steel–timber hybrid shear wall with slip friction dampers: theoretical analysis and experimental investigation. Struct. Design Tall Spec. Build. **29**(15), e1789 (2020)
20. Li, J., Wang, W.: Assessments on seismic performance of self-centering hybrid damping systems under far-field and near-field ground motions. J. Constr. Steel Res. **192**, 107209 (2022)
21. Pan, Y., Shahnewaz, M., Dickof, C., Tannert, T.: Seismic fragility assessment of a balloon-framed CLT building with self-centering hold-down. In: World Conference on Timber Engineering 2023 (WCTE2023)
22. Granello, G., Palermo, A., Pampanin, S., Pei, S., Van De Lindt, J.: Pres-lam buildings: state-of-the-art. J. Struct. Eng. **146**(6), 04020085 (2020)

23. Newcombe, M.P., Pampanin, S., Buchanan, A.H.: Design, fabrication and assembly of a two-storey post-tensioned timber building. In: Proc. World Conference on Timber Engineering, pp. 3092–3100 (2010)
24. Pei, S., Van De Lindt, J.W., Berman, J., Ryan, K., Dolan, J.D., Pryor, S., Wichman, S., Busch, A., Zimmerman, R.: Full-scale 3-D shake table test of a ten-story mass timber building. In: Proc., World Conf. on Timber Engineering (WCTE 2023), pp. 2084–2089 (2023)
25. Pampanin, S., Ciurlanti, J., Bianchi, S., Perrone, D., Palmieri, M., Grant, D.N., Granello, G., Palermo, A., Filiatrault, A., Costa, A.C., Candeias, P.X.: Enhancing seismic safety and reducing seismic losses: overview and preliminary results of SERA Project-3D shaking table tests on an integrated low-damage building system. In: 17th World Conference on Earthquake Engineering (17WCEE), pp. 13–18 (2020)
26. Pei, S., Ryan, K.L., Berman, J.W., van de Lindt, J.W., Pryor, S., Huang, D., Wichman, S., Busch, A., Roser, W., Wynn, S.L., Ji, Y.E.: Shake-table testing of a full-scale 10-story resilient mass timber building. J. Struct. Eng. **150**(12), 04024183 (2024)
27. Brown, J.R., Li, M., Palermo, A., Pampanin, S., Sarti, F., Nokes, R.: Experimental testing and analytical modelling of single and double post-tensioned CLT shear walls. Eng. Struct. **256**, 114065 (2022)
28. Pieroni, L., Araujo, R.G.A., Simpson, B.G., Freddi, F., Mishra, P., Barbosa, A.R., Sinha, A., van de Lindt, J., Brown, N.: Estimating first and higher-mode effects for the design of rocking mass timber walls with controlled overturning moments. In: Proceedings of the 18th World Conference on Earthquake Engineering (WCEE 2024) (2024)
29. Newcombe, M.P., Pampanin, S., Buchanan, A.H.: Experimental testing of a two-storey post-tensioned timber building. In: 9th US National & 10th Canadian Conference on Earthquake Engineering, Toronto, Canada, vol. 8, (2010)
30. Palermo, A., Pampanin, S., Buchanan, A., Newcombe, M.: Seismic design of multi-storey buildings using laminated veneer lumber (LVL) (2005)
31. Pei, S., van de Lindt, J.W., Barbosa, A.R., Berman, J.W., McDonnell, E., Daniel Dolan, J., Blomgren, H.E., Zimmerman, R.B., Huang, D., Wichman, S.: Experimental seismic response of a resilient 2-story mass-timber building with post-tensioned rocking walls. J. Struct. Eng. **145**(11), 04019120 (2019)
32. Akbas, T., Sause, R., Ricles, J.M., Ganey, R., Berman, J., Loftus, S., Dolan, J.D., Pei, S., van de Lindt, J.W., Blomgren, H.E.: Analytical and experimental lateral-load response of self-centering posttensioned CLT walls. J. Struct. Eng. **143**(6), 04017019 (2017)
33. Chen, Z., Popovski, M., Iqbal, A.: Structural performance of post-tensioned CLT shear walls with energy dissipators. J. Struct. Eng. **146**(4), 04020035 (2020)
34. Iqbal, A., Pampanin, S., Buchanan, A.H.: Seismic performance of full-scale post-tensioned timber beam-column connections. J. Earthq. Eng. **20**(3), 383–405 (2016)
35. Di Cesare, A., Ponzo, F.C., Lamarucciola, N., Nigro, D.: Experimental seismic response of a resilient 3-storey post-tensioned timber framed building with dissipative braces. Bull. Earthq. Eng. **18**(15), 6825–6848 (2020)
36. Ding, Y., Zhou, Z., Huang, L., Si, Y.: Seismic performance of self-centering glulam frame with friction damper. Eng. Struct. **245**, 112857 (2021)
37. Matteoni, M., Ciurlanti, J., Bianchi, S., Pampanin, S.: Fragility functions for low-damage post-tensioned timber frames. Earthq. Eng. Struct. Dynam. **53**(15), 4741–4762 (2024)
38. Pacchioli, S., Pozza, L., Trutalli, D., Polastri, A.: Earthquake-resistant CLT buildings stiffened with vertical steel ties. J. Build. Eng. **40**, 102334 (2021)
39. Dowden, D.M., Tatar, A.: Shake table test of a full-scale 10-story mass timber building with uplift friction dampers. In: World Conference on Earthquake Engineering (2024)
40. Keskisalo, M.: Use of tension rods in wood construction–14 storeys with laminated veneer lumber as shear walls: Lighthouse Joensuu. In: Internationales Holzbau-Forum IHF 2018, pp. 6–7 (2018)

41. Gräfe, M.: Vorgespannte Konstruktionen aus Brettsperrholz: Entwicklung, experimentelle und theoretische Untersuchungen, Entwurf und Bemessung. Doctoral dissertation, Technische Universität München (2020)
42. Ugalde, D., Almazán, J.L., Santa María, H., Guindos, P.: Seismic protection technologies for timber structures: a review. Eur. J. Wood Wood Prod. **77**, 173–194 (2019)
43. Chen, F., Li, M., Li, Z.: Self-centering mass timber structures: a review on recent research progress. Eng. Struct. **303**, 117474 (2024)
44. Filiatrault, A.: Analytical predictions of the seismic response of friction damped timber shear walls. Earthq. Eng. Struct. Dynam. **19**(2), 259–273 (1990)
45. Wakashima, Y., Ishikawa, K., Shimizu, H., Kitamori, A., Matsubara, D., Hanai, S., Tesfamariam, S.: Friction-based connectors with wooden dowels for timber shear walls: shake table and time-dependent test. Eng. Struct. **307**, 117838 (2024)
46. Fang, C., Wang, W., Qiu, C., Hu, S., MacRae, G.A., Eatherton, M.R.: Seismic resilient steel structures: a review of research, practice, challenges and opportunities. J. Constr. Steel Res. **191**, 107172 (2022)
47. Li, Z., Dong, H., Wang, X., He, M.: Experimental and numerical investigations into seismic performance of timber-steel hybrid structure with supplemental dampers. Eng. Struct. **151**, 33–43 (2017)
48. Kam, W.Y., Pampanin, S., Carr, A.J., Palermo, A.: Advanced Flag-Shape Systems for High Seismic Performance Including Near-Fault effects. NZSEE (2007)
49. Dinehart, D.W., Shenton, H.W.: Comparison of the response of timber shear walls with and without passive dampers. In: Proc. of Structural Engineers World Congress, San Francisco, CA, July, Paper T207-5 (1998)
50. Dinehart, D.W., Shenton, H.W., Elliott, T.E.: The dynamic response of wood-frame shear walls with viscoelastic dampers. Earthq. Spectra. **15**(1), 67–86 (1999)
51. Chen, J., Wang, R., Furuta, T., Xiong, H.: Experimental research on lateral performance of CLT shear walls with novel dissipative angle brackets and hold-downs. J. Build. Eng. **86**, 108929 (2024)
52. Chen, J., Wu, Z., Zheng, Y., Furuta, T., Xiong, H.: Full-scale shaking table tests of cross-laminated timber structures adopting dissipative angle brackets and hold-downs with soft-steel and rubber. Eng. Struct. **313**, 118292 (2024)
53. Yi, D., Yuan, J., Pan, Y., Fan, Y.: Full-scale seismic performance testing of a predamaged two-story traditional timber frame on a slope reinforced using viscoelastic dampers. J. Struct. Eng. **150**(4), 04024026 (2024)
54. Tsiavos, A., Kolyfetis, D., Panzarasa, G., Burgert, I., Stojadinovic, B.: Shaking table investigation of a low-cost and sustainable timber-based energy dissipation system with recentering ability. Bull. Earthq. Eng. **21**(8), 3949–3968 (2023)
55. Chapain, S., Aly, A.M.: Vibration attenuation in a high-rise hybrid-timber building: a comparative study. Appl. Sci. **13**(4), 2230 (2023)
56. Celebi, E.: Tuned Liquid Column Damper (TLCD) for Taller Timber Buildings (2024)
57. Daneshvar, H., Dickof, C., Tannert, T., Hei, C.Y.: Experimental investigation of performance of perforated steel plate as structural fuse for mass timber seismic force resisting systems. In: Canadian Society of Civil Engineering Annual Conference, pp. 317–332. Springer Nature, Cham (2022)
58. Daneshvar, H., Niederwestberg, J., Dickof, C., Jackson, R., Chui, Y.H.: Perforated steel structural fuses in mass timber lateral load resisting systems. Eng. Struct. **257**, 114097 (2022)
59. Maître, K., Lestuzzi, P., Geiser, M.: Experimental investigations of a new highly ductile hold-down with adaptive stiffness for timber seismic bracing walls. Bull. Earthq. Eng. **21**(5), 2603–2634 (2023)
60. Dang, D., Yuan, K., Liu, Y., Liu, Y.: Experimental study on seismic performance of timber frame with novel energy-dissipation joints. Eng. Struct. **300**, 117137 (2024)

61. Dong, H., Tao, J., Wang, X., Luo, J.: Enhancing seismic resilience in hybrid steel frame–Light wood shear wall structures with slotted-bolted dampers: design parameters and performance evaluation using "DowelType" hysteretic model. In: Structures, vol. 68, p. 107229. Elsevier, Amsterdam (2024)
62. Cléroux, M., Kim, E., Lacroix, D.: Cyclic behavior and damage assessment of superelastic SMA dowel connections for braced timber frame applications. J. Perform. Constr. Facil. **38**(5), 04024027 (2024)
63. Mowafy, A., Imanpour, A., Chui, Y.H., Daneshvar, H.: Development and experimental validation of a timber beam-to-steel column connection with replaceable U-shaped fuses. J. Struct. Eng. **150**(11), 04024171 (2024)
64. Salvalaggio, M., Lorenzoni, F., Valluzzi, M.R.: Impact of sound-insulated joints in the dynamic behavior of cross-laminated timber structures. J. Build. Eng. **91**, 109525 (2024)
65. Thiers-Moggia, R., Málaga-Chuquitaype, C.: Performance-based seismic design and assessment of rocking timber buildings equipped with inerters. Eng. Struct. **248**, 113164 (2021)
66. Thiers-Moggia, R., Málaga-Chuquitaype, C.: Recent studies on inerter-equipped rocking structures. In: 17th World Conference on Earthquake Engineering (17WCEE)
67. Chillemi, M., Furtmüller, T., Adam, C., Pirrotta, A.: Fluid inerter-based vibration control of multi-modal structures subjected to vertical vibration. Eng. Struct. **307**, 117938 (2024)
68. Franco-Anaya, R., Iqbal, A.: Seismic Response Reduction of a Timber Frame using Semi-Active Resettable Devices (2007)
69. Das, S., Tesfamariam, S.: Multiobjective design optimization of multi-outrigger tall-timber building: Using SMA-based damper and Lagrangian model. J. Build. Eng. **51**, 104358 (2022)
70. Yan, L., Li, Y., Chang, W.S., Huang, H.: Seismic control of cross laminated timber (CLT) structure with shape memory alloy-based semi-active tuned mass damper (SMA-STMD). In: Structures, vol. 57, p. 105093. Elsevier, Amsterdam (2023)
71. Jiang, Z.Q., Liu, L.T., Chang, W.S., Wang, W., Wang, Y.H., Huang, H.: Enhancing CLT floor vibration mitigation with pre-strained shape memory alloy-tuned mass dampers. In: Structures, vol. 67, p. 106980. Elsevier, Amsterdam (2024)
72. Junda, E., Málaga-Chuquitaype, C.: Seismic acceleration demands in tall CLT buildings, predictive models and intensity measures. Eng. Struct. **298**, 117024 (2024)
73. Demirci, C., Málaga-Chuquitaype, C., Macorini, L.: Seismic drift demands in multi-storey cross-laminated timber buildings. Earthq. Eng. Struct. Dynam. **47**(4), 1014–1031 (2018)
74. Bahar, A., Pozo, F., Meybodi, M.R., Karami, S.: Magnetorheological fluid dampers: a close look at efficient parametric models. Struct. Control Health Monit. **2024**(1), 6860185 (2024)
75. Koutsoloukas, L., Nikitas, N., Aristidou, P.: Passive, semi-active, active and hybrid mass dampers: a literature review with associated applications on building-like structures. Dev. Built Environ. **12**, 100094 (2022)
76. Symans, M.D., Cofer, W.F., Fridley, K.J.: Base isolation and supplemental damping systems for seismic protection of wood structures: literature review. Earthq. Spectra. **18**(3), 549–572 (2002)
77. Filiatrault, A., Folz, B.: Performance-based seismic design of wood framed buildings. J. Struct. Eng. **128**(1), 39–47 (2002)
78. van de Lindt, J.W., Liu, H., Symans, M.D., Shinde, J.K.: Seismic performance and modeling of a half-scale base-isolated wood frame building. J. Earthq. Eng. **15**(3), 469–490 (2011)
79. Delfosse, G.C.: Wood framed individual houses on seismic isolators. In: Proceedings of the International Conf. on Natural Rubber for Earthquake Protection of Buildings and Vibration Isolation, Kuala Lumpur, Malaysia, February, pp. 104–111 (1982)
80. Reed, J.W., Kircher, C.A.: Base isolation of a five-story wood-frame building. In: Proceedings of a Seminar and Workshop on Base Isolation and Passive Energy Dissipation (ATC-17), pp. 133–142 (1986)

81. Sakamoto, I., Ohashi, Y., Fujii, Y.: Seismic behavior of base isolated two-storied wooden building. In: Proceedings of the 1990 International Timber Engineering Conference, Tokyo, Japan, vol. 3, pp. 938–945 (1990)
82. Pall, A.S., Pall, R.: Seismic response of a friction-base-isolated house in Montreal. In: 6th Canadian Conference on Earthquake Engineering, pp. 375–382 (1991)
83. Zayas, V.A., Low, S.S.: Seismic isolation of a four-story wood building. In: Earthquake Performance and Safety of Timber Structures, Forest Products Society, Proc. No. 7289, pp. 83–91 (1997)
84. Ou, T., Wang, D.: Study on seismic performance of base-isolated and base-fixed ancient timber buildings in hanging-wall/footwall Earthquakes. J. Asian Archit. Build. Eng. **22**(2), 374–389 (2023)
85. Quizanga, D., Almazán, J., Valdivieso, D., Guindos, P., Hernández, F., Lopez-Garcia, D.: Shaking table test of a cost-efficient isolation system for light frame timber buildings. J. Build. Eng. **82**, 108402
86. Quizanga, D., Almazán, J.L., Valdivieso, D., López-García, D., Guindos, P.: Shaking table test of a timber building equipped with a novel cost-effective, impact-resilient seismic isolation system. J. Build. Eng. **82**, 108402 (2024)
87. Medel-Vera, C., Contreras, M.C.: Resilience-based predictive models for the seismic behaviour of mid-rise, base-isolated CLT buildings for social housing applications in Chile. J. Build. Eng. **44**, 103397 (2021)
88. Poshtiban, P.: Seismic Performance of CLT Structures with Base Isolation—Dissertation (2022)

Compartment Fire Dynamics in Taller Timber Buildings: Guidance for Performance-Based Fire Safety Engineering

Ian Pope[1(✉)], Antonela Čolić[2], Chamith Karannagodage[3], Ahmed Ahmed Ali Awadallah[1,2], and Andrea Lucherini[4]

[1] The Danish Institute of Fire and Security Technology (DBI), Hvidovre, Denmark
{ipo,aaa}@dbigroup.dk

[2] School of Engineering, The University of Edinburgh, Edinburgh, UK
a.colic@sms.ed.ac.uk

[3] Institute of Structural Engineering, ETH Zurich, Zurich, Switzerland
karannagodage@ibk.baug.ethz.ch

[4] Department for Fire-Safe Sustainable Built Environment (FRISSBE), Slovenian National Building and Civil Engineering Institute (ZAG), Ljubljana, Slovenia
andrea.lucherini@zag.si

Abstract. In comparison to non-combustible construction materials commonly used for taller buildings, timber elements can significantly alter the fire dynamics in a compartment. This fundamentally challenges many of the conventional fire safety strategies and design approaches for mid-rise and high-rise buildings. Consequently, many building industry practitioners are questioning the limitations of existing methodologies, while searching for additional ways to account for this different fire behaviour in the design, construction, and operation of timber buildings. In seeking to address these questions, this chapter describes the state-of-the-art and recent advances in understanding the fire behaviour in compartments with areas of exposed timber (e.g., engineered wood products), and protected timber elements that may contribute to the fire if their encapsulation fails. Relevant experimental findings and engineering approaches to date are summarised and discussed, and design guidance is provided in relation to the typical phases of realistic or 'natural' fires, namely the growth phase, the fully-developed phase, the fire decay, and the cooling phase. Critical fire phenomena and their impacts on the fire safety strategy are addressed, such as fire spread; active fire suppression; heat induced delamination and char fall-off; and self-extinguishment.

Keywords: Timber · Fire dynamics · Fire safety · Self-extinguishment · Heat induced delamination · Char fall-off

1 Introduction

Mid- and high-rise buildings present fundamentally different fire safety challenges than low-rise buildings due to the higher consequences of external flame spread, compartmentation failure, and potential collapse, as well as significantly longer egress times. These

G. Fink et al. (Eds.): *Holistic Design of Taller Timber Buildings*, SPRTRCIENG, pp. 272–287, 2025.
https://doi.org/10.1007/978-3-032-02098-7_21

challenges have generally been addressed through prescriptive and/or performance-based design approaches that have been developed for buildings that are predominantly constructed from non-combustible materials (e.g., concrete, steel, masonry). In prescriptive approaches, an implicit level of safety is achieved for 'conventional' buildings by setting requirements that have been deemed adequate on the basis of years of experience and tradition. Examples of this include imposing restrictions on egress distances and mandating minimum separation distances between buildings. In contrast, performance-based fire engineering requires the explicit quantification of safety [1, 2]. This generally involves analysis of the fire dynamics in a building and the development of a fire safety strategy that quantifiably ensures that building occupants will not be exposed to untenable conditions, along with other objectives, such as preventing structural collapse or damage to neighbouring property. A holistic approach is critical in both prescriptive and performance-based design to prevent a single-point failure from invalidating the entire strategy for a building.

Understanding and reliably predicting the fire dynamics in a building is, therefore, fundamental to performance-based fire safety engineering. Over the past century, extensive research and development efforts have provided engineering tools that have become widespread and generally accepted for conventional buildings predominantly composed of non-combustible materials. As illustrated in Fig. 1, the timeline of a compartment fire can typically be divided into distinct phases that are addressed separately in fire safety strategies [3]. The initial 'growth phase' of the fire incorporates the period from ignition until a relatively steady peak fire size is reached—often following a rapid transition or 'flashover'. At the end of the growth phase, the fire enters the 'fully-developed phase', which generally continues until the available fuel has been consumed to an extent that the fire size begins to decrease. This initiates the 'decay phase', during which the fire continues to decrease in intensity until all the available fuel has been consumed. Finally, the 'cooling phase' incorporates the remaining time until the entire structure has returned to ambient temperature.

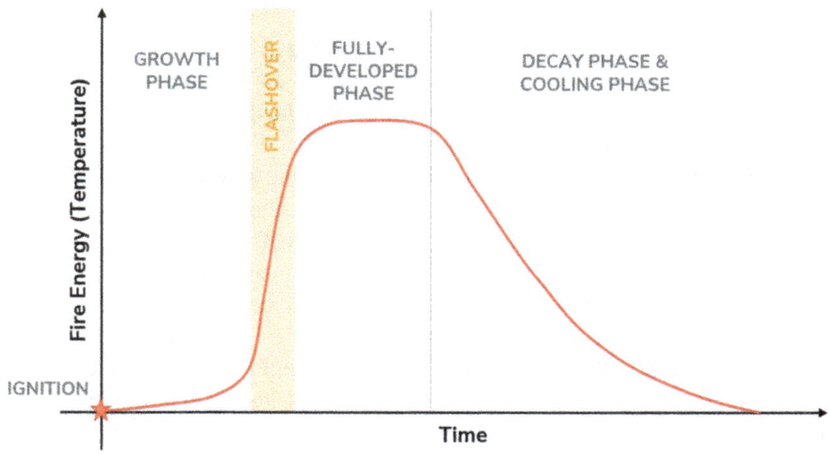

Fig. 1. Timeline of a typical compartment fire with distinct phases.

Compared to a conventional building, the additional combustible surfaces in a building with exposed timber elements—or with initially protected timber surfaces that can become exposed during fire—can drastically alter the compartment fire dynamics in every phase, to the extent that existing prescriptive and performance-based methodologies may not be applicable. This chapter describes the effects of timber elements on the compartment fire dynamics in each phase and provides guidance on how to account for this in performance-based design.

2 Growth Phase

After ignition, a compartment fire typically grows according to the available oxygen and the intrinsic fuel characteristics: physical and chemical properties, geometry, and distribution. This causes a gradual increment of the fire heat release rate (HRR [W]), which is typically approximated by an $\alpha \cdot t^2$ relationship in fire safety engineering applications (α [W/s^2] is the fuel-specific fire growth rate and t [s] is the time from ignition). A smoke layer usually descends within the enclosure, compartment temperatures increase, and flashover conditions may be attained once the heat flux from the smoke layer is sufficient to ignite all combustible surfaces and items in the compartment. A core part of performance-based design related to this phase is to ensure that the Available Safe Egress Time (ASET) is greater than the Required Safe Egress Time (RSET)—the time necessary to allow for building occupants to reach a place of safety. The extent of the ASET is delimited by the time to onset of untenable conditions in the vicinity of occupants, due to unacceptably reduced visibility, exposure to smoke, and/or exposure to heat.

Recent research has highlighted that faster fire growth may be observed in timber-lined compartments. Flashover in exposed timber compartments has been observed to occur earlier than in non-combustible compartments [4, 5], exacerbated by the area, number, and configuration of exposed surfaces. Similarly, exposed timber ceilings in large open-plan compartments can promote faster fire spread rates and transition to fully-developed fire conditions once the combustible ceiling ignites [6–9]. The time to untenable conditions in a timber compartment can be shortened if timber becomes involved early in the fire, and this concerns the development of compartment temperatures [4, 10], smoke toxicity [11], and visibility. Consequently, the potential for accelerated fire growth in compartments with exposed timber can affect evacuation strategies and the design of active fire protection measures (e.g., smoke and heat extraction). To address these risks, it is crucial to account for ignition and flame spread over combustible timber surfaces and the energy contribution from burning timber. These challenges can be managed through encapsulation, timber treatments and well-maintained coatings, installing automatic suppression systems, or changing compartment geometries, for instance interrupting the continuity of a timber ceiling using beams or non-combustible elements. However, the impacts of these mitigation measures should be explicitly quantified to verify that an acceptable level of safety is still achieved, rather than relying on 'trade-offs' or comparative methods that may not account for the different fire dynamics in timber buildings.

2.1 Thermal Degradation and Ignition of Timber

Depending on the specific species, wood is generally composed of cellulose (40–55%), hemicellulose (15–35%), lignin (18–40%) and other minerals and extracts [12]. Like any construction material, wood is subjected to different thermal decomposition processes at elevated temperatures, which lead to both physical and chemical changes. As temperatures rise above ambient temperature, the moisture within the wood is displaced through diffusion and/or evaporation, until the wood is completely dried above 100 °C. For temperatures above 200 °C, wood typically starts decomposing (pyrolyzing) rapidly, releasing combustible gases and forming a residual carbonaceous char [13]. Above 300 °C, this char formation process is typically assumed to be complete, and the location of the 300 °C isotherm is commonly used to represent the base of the char layer as a simplified approximation for structural design. In the presence of oxygen, charred wood further degrades due to oxidation reactions (smouldering), typically initiated in the temperature range of 400–500 °C, until the char itself has been consumed [14]. In addition to the combustion of flammable gases produced through pyrolysis, oxidation of the char layer also releases heat (exothermic reaction), particularly during the decay phase when oxygen can more easily reach the char layer (due to the reduction of combustible pyrolysis gases released from within the timber members) [15].

During the growth phase, the production of pyrolysis gases creates a risk of ignition and flame spread over exposed timber surfaces that can increase the growth rate of a compartment fire. A value of 12 kW/m^2 is commonly used for the critical heat flux for piloted ignition of timber [3, 13]. Designers may prevent or delay the time at which this threshold is breached through adequate separation of timber elements from heat sources by, for example, encapsulating walls or increasing the height of an exposed timber ceiling above combustibles on the floor. This can be verified through heat transfer and/or fluid dynamics models that can estimate the temperature and/or heat flux to the timber during the pre-flashover stage. However, flame spread is an extremely complex process, particularly for vertical surfaces or beneath ceilings, and there are currently no computational models that have been validated for these scenarios. Consequently, delaying or preventing ignition is a more robust solution, particularly as research suggests that flame spread over exposed timber ceilings can be extremely rapid and lead to almost immediate loss of tenable conditions in a compartment [7, 9].

2.2 Fire Retardancy and Fire Protection

In order to reduce the ignitability of timber and its contribution to the fire growth phase (reaction to fire), fire retardants represent a common solution to delay (or prevent) ignition, and reduce flame spread and the energy (heat) release by the timber combustion [14].

Over decades, many different types of fire retardants have been developed and extensively used. However, while many fire retardants have shown effective performance, they are currently challenged due to their toxicity, volatility and leaching problems. Concerns also arise around their long-term durability and their impact on other timber characteristics, such as moisture sensitivity, appearance, weathering resistance, strength and stiffness [16]. Moreover, since fire retardants have been mainly developed to reduce the timber ignitability and flame spread during the fire growth phase, they are typically effective for relatively mild fire conditions (e.g., 30 kW corner fire in the EN 13823 Single Burning Item test) and their effectiveness can considerably vary depending on the heating conditions. This is evident in the case of intumescent coatings—swelling char-forming coatings that react upon heating—whose fire protection effectiveness largely depends on many factors, such as fire exposure and substrate conditions [17, 18]. Therefore, for a performance-based design, it is important that the designer determines the actual performance of the chosen solution (e.g., critical heat flux for ignition), rather than relying on standard classification and implicit assessments.

The impact of timber elements in the growth phase can also be limited through the strategic use of different types of encapsulation, such as calcium silicate and gypsum plasterboards. While in many cases it is preferable to expose timber surfaces for their aesthetic qualities, encapsulation materials are often required in some areas for other reasons, such as acoustic performance. Encapsulation materials provide a physical barrier that insulates the timber, slowing down the rate of temperature rise at the timber surface and thereby delaying pyrolysis and ignition. The effectiveness of these systems in the growth phase depends on their detailing (e.g., thickness, fixation and joints) and placement, and it is often important to ensure that they are also robust enough to function through the severe exposures in the fully-developed phase (see Sect. 3.2).

2.3 Active Fire Suppression

Active fire suppression systems are often applied to mitigate the fire risk when designing buildings with high fuel loads, to protect property from fire damage, and to prevent structural failure. The logic behind the use of active suppression systems is that they can limit and possibly suppress the growth of the fire. If designed and functioning correctly, these systems may provide a means by which the relatively higher heat release rates in timber compartments can be avoided. However, very little experimentation has been conducted where these systems were functional and used in the context of timber compartments. For example, Zelinka et al. 2018 [19] showed that sprinklers, when properly operational, can suppress the fire in a timber compartment, while another test showed the ability of a water mist system to control the fire [20].

It is important to recognise and account for the fact that active suppression systems are not 100% reliable and may fail in part or whole. Non-activation or delayed activation may result in the fire growing beyond the ability of the system to control it. Indeed, this was shown to be the case in the recent experimental work by Bøe et al. 2024 [21] involving a compartment where, as part of the study, the sprinkler nozzles were intentionally turned off and an adjacent corridor where the sprinklers were activated. The fire grew to flashover in the compartment and extended into the corridor. The sprinklers in the corridor had no effect on the fire, and the upper parts of the corridor walls and the ceiling ignited, with flames spreading along the corridor. This reveals an additional vulnerability of active suppression systems: fire spread on ceilings and upper parts of walls may not be suppressed if ignited, and conventional suppression systems are not designed to suppress fires in compartments with combustible ceilings. It is therefore important for the designer to recognise these issues and explicitly account for the potential failure of the suppression system in the fire safety strategy.

3 Fully-Developed Phase

A fire may be defined as fully-developed once it reaches a relatively steady maximum size and intensity, where the heat release rate inside the compartment is constrained by either the availability of oxygen (ventilation-controlled fires) or by the availability and configuration of fuel (fuel-controlled fires) [22, 23]. The fully-developed phase is generally initiated by a flashover, when all the exposed combustible surfaces and items in the compartment are almost simultaneously ignited by the intense heat flux provided by the smoke layer. The combusting fuel can be separated into the 'moveable fuel' (furniture and other temporary contents) and the 'permanent fuel' (integral building elements, e.g. walls and ceilings). The high heat fluxes in a fully-developed compartment fire will heat up and damage protection materials and structural elements, so the severity and duration of this phase are crucial factors in determining the performance of the structure and compartmentation. In taller buildings, the fire safety strategy usually requires that the structure must not collapse. This is essential for the safety of occupants and firefighters, and to prevent damage to neighbouring buildings and impairment of public infrastructure. Accordingly, in a performance-based design, the designer must ensure that the structure will maintain sufficient integrity and stability throughout the fully-developed phase, as well as the subsequent decay and cooling phases. This is known as 'design for burnout' [24].

During this phase, the additional fuel load from burning timber elements will increase the total heat release rate of the fire [4, 10, 25–27], which depends on the amount of exposed timber, the compartment geometry, and available ventilation [28]. In ventilation-controlled compartment fires, the excess fuel will combust outside the compartment, having minimal impact on the rate of heat generation internally. In fuel-controlled compartment fires, ignition of wooden surfaces will increase the internal HRR, raising the temperatures and heat fluxes within the compartment [29] and resulting in faster fire spread and a larger fire size in general [6–8]. Timber compartment linings also have better insulative properties compared to concrete or masonry, resulting in relatively lower heat losses and potentially higher compartment temperatures [30].

In both ventilation-controlled and fuel-controlled fires, the configuration of exposed timber surfaces can also alter the internal flow fields, resulting in highly non-uniform temperatures or heat flux distributions [26, 31]. In particular, burning walls may create higher local heat fluxes and momentum-driven flows that can invalidate design assumptions of uniform conditions in a compartment [31]. This is particularly problematic for 'zone models' and parametric fire curves that are commonly used in design, because the assumption of uniform conditions is fundamental to these models. Furthermore, higher oxygen concentrations near the floor of a compartment can support enhanced combustion and oxidation of charred material in this region, leading to greater damage to elements locally. Consequently, these methods should not be applied to compartments with exposed timber without consideration of the potential and consequences of non-uniform conditions.

The duration of the fully-developed phase may also be extended significantly if self-extinguishment of exposed timber surfaces is not ensured. Char fall-off or heat induced delamination, failure of encapsulation, or excessive heat feedback may all result in continued burning until failure of the structure or compartmentation [27, 29, 32].

3.1 Exposed Wood—Heat Induced Delamination and Char Fall-Off

As engineered wood products (EWPs) are composites with multiple elements glued together, their structural response relies on sufficient composite action, meaning that the bonded elements (e.g., lamellae or veneers) behave as a single unit. Debonding is a general term used for any loss of composite action. In ambient conditions, this failure can occur in timber, as timber is a natural composite; in the adhesive bulk; and at the adhesive-timber interphase. When exposed to fire, EWPs usually experience debonding in either the timber or the adhesive-timber interphase (i.e., bond line). This debonding in fire conditions is further divided into two types of failure modes: (i) char fall-off and (ii) heat induced delamination (HID), as presented in Fig. 2. Char fall-off may appear in any part of the charred cross-section. Heat induced delamination—i.e., delamination, glue/bond line integrity failure (GLIF) [33], and premature char fall-off [34]—appears in the proximity of the bond line.

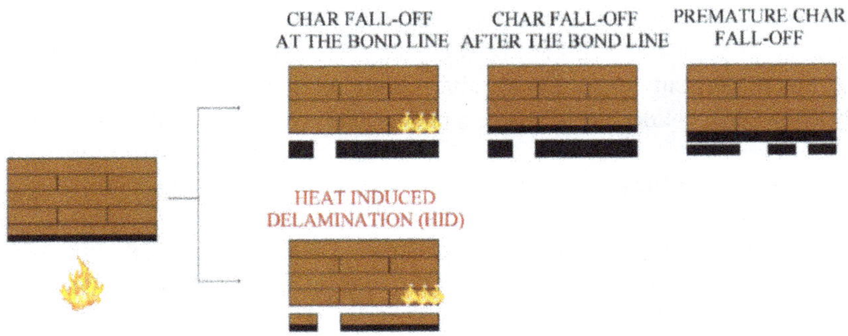

Fig. 2. Schematic representation of char fall-off and heat induced delamination from [38]. Reproduced under the CC BY license (http://creativecommons.org/licenses/by/4.0/).

Both failures initiate a change in fire dynamics and heat transfer [10, 27, 35, 36] because the loss of outer layers increases the heat transfer to the residual section (accelerating further degradation and pyrolysis), while the fallen pieces continue to combust and radiate heat back to the structural elements. This is particularly severe when burning pieces accumulate at the base of exposed timber walls, where oxygen concentrations are high, and the heat feedback can sustain flaming over the entire wall. The result is a cyclic phenomenon where the loss of the structural element's cross-section contributes to the ongoing burning of that same element.

In addition to the fire dynamics changes, HID may lead to the progressive loss of structural capacity [37] due to the loss of composite action and/or the premature loss of cross-section which has not completely charred.

When a large enough area is affected by HID or char fall-off, and the energy released by burning timber elements is sufficient, the compartment heat fluxes can be sufficient to sustain flaming combustion of the timber, even after the moveable fuel load is depleted. Similarly, if this occurs during the decay phase, a second flashover can occur. Both scenarios can extend the fire's duration and compromise the intended fire strategy, e.g., through failure of compartmentation or structural collapse.

Bond line response is dependent on the product (e.g. choice of adhesive chemistry, timber species, and manufacturing) and external conditions (e.g. heating conditions, orientation, mechanical stresses, and pressure induced by the moisture movement). To date, researchers have focused mostly on heating conditions, whereby the commonly accepted performance criterion is a generalised temperature. Thermal decomposition temperatures, such as glass transition, phase transition, or pyrolysis for adhesive and/or timber, can all compromise bond line mechanical response. In terms of the thermo-mechanical properties, both timber and adhesive film stiffness and strength start decreasing at temperatures lower than 100 °C [38, 39]. However, they do not deteriorate at the same rate, which can lead to unpredictable failure modes in both the timber and the bond line region [40, 41]. It was noted that HID can occur at bond line temperatures as low as 100 °C when the heating conditions are high enough to create a high moisture-related pressure gradient in the element and the load is sufficiently high [42]. However, the implications of mechanical stresses and moisture front position on the bond line response are still under research.

Current solutions for prediction and prevention of HID include careful manufacturing specification of the final product (i.e., thicker first lamella, edge bonding, and adhesive choice [25, 34, 37, 41, 43]). Designers are encouraged to impose a critical bond line temperature to account for HID depending on the adhesive and timber choice [38], at which the preceding lamella is considered as detached, and thus the layer is no longer accounted for thermally or structurally and is added to the fuel load.

3.2 Encapsulated Wood—Encapsulation Failure

Encapsulating some (if not all) timber elements is essential in higher consequence class buildings (e.g., buildings with vulnerable occupants, or where structural collapse would be too detrimental) because it directly impacts the life safety objective in the early stages of fire, as well as structural performance, compartmentation, and property protection during the fully-developed and decay phases.

For structures designed to survive burnout using encapsulation as a protection strategy, it must be demonstrated that the encapsulation will prevent the involvement of timber structural elements throughout the expected fire duration. This is relevant to both mass timber buildings and light-frame timber buildings. Designers should choose a challenging fire scenario, ignition threshold, and expected thermal and mechanical properties for the specific duration, and the integrity and insulation performance of the encapsulation should be quantified under those conditions. Additionally, to address the risk of continued smouldering after a fire, a management plan should be developed with the local fire service to remove the encapsulation in fire-affected compartments and check for hot-spots or other signs of combustion in gaps and concealed spaces [10, 44].

If the encapsulation fails due to loss of integrity (e.g., fall-off) or the wood behind exceeds the ignition threshold, the initially protected wood will begin to contribute additional fuel to the fire and mechanically degrade. This can extend the fire duration and lead to failure of the structure or compartmentation. Failure of encapsulation can occur due to insufficient thickness or fixing [45, 46], particularly if there is a higher fire load than the product was tested for (due to increased exposed timber surfaces and/or changes in fire dynamics) [47]. Due to the inherent uncertainties in encapsulation performance, safety factors and redundancies are essential.

3.3 Compartmentation Failure

Compartmentation is a fundamental aspect of the fire safety strategy, as it isolates the fire and stops it from spreading from its origin to other parts of the building (vertically or horizontally). This is crucial in taller buildings where fire spread (particularly vertically to other floors) may compromise many aspects of the overall strategy, such as egress, suppression, fire service intervention, and even the structural integrity. In a tall timber building with an inherently higher fuel load, the potential consequences of compartmentation failure are even greater.

3.3.1 External Fire Spread—Exposure to Facades and Neighbouring Buildings

The additional fuel load from burning timber elements can lead to increased exposure to facades and nearby buildings or objects, especially in ventilation-controlled fires where the excess fuel combusts in an external spilled plume [26, 48, 49]. However, significant external flaming has also been observed in experiments with relatively well-ventilated timber compartments [9]. The flame height and thermal exposure to the facade are, among other things, dependent on the surface area of exposed timber, and the opening dimensions [48–50]. While models for predicting external flame height in exposed timber compartments have been proposed [49, 51], these have not yet been systematically validated in large-scale experiments, so they should be applied with a high degree of conservatism. Furthermore, many national standard façade fire tests do not induce the same level and duration of exposure as can be expected in severe fire scenarios [50].

Compounding the effects of greater external flaming in timber compartments, the exposure to neighbouring buildings will be even more challenging if the fire spreads to multiple floors. Currently, fires spreading externally from floor to floor on a building can be prevented by architectural constraints or fire-resistant glazing, which must account for the increased severity of external flaming due to exposed timber.

3.3.2 Internal Fire Spread

The risk of internal fire spread may be increased if the elements contributing to compartmentation are not designed with consideration of the real fire exposure and duration in compartments with exposed timber. This includes the connections between elements, which generally require tolerance gaps and contain highly conductive metallic components. Furthermore, separating elements made from timber may promote the spread of fire or smoke through localised smouldering, or combustion of the timber in cavities between compartments. These risks can be reduced by preventing the timber separating elements from heating up significantly (e.g., by encapsulating them) and by avoiding or blocking cavities. Due to the persistent and highly localised nature of smouldering combustion [44], it may not be possible to entirely eliminate this risk through design. However, the relatively slow rate of smouldering combustion provides the opportunity for fire service intervention, and the fire service should be consulted and involved in developing a strategy to address this challenge. Overall, the robustness of design against internal fire spread can be improved through a multilayer approach to fire protection that avoids the potential for single-point failures.

4 Decay Phase

The fully-developed phase of a typical fire event lasts until the remaining fuel is no longer able to sustain the steady fuel- or ventilation-controlled heat release rate. This instant represents the beginning of the fire decay phase. During this phase, the conditions within the compartment evolve towards fuel burnout. As in the pre-flashover stages, the fire heat release rate is directly dependent on the fuel characteristics, and the compartment gas temperature tends to follow the fire heat release rate trend. The reduced soot production and increased intake of cold air through the opening results in a decrease in smoke density (increase in optical depth). The heat fluxes to the compartment elements significantly decrease, and the heat transfer becomes more complex and non-uniform [15].

While these phenomena are typical for traditional compartment fires with defined movable fuel load and non-combustible compartment elements, timber compartment elements can greatly affect the fire decay phase (i.e., onset, duration, etc.) [25, 35]. If flaming self-extinguishment of timber elements is achieved (see Sect. 4.1), the compartment will transition into the cooling phase. Conversely, under certain conditions, the combustible compartment elements can continue to burn and sustain the fire development, and timber compartments might never achieve flame extinction and burnout. Burnout of timber compartments is commonly considered to occur once the moveable fuel is consumed and flaming extinction of timber elements is achieved. This phenomenon is affected by many discussed factors, including the area and configuration of exposed timber, ventilation conditions, and compartment geometry [28, 32, 46].

4.1 Self-Extinguishment

During the decay phase, once the heat release rate from the remaining moveable fuel has decreased sufficiently, the incident heat flux on exposed timber surfaces may become low enough that the resulting pyrolysis rate of the timber can no longer sustain flaming combustion. Consequently, flaming on these surfaces will cease and their energy contribution to the compartment will reduce drastically (apart from residual smouldering). This process is known as flaming 'self-extinguishment' (also commonly known as 'self-extinction' or 'auto-extinction'), although some use self-extinguishment to mean the end of both flaming and smouldering combustion [25]. Ensuring the conditions for flaming self-extinguishment of exposed timber surfaces is a crucial step in designing for the eventual burnout of a compartment fire.

Flaming self-extinguishment in timber compartments is a complex phenomenon that requires several conditions to occur [32, 46]. Fundamentally, the amount of heat from the compartment being transferred through the insulating char layer to the pyrolysing timber must fall below a critical value. Depending on the local oxygen concentration and airflow velocity, this can be achieved when the external heat flux onto the surface of the charring timber falls below 30–45 kW/m^2 [35, 52–54]. This critical heat flux criterion applies once the char layer has reached a sufficient thickness, so it can be invalidated by char fall-off or heat induced delamination that increases the heat transfer to the underlying timber [10, 35, 40].

Once the moveable fuel load has substantially been consumed, the residual thermal exposure to the exposed timber will be dominated by radiative and convective heat feedback from the other compartment surfaces. Consequently, the area and configuration of exposed timber surfaces must allow for the residual heat fluxes in the compartment to fall below the critical value. This also requires that the performance of encapsulation is maintained for protected timber elements, so that they do not contribute additional fuel and heat [10, 26, 27, 32].

Char fall-off or heat induced delamination and encapsulation failure all depend on the severity and duration of the fire. Therefore, controlling or accounting for the fuel load and ventilation characteristics in a compartment is equally as important as optimising the timber elements, adhesive, and encapsulation system (see Sects. 3.1 and 3.2).

In some cases, the configuration of exposed timber surfaces can have a greater influence on the potential for flaming self-extinguishment than the total exposed area, since this will control the heat transfer between surfaces. For example, the heat feedback between two or more adjacent burning timber surfaces may be enough to sustain flaming until the structure or compartmentation fail, or for long enough to induce encapsulation failure or delamination that subsequently causes a secondary flashover [10, 27, 32]. Furthermore, it may be safer to expose a ceiling than a wall, because the burning rate of the ceiling is often lower due to the lack of oxygen in the upper region of a compartment, and the optically thick smoke layer below the ceiling can reduce the radiation received from other surfaces [26]. As a limitation to this, it should be noted that most compartment fire experiments have not included mechanical loading of exposed horizontal elements, so they may not be representative of the effects of realistic stresses on the bond lines. Nonetheless, the total percentage of exposed area has been found to be strongly correlated to the incidence of self-extinguishment in compartment fire experiments, with a high occurrence of self-extinguishment for compartments with less than 20% of the total surface area exposed [28].

Considering the complexity and uncertainties involved in designing for self-extinguishment in timber buildings, it is essential to approach the design in a proportionately conservative way. As part of a holistic fire safety strategy, designing for self-extinguishment can reduce the risk of compartmentation failure and structural collapse. Nonetheless, the other elements of the strategy—e.g., providing safe egress routes and facilitating fire service intervention—should be robust and provide sufficient redundancy to avoid the possibility of cascading failures.

5 Cooling Phase

Once the movable fuel in a compartment has been combusted and flaming extinction has occurred on all the timber elements, the fire heat release rate tends to become negligible, and the compartment volume enters a pure cooling phase [15]. This phase is characterized by the two significantly different modes of cooling taking place in the gas-phase and the solid-phase. The smoke inside the compartment is rapidly evacuated and cold air flows continuously into the enclosure through openings. Consequently, the gases inside the compartment quickly become optically thin and tend to ambient temperature. In contrast, the solid compartment elements (e.g. linings and structural elements), slowly cool down through surface convective cooling, and radiative heat exchange occurs between the exposed solid surfaces. These cooling phenomena primarily depend on the characteristics of the compartment (e.g., geometry and opening) and its elements (e.g. thermal inertia of compartment elements), and they can be modelled using simplified numerical heat transfer models [55].

As regards timber structures, the relevance of considering the thermal conditions during the cooling phase has been highlighted by several researchers, particularly in light of the continuous wood degradation due to the in-depth penetration of the heat wave [15, 37, 55–59]. In addition, during this phase, timber compartment elements may suffer from significant smouldering, which typically acts over longer timescales but can eventually lead to failure of the structure or compartmentation without intervention [44, 59, 60].

6 Conclusions

Performance-based design of timber buildings requires explicit consideration of the realistic compartment fire dynamics and how timber elements can influence this. The combustible nature of timber has the potential to fundamentally alter the fire behaviour in comparison to a non-combustible building, and conventional design approaches are often incapable of accounting for this. By examining the current state-of-the-art research on timber compartment fire dynamics, the critical phenomena in each phase of a fire can be explicitly quantified and addressed in a holistic fire safety strategy.

References

1. Cowlard, A., Bittern, A., Abecassis-Empis, C., Torero, J.L.: Some considerations for the fire safe design of tall buildings. Int. J. High Rise Build. **2**, 63–77 (2013)
2. Schmidt, L., Hilditch, R., Ervine, A., Madden, J.: Explicit fire safety for modern mass timber structures—from theory to practice. In: World Conference on Timber Engineering (WCTE 2023), Oslo, Norway, pp. 1738–1747 (2023)
3. Drysdale, D.: An Introduction to Fire Dynamics, 3rd edn. John Wiley & Sons, Hoboken, NJ (2011)
4. Li, X., Zhang, X., Hadjisophocleous, G., McGregor, C.: Experimental study of combustible and non-combustible construction in a natural fire. Fire Technol. **51**, 1447–1474 (2015). https://doi.org/10.1007/s10694-014-0407-4
5. Su, J., Leroux, P., Lafrance, P.-S., et al.: Fire Testing of Rooms with Exposed Wood Surfaces in Encapsulated Mass Timber Construction. National Research Council of Canada, Construction Research Centre (2018)
6. Gupta, V., Osorio, A.F., Torero, J.L., Hidalgo, J.P.: Mechanisms of flame spread and burnout in large enclosure fires. Proc. Combust. Inst. **38**, 4525–4533 (2021). https://doi.org/10.1016/j.proci.2020.07.074
7. Kotsovinos, P., Rackauskaite, E., Christensen, E., et al.: Fire dynamics inside a large and open-plan compartment with exposed timber ceiling and columns: CodeRed #01. Fire Mater. (2022). https://doi.org/10.1002/fam.3049

8. Nothard, S., Lange, D., Hidalgo, J.P., et al.: Factors influencing the fire dynamics in open-plan compartments with an exposed timber ceiling. Fire Saf. J. **129**, 103564 (2022). https://doi.org/10.1016/j.firesaf.2022.103564
9. Bøe, A.S., Friquin, K.L., Brandon, D., et al.: Fire spread in a large compartment with exposed cross-laminated timber and open ventilation conditions: #FRIC-02 - exposed wall and ceiling. Fire Saf. J. **141**, 103986 (2023). https://doi.org/10.1016/j.firesaf.2023.103986
10. Su, J., Lafrance, P.-S., Hoehler, M., Bundy, M.: Fire Safety Challenges of Tall Wood Buildings - Phase 2: Task 3 - Cross Laminated Timber Compartment Fire Tests. National Institute of Standards and Technology (2018)
11. Hull, T.R., Brein, D., Stec, A.A.: Quantification of toxic hazard from fires in buildings. J. Build. Eng. **8**, 313–318 (2016). https://doi.org/10.1016/j.jobe.2016.02.014
12. Neugebauer, R.: Biologische Transformation. Springer, Berlin, Heidelberg (2019)
13. Bartlett, A.I., Hadden, R.M., Bisby, L.A.: A review of factors affecting the burning behaviour of wood for application to tall timber construction. Fire Technol. **55** (2019). https://doi.org/10.1007/s10694-018-0787-y
14. Browne, F.L.: Theories of the Combustion of Wood and its Control. Forest Products Laboratory, Forest Service U. S. Department of Agriculture, USA (1958)
15. Lucherini, A., Torero, J.L.: Defining the fire decay and the cooling phase of post-flashover compartment fires. Fire Saf. J. **141**, 103965 (2023). https://doi.org/10.1016/j.firesaf.2023.103965
16. LeVan, S.L., Winandy, J.E.: Effects of fire retardant treatments on wood strength: a review. Wood Fiber Sci. **22**, 113–131 (1990)
17. Hartl, A., Razzaque, Q.S., Lucherini, A., Maluk, C.: Comparative study on the fire behaviour of fire-rated gypsum plasterboards vs. thin intumescent coatings used in mass timber structures. In: Proceedings of the 11th International Conference on Structures in Fire (SiF2020). The University of Queensland, online (2020)
18. Lucherini, A., Razzaque, Q.S., Maluk, C.: Exploring the fire behaviour of thin intumescent coatings used on timber. Fire Saf. J. **109**, 102887 (2019). https://doi.org/10.1016/j.firesaf.2019.102887
19. Zelinka, S.L., Hasburgh, L.E., Bourne, K.J., et al.: Compartment Fire Testing of a Two-Story Mass Timber Building. U.S. Department of Agriculture, Forest Service, Forest Products Laboratory, Madison, WI (2018)
20. Kotsovinos, P., Christensen, E.G., Gale, J., et al.: The effectiveness of a water mist system in an open-plan compartment with an exposed timber ceiling: CodeRed# 03. SPFE Extra Magazine (2022)
21. Bøe, A.S., Hox, K., Mikalsen, R.F., Friquin, K.L.: Large-scale fire experiments in a cross-laminated timber compartment with an adjacent corridor—partly and fully protected with a water sprinkler system. Fire Saf. J. **148**, 104212 (2024). https://doi.org/10.1016/j.firesaf.2024.104212
22. Thomas, P.H., Heselden, A.J.M., Law, M.: Fully-Developed Compartment Fires - Two Kinds of Behaviour. Ministry of Technology and Fire Offices' Committee - Joint Fire Research Organization (1967)
23. Torero, J.L., Majdalani, A.H., Abecassis-Empis, C., Cowlard, A.: Revisiting the compartment fire. Fire Saf. Sci. **11**, 28–45 (2014). https://doi.org/10.3801/IAFSS.FSS.11-28
24. Law, A., Bisby, L.: The rise and rise of fire resistance. Fire Saf. J. **116**, 103188 (2020). https://doi.org/10.1016/j.firesaf.2020.103188
25. Crielaard, R., van de Kuilen, J.-W., Terwel, K., et al.: Self-extinguishment of cross-laminated timber. Fire Saf. J. **105**, 244–260 (2019). https://doi.org/10.1016/j.firesaf.2019.01.008
26. Gorska, C., Hidalgo, J.P., Torero, J.L.: Fire dynamics in mass timber compartments. Fire Saf. J. **120**, 103098 (2020). https://doi.org/10.1016/j.firesaf.2020.103098

27. Hadden, R.M., Bartlett, A.I., Hidalgo, J.P., et al.: Effects of exposed cross laminated timber on compartment fire dynamics. Fire Saf. J. **91**, 480–489 (2017). https://doi.org/10.1016/j.firesaf.2017.03.074
28. Ali Awadallah, A.A., Hadden, R.M., Law, A.: Meta-analysis of temperature, heat release rate, delamination and auto-extinction of timber compartments. Fire Saf. J. **146**, 104164 (2024). https://doi.org/10.1016/j.firesaf.2024.104164
29. McNamee, R., Zehfuss, J., Bartlett, A.I., et al.: Enclosure fire dynamics with a cross-laminated timber ceiling. Fire Mater. **45**, 847–857 (2021). https://doi.org/10.1002/fam.2904
30. Lange, D., Sjöström, J., Schmid, J., et al.: A comparison of the conditions in a fire resistance furnace when testing combustible and non-combustible construction. Fire Technol. **56**, 1621–1654 (2020). https://doi.org/10.1007/s10694-020-00946-6
31. Pope, I., Gupta, V., Xu, H., et al.: Fully-developed compartment fire dynamics in large-scale mass timber compartments. Fire Saf. J. **141**, 104022 (2023). https://doi.org/10.1016/j.firesaf.2023.104022
32. Wiesner, F., Xu, H., Lange, D., et al.: Large-scale compartment fires to develop a self-extinction design framework for mass timber-part 2: results, analysis and design implications. Fire Saf. J. **152**, 104346 (2025). https://doi.org/10.1016/j.firesaf.2025.104346
33. Klippel, M., Schmid, J., Fahrni, R., Frangi, A.: Assessing the adhesive performance in CLT exposed to fire. In: WCTE 2018 - World Conference on Timber Engineering. Seoul, Republic of Korea (2018)
34. Hopkin, D., Węgrzyński, W., Spearpoint, M., et al.: Large-scale enclosure fire experiments adopting CLT slabs with different types of polyurethane adhesives: genesis and preliminary findings. Fire. **5**, 39 (2022). https://doi.org/10.3390/fire5020039
35. Bartlett, A.I., Hadden, R.M., Hidalgo, J.P., et al.: Auto-extinction of engineered timber: application to compartment fires with exposed timber surfaces. Fire Saf. J. **91**, 407–413 (2017). https://doi.org/10.1016/j.firesaf.2017.03.050
36. Emberley, R., Inghelbrecht, A., Yu, Z., Torero, J.L.: Self-extinction of timber. Proc. Combust. Inst. **36**, 3055–3062 (2017). https://doi.org/10.1016/j.proci.2016.07.077
37. Wiesner, F., Deeny, S., Bisby, L.A.: Influence of ply configuration and adhesive type on cross-laminated timber in flexure at elevated temperatures. Fire Saf. J. **120**, 103073 (2021). https://doi.org/10.1016/j.firesaf.2020.103073
38. Čolić, A., Wiesner, F., Hopkin, D., et al.: Application of microscale methods to study the heat-induced delamination in engineered wood products bonded with one-component polyurethane adhesives. Int. J. Adhes. Adhes. **135**, 103834 (2024). https://doi.org/10.1016/j.ijadhadh.2024.103834
39. Clauβ, S., Dijkstra, D.J., Gabriel, J., et al.: Influence of the chemical structure of PUR prepolymers on thermal stability. Int. J. Adhes. Adhes. **31**, 513–523 (2011). https://doi.org/10.1016/j.ijadhadh.2011.05.005
40. Emberley, R., Inghelbrecht, A., Doyle, N., Torero, J.L.: Components and consequences of cross-laminated timber delamination. In: Harada, K., Matsuyama, K., Himoto, K., et al. (eds.) Fire Science and Technology 2015, pp. 273–279. Springer, Singapore (2017)
41. Frangi, A., Fontana, M., Mischler, A.: Shear behaviour of bond lines in glued laminated timber beams at high temperatures. Wood Sci. Technol. **38**, 119–126 (2004). https://doi.org/10.1007/s00226-004-0223-y
42. Čolić, A., Wiesner, F., Spearpoint, M., et al.: Influence of adhesive type on heat induced delamination in tension loaded shear lap cross laminated timber samples exposed to radiant heat flux. In: Proceedings of the 13th International Conference on Structures in Fire (SiF 2024), pp. 1405–1416. University of Coimbra, Portugal (2024)
43. Brandon, D., Dagenais, C.: Fire Safety Challenges of Tall Wood Buildings—Phase 2: Task 5—Experimental Study of Delamination of Cross Laminated Timber (CLT) in Fire. National Fire Protection Association (2018)

44. Mitchell, H., Amin, R., Heidari, M., et al.: Structural hazards of smouldering fires in timber buildings. Fire Saf. J. **140**, 103861 (2023). https://doi.org/10.1016/j.firesaf.2023.103861
45. Chorlton, B., Forrest, B., Gales, J., Weckman, B.: Performance of type X gypsum board on timber to non-standard fire exposure. Fire Mater. **45**, 885–900 (2021). https://doi.org/10.1002/fam.2822
46. Xu, H., Pope, I., Gupta, V., et al.: Large-scale compartment fires to develop a self-extinction design framework for mass timber—part 1: literature review and methodology. Fire Saf. J. **128**, 103523 (2022). https://doi.org/10.1016/j.firesaf.2022.103523
47. Brandon, D.: Fire Safety Challenges of Tall Wood Buildings—Phase 2: Task 4—Engineering Methods. National Fire Protection Association (2018)
48. Bartlett, A., Law, A.: Influence of excess fuel from timber lined compartments. Construct. Build Mater. **235**, 117355 (2020). https://doi.org/10.1016/j.conbuildmat.2019.117355
49. Gorska, C.: Fire Dynamics in Multi-Scale Timber Compartments. PhD Thesis, The University of Queensland (2020)
50. Sjöström, J., Brandon, D., Temple, A., et al.: Exposure from Mass Timber Compartment Fires to Facades. RISE (2021)
51. Hopkin, D., Spearpoint, M.: A Calculation Method for Flame Extension from Partially Exposed Mass Timber Enclosures, Ljubljana, Slovenia (2021)
52. Emberley, R., Putynska, C.G., Bolanos, A., et al.: Description of small and large-scale cross laminated timber fire tests. Fire Saf. J. **91**, 327–335 (2017). https://doi.org/10.1016/j.firesaf.2017.03.024
53. Cuevas, J., Torero, J.L., Maluk, C.: Flame extinction and burning behaviour of timber under varied oxygen concentrations. Fire Saf. J. **120**, 103087 (2020). https://doi.org/10.1016/j.firesaf.2020.103087
54. Emberley, R., Do, T., Yim, J., Torero, J.L.: Critical heat flux and mass loss rate for extinction of flaming combustion of timber. Fire Saf. J. **91**, 252–258 (2017). https://doi.org/10.1016/j.firesaf.2017.03.008
55. Lucherini, A., Jovanović, B., Torero, J.L., et al.: Thermal characterisation of the cooling phase of post-flashover compartment fires. Int J Therm Sci. **199**, 108933 (2024). https://doi.org/10.1016/j.ijthermalsci.2024.108933
56. Gernay, T., Zehfuß, J., Brunkhorst, S., et al.: Experimental investigation of structural failure during the cooling phase of a fire: timber columns. Fire Mater. (2022). https://doi.org/10.1002/fam.3110
57. Awadallah, A.A.A., Bisby, L.: Predicting the structural capacity of fire-exposed CLT compression members. In: Proceedings of the 12th International Conference on Structures in Fire (SiF 2022), pp. 711–721. The Hong Kong Polytechnic University, Hong Kong (2022)
58. Gernay, T.: Fire resistance and burnout resistance of timber columns. Fire Saf. J. **122**, 103350 (2021). https://doi.org/10.1016/j.firesaf.2021.103350
59. Wiesner, F., Hadden, R., Deeny, S., Bisby, L.: Structural fire engineering considerations for cross-laminated timber walls. Construct. Build Mater. **323**, 126605 (2022). https://doi.org/10.1016/j.conbuildmat.2022.126605
60. Wiesner, F., Bartlett, A., Mohaine, S., et al.: Structural capacity of one-way spanning large-scale cross-laminated timber slabs in standard and natural fires. Fire Technol. **57**, 291–311 (2021). https://doi.org/10.1007/s10694-020-01003-y

Structural Fire Behaviour

Felix Wiesner[1] (✉), Daniel Brandon[2], Andrea Lucherini[3], and Pedro Palma[4]

[1] Department of Wood Science, Faculty of Forestry, University of British Columbia, Vancouver, BC, Canada
felix.wiesner@ubc.ca

[2] Division of Fire Safety Engineering, Lund University, Lund, Sweden

[3] Department for Fire-Safe Sustainable Built Environment (FRISSBE), Slovenian National Building and Civil Engineering Institute (ZAG), Ljubljana, Slovenia

[4] Structural Engineering Research Laboratory, Empa—Swiss Federal Laboratories for Materials Science and Technology, Dübendorf, Switzerland

Abstract. Fire safety and fire protection objectives require that buildings and parts of buildings do not collapse during a fire. This requires that the load-carrying capacity is maintained to a minimum acceptable level during a fire. This chapter briefly describes the historical background and state of the art of fire resistance and its determination for timber members through testing or calculations. The thermal and mechanical principles that underpin structural behaviour of wood at elevated temperatures are explained in the context of explicit calculation methods that enable explicit evaluation of the structural capacity beyond fire resistance, which is a formalised and codified assessment of structural elements against a standard fire. The importance of connections to the overall structure in fire is explained along with suitable design considerations. Ultimately, knowledge gaps with respect to novel and more complex engineered timber products for taller timber buildings are highlighted alongside potential limitations of established design parameters.

Keywords: fire resistance · fire safety · structures · load-carrying capacity · timber · connections

1 Fire Resistance

In the past, catastrophic fires were often driven by the spread of fire, and structural collapse was considered a consequence of out-of-control fires. Thus, structural fire safety was regulated implicitly through the imposition of non-combustible materials, for example in the rebuilding of Rome and London after their great fires in 64 AD and 1666, respectively.

The increased awareness of the importance of individual buildings, paired with an increased societal valuation of life led to the recognition that structures and their individual structural elements play important roles to achieve life safety and to prevent catastrophic fire spread. Preventing structural collapse and constraining a fire to its compartment of origin can reduce deaths among occupants and emergency personnel and limit material losses. This led to the development of fire resistance [1, 2] to provide a

G. Fink et al. (Eds.): *Holistic Design of Taller Timber Buildings*, SPRTRCIENG, pp. 288–299, 2025.
https://doi.org/10.1007/978-3-032-02098-7_22

means to compare different building assemblies measured against criteria to limit the conduction of heat, prevent the passage of smoke or flame, and maintain load-carrying capacity throughout a fire, which this chapter focuses on. The former two may also be fulfilled by non-load-carrying building elements.

The initial concept of fire resistance was to assess performance against the three criteria above (load-carrying capacity, integrity, and insulation) to the equivalent duration of a full burn-out fire [3] this means that collapse should never occur in a fire with a finite specified fuel load. Modern codified guidelines and regulations specify fire resistance not only with respect to the fuel load but also accommodate the risk profile of a building. For example, even if a fire has equal physical consequences for a structural element, that element would require a higher fire resistance rating in a high-rise hotel, compared to a single-family house. This takes the expected direct and often indirect losses into account, as well as the increased difficulty to fire-fighting operations that arise in taller buildings.

The assessment of fire resistance is done against a so-called 'standard fire'. This can be done experimentally, in a furnace, or using current modelling and design procedures. Multiple variants of standard fires exist, although the most common one is described as the standard temperature-time curve, ISO 834, that represents the evolution of the temperatures in a compartment engulfed in a fully developed fire. It prescribes a defined rise in temperature and pressure conditions, although the means of measuring this rise vary between jurisdictions. For example, in North America, thermocouples measure the apparent gas phase temperature, while in Europe, plate thermometers are prescribed to measure the adiabatic surface temperature. Thus, while the concept is standardised, the outcomes of fire resistance tests in one jurisdiction do not guarantee acceptance in others.

Standard fires are not supposed to replicate real fires but rather provide a standardised exposure against which fire resistance can be assessed. It is important to remember that fire resistance duration is decoupled from time in a real fire. Fire resistance and heat exposure in standard fires are only comparable to the conditions in a fully developed, under-ventilated fire. Alternative time-temperature curves may be used to better reflect specific compartments [4], for example, parametric fire curves account for ventilation and heat losses through compartment geometries to account for different compartment fire heating rates and a fire decay phase as fuel burns out.

2 Effect of Temperature on Mechanical Properties

All materials experience a change in their mechanical properties in response to a change in temperature. Elevated temperatures cause a deterioration of mechanical properties through a variety of mechanisms. In wood, both strength and stiffness are affected by physical changes, often associated with the movement and storage of water, and through the thermo-chemical decomposition of the polymers that make up the majority of its biomass. Depending on the specific species, wood is generally composed primarily of cellulose (40–50%), hemicelluloses (20–30%), lignin (20–30%), and other minerals and extracts [5]. Like other construction materials, wood undergoes various thermal decomposition processes at elevated temperatures, resulting in both physical and chemical changes. At temperatures up to 100 °C, the water within wood is displaced, diffuses, and/or evaporates, causing shrinkage. At temperatures exceeding 200 °C, the hemicelluloses and lignin begin to decompose and release combustible gases; this thermal decomposition process is known as pyrolysis. Above 280–300 °C, the cellulose compound starts to decompose and residual carbonaceous char is formed during pyrolysis. In the presence of sufficient oxygen, charred wood undergoes further degradation due to oxidation reactions, typically occurring in the temperature range of 400–500 °C [6].

From a structural perspective, the cross-sections of fire-exposed timber elements show reduced mechanical properties in the areas under elevated temperatures and no load-carrying capacity once it has charred. The structural performance of timber elements in fire is most commonly and most directly linked to the concept of charring. Charring is the thermal decomposition process that forms residual high-carbon char upon pyrolysis. This process generally occurs at temperatures above 280–300 °C and significantly affects the mechanical properties of timber. Beyond 300 °C, the conversion of wood to char is typically assumed to be complete, and the resulting charred wood is considered to have negligible strength and stiffness [7]. In particular, the 300 °C isotherm is often used to define the location of the char front. By tracking its progression, it is possible to estimate the char depth and charring rate, which are crucial parameters for assessing the load-carrying capacity of timber elements exposed to fire.

After estimating the effect of charring, it is essential to understand the impact of heat penetration in uncharred heated wood to evaluate the actual load-carrying capacity of timber structural members exposed to fire. As mentioned, timber decomposition processes at temperatures below the charring threshold also lead to significant reductions in mechanical properties. Figure 1 illustrates typically assumed reduction factors for strength and modulus of elasticity parallel to the grain for typical softwood exposed to a

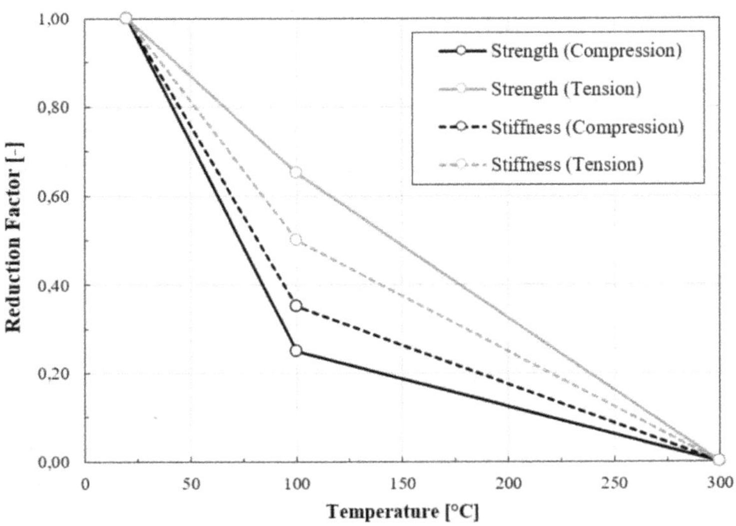

Fig. 1. Reduction factor for strength and stiffness (modulus of elasticity) parallel to grain of softwood exposed to the standard fire curve (after [8]).

standard fire, according to Eurocode 5 Part 1-2 [8]. For example, at 100 °C, the compression strength of timber is assumed to be reduced by 75% compared to its value at normal ambient temperature, while its tensile strength is reduced by 35%. However, given the low thermal conductivity of timber, the temperature gradient within fire-exposed cross-sections is very steep, at least during the fully developed phase of a fire. This means that underneath the external charred layer, the temperatures drop in short distance, often keeping the core of the cross-section at relatively low temperatures. In the cooling phase of a real fire, however, this gradient tends to flatten and elevated temperatures tend to homogenise inside the cross-section.

3 Current Assessments of Fire Resistance and Structural Fire Capacity

Fire resistance requirements can be met either by the use of standard constructive solutions, through testing, through established design models, or more complex simulations. These design models consider the reduced mechanical properties of the char layer and heated layer of uncharred timber. Under standard fire exposure, the charring rate of timber cross-sections converges to a quasi-constant value once a stable char layer is formed after approximately 20 minutes and the temperature profile beyond the char stabilizes [9]. This nearly constant char rate and steady-state temperature profile underneath the char front is the basis of the effective (or reduced) cross-section method, which subtracts a char layer and a so-called zero-strength layer from the initial cross-section. Variations of this method are implemented in design standards globally.

For a calculation of the char layer thickness for wood under standard fire exposure, a one-dimensional charring rate of 0.65 mm/min is generally assumed for softwoods; this is increased to account for corner rounding in multidimensional exposure conditions. For hardwoods or high-density softwoods, lower rates may be expected. As the strength of the char layer is negligible, it should be fully subtracted from the cross-section for structural calculations. This, however, does not hold for the heat-affected zone. Instead of calculating the structural capacity using the steep temperature gradient and the corresponding temperature-adjusted mechanical properties, which are lowest near the char front and highest deeper into the member, the use of the zero-strength layer [10] simplifies calculations and assumes a thinner uniform layer with zero strength underneath the charred layer, while the remaining cross-section is assumed to maintain the full initial capacity (Fig. 2). Several design guidance documents assume a constant zero-strength layer of 7 mm, which is based on the work of Schaffer on bending members [10]. Although rarely implemented in practice, numerical methods can be used to determine structural fire resistance, where simpler methods are not sufficient. Such numerical methods are usually based on uncoupled thermal-mechanical analyses, in which the temperature fields are calculated at successive fire exposure times and the temperature-dependent mechanical properties are adjusted accordingly. This approach allows for determining the stress fields and the mechanical behaviour during the fire exposure and assumes that the stress fields do not influence the temperature distribution, which is not always valid, particularly in the case of timber connections.

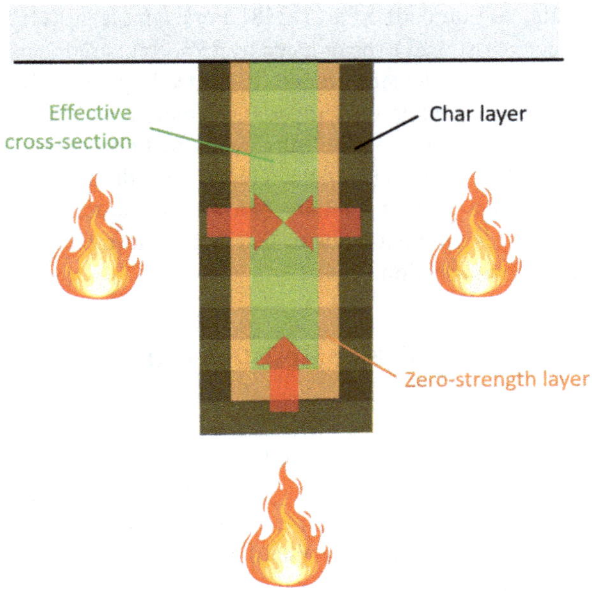

Fig. 2. Schematic showing char and zero-strength layers (ZSL) defining the effective cross-section in a mass timber beam subjected to fire from three sides. Arrows denote progression of char and ZSL fronts.

The temperature calculations can be done using, for example, non-linear finite difference or finite element heat transfer calculation methods. Most heat transfer calculations rely on effective thermal properties that implicitly take into account multiple complex phenomena, such as pyrolysis, fissures and associated mass transfer. These effective thermal properties are intended to give good approximations of the temperature fields of timber exposed to the standard fire exposure.

For structural calculations, numerous proposed relationships link temperature with the strength and stiffness of wood under compression, tension, and shear loading [11–13]. These relationships vary significantly among authors, which can be attributed to the differences in outcomes that arise between tests in steady-state or transient temperature conditions [7, 14]. Relationships provided by Eurocode 5 [8], shown in Fig. 1, were determined from transient conditions in standard fire exposure; while steady state conditions result in less onerous reductions, most fire situations will induce transient conditions. The use of advanced thermal or thermal-mechanical simulations does not guarantee conservative designs, unless the scope of application and limitations of such models are strictly considered.

4 Connections

Connections often govern the design of timber structures at normal ambient temperature, namely because of the limited load-carrying capacity and stiffness of common connections compared with the timber members being joined. Connections with metallic dowel-type fasteners (nails, staples, dowels, bolts, screws) and plates are efficient and widely used in timber structures. At normal ambient temperature, the serviceability and load-carrying capacity of timber connections depends on the type, geometry, and layout of the fasteners, on the angle between the loading direction and the axis of the fastener, and on the embedment, withdrawal, shear, and splitting properties of the connected timber members.

In fire, the performance of timber connections is even more critical for the behaviour of timber structures. This is namely due to charring of fire-exposed timber members, increased heat transfer into the timber members through the metallic fasteners and plates, high localized stresses close to the fasteners, heat transfer through gaps and slots, and the already mentioned temperature dependency of mechanical properties.

Fig. 3 Beam-to-column connections after fire resistance testing [15].

Given the complex mechanical and thermal interactions that take place in timber connections during fire exposure, the study of their structural fire behaviour, particularly the prediction of their fire resistance, has been heavily based on experimental research [15–22](Fig. 3). Even though many experiments have been performed on connection configurations that are not typical in taller timber buildings, many findings still apply. The most important parameters that influence the fire resistance of timber connections are the load level (ratio between the load applied in the fire situation and the load-carrying capacity at normal ambient temperature), the thickness of the timber side members, the width of gaps and slots in the connection area, the type of fastener and how much the steel parts protrude from the timber members, the end and edge distances of the fasteners, and the fastener spacing. Metallic parts and wide gaps or slots increase the heat transfer into the timber members, leading to increased charring in the connection zone. The most common design methods are based on protecting the connection through increased member cross-section and fastener end and edge distances and spacing or by attaching panels to cover the connection zone (the latter being aesthetically not always very appealing). Gaps and slots can also be filled or covered to limit heat transfer. This means that most fire resistance requirements can the fulfilled through an adequate choice of connection geometry based on the imposed load level. The upcoming version of Eurocode 5 provides tabulated data for this purpose. Empirical or semi-empirical fire resistance models (function of the type of connection, time of fire exposure or load level, and side-member thickness) are also available, even for fire resistances up to 120 min. Advanced thermal and thermal-mechanical modelling of connections require additional considerations, namely due to the abovementioned complex heat transfer phenomena and because of the influence that the large deformations have on the temperature distribution, as the surfaces between members become progressively more exposed to fire.

5 Challenges to Existing Methods and Knowledge

5.1 Fire Decay and Cooling Phases

Most of the discussed assessment methods aimed at quantifying the load-carrying capacity of timber structural elements exposed to fire traditionally focus on the fully developed phase of post-flashover compartment fires, usually known as the "heating phase." Nevertheless, after this phase, a compartment fire within a building enclosure is followed by decay and cooling phases [23]. Current standardised fire testing and structural fire assessment methodologies do not require investigating the effects of the fire decay and cooling phases because they are traditionally considered less critical due to their lower temperatures and because the fire resistance prescribed by building authorities is assumed to implicitly account for these phases.

During the fire decay and cooling phases heat continues to diffuse within structural materials through conduction. In particular, even if the temperature at the exposed surface is decreasing, the heat wave continues to diffuse and penetrate the structural elements, leading to an internal temperature rise and, consequently, a reduction in the load-carrying capacity of structural elements [24]. Since timber undergoes an irreversible loss of mechanical properties at relatively low temperatures compared to traditional construction materials like steel and concrete (refer to Fig. 1), the penetration of the thermal wave during the fire decay and cooling phases poses a serious challenge for timber structures. The relevance of considering this phenomenon has been recently highlighted by several researchers [25–27], and experimental evidence has shown timber structural elements failing during the cooling phase [28–30]. Research has underlined that in-depth penetration of the heat wave is highly affected by the thermal boundary conditions during the fire decay and cooling phases, as well as the duration of the various fire phases [31]. Typically, the heating phase dominates the development of the char depth, but the cooling phase has a non-negligible contribution. In addition, the uncharred heated timber increases throughout the entire fire scenario, particularly during cooling and possibly even after the end of the cooling phase. Due to the four-sided fire exposure, timber columns are particularly susceptible to this phenomenon [29].

Current structural fire engineering approaches do not address this issue explicitly, and recent research outcomes have highlighted the necessity of incorporating the fire decay and cooling phases in modern performance-based methodologies for fire-safe timber structures. One way to address this issue, at least partially, is to fully encapsulate the structure, however, this can stifle the full potential or cause overly conservative designs in the absence of fully engineered solutions.

5.2 Different Engineered Timber Products

5.2.1 Stress Distribution

Existing structural fire design guides mainly utilise the concept of a zero-strength layer to perform structural calculations for the determination of the fire resistance. This concept originated from work by Schaffer et al. [10] for simply-supported glulam beams in bending. Recent studies indicated that the generally implemented zero-strength layer of 7 mm does not reliably result in conservative predictions of the fire resistance [32]. More significant overestimations of the capacity are made for:

- Timber members that are loaded in compression and prone to buckling
- Non-homogeneous timber members

This is because the mode of loading determines the strain and stress distributions and the stresses near the char front. The capacity of a column that is prone to buckling is more dependent on the strength and stiffness of the wood material close to the char front than beams that are loaded in bending. The upcoming version of Eurocode 5 prescribes different zero-strength layer values for different types of loading.

The simplification implemented by the zero-strength layer can lead to non-conservative errors for engineered timber materials with non-homogeneous cross-sections, such as CLT. This can be the case when the zero-strength layer partially falls within non-load-carrying cross-layers. The zero-strength layer was originally proposed due to its ease of use; if it requires a wide range of values to account for varying scenarios that arise in more complex timber buildings it may be preferable to calculate the load-carrying capacity directly from the temperature gradient and the relationships of temperature and mechanical properties that are shown in Fig. 1.

5.2.2 Adhesive Performance and Composite Action

Recent increased interest and implementation of timber structural elements for tall timber structures have mainly been driven by engineered timber products, of which the most popular types, cross-laminated timber (CLT) and glue-laminated timber, utilize adhesives to form structural elements out of multiple boards.

At normal ambient temperatures, adhesives are expected to provide the same or more shear strength than the wood itself. This ensures composite action so that the assembly of individual timber boards act as one composite structural element to greatly increase the second moment of area and thus the flexural stiffness.

Different performance of adhesives at elevated temperatures have been an important consideration for the fire dynamics in a compartment, especially for CLT, as highlighted in Chap. 21. Even before fall-off of charred wood, elevated temperatures and moisture levels can weaken mechanical properties of adhesives [33–35]. When this weakening reduces composite action, significant reductions in overall structural capacity are possible [28, 36–38].

Different adhesive formulations exhibit different mechanical performance at elevated temperatures and their selection is dependent on the balance of mechanical performance, costs and environmental considerations. In North America, a minimum level of performance of adhesive in fire situations is assessed in PRG 320 [39]. While this test applies loads, the performance requirements for an adhesive to pass are solely based on the occurrence of char fall-off and not on structural performance. Thus, while a product might be suitable from a fire dynamics perspective, the adhesive may still cause a product to have an inferior flexural capacity during a fire.

5.2.3 Smouldering

Smouldering, also sometimes referred to as glowing, describes the oxidation of solid-phase carbon. It is an exothermic reaction and can under some conditions self-sustain. It occurs where airflow can interact with sufficiently hot char and heat losses are reduced, for example near insulation or in corners and crevices. If smouldering is self-sustained it will eventually lead to structural collapse, unless extinguished by fire service intervention. Smouldering progression is markedly slower than surface flame spread and thus fire service intervention after cessation of flaming is possible if smouldering is detected; this is usually achieved from infrared camera inspections and the potential removal of insulation and fire protection. Consequently, most codified guidance documents do not currently consider smouldering explicitly for structural design requirements. An exception is Japan, where for some jurisdictions and building heights, fire resistance tests must be continued after the prescribed fire resistance time has been achieved. Tested building elements must achieve self-extinguishment of smouldering to pass the fire resistance test.

6 Concluding Remarks

The structural fire safety of timber structures has long been investigated and successfully implemented in buildings with a timber frame structural system, mainly via standard fire resistance testing and associated calculation methods. These designs are underpinned by knowledge of the relationship between mechanical strength and temperature and expected temperature profiles in standard fires. However, knowledge gaps exist for fires beyond standard fires. Recent research outcomes emphasise that both the heating and cooling phases play a crucial role in assessing the heat penetration and the consequent reduction of mechanical properties in timber structural elements exposed to fire. Indeed, during the fire decay and cooling phases, the load-carrying capacity of timber elements can significantly reduce due to heat wave penetration (further charring and heated timber), and ignoring this phenomenon can lead to substantial underestimation, resulting in unsafe assessments. In addition, the constant development of novel timber products requires consideration regarding their link to existing design solutions.

References

1. Law, A., Bisby, L.: The rise and rise of fire resistance. Fire Saf. J. **116**, 103188 (2020). https://doi.org/10.1016/j.firesaf.2020.103188

2. Babrauskas, V., Williamson, R.B.: The historical basis of fire resistance testing—Part I. Fire Technol. **14**(3), 184–194 (1978). https://doi.org/10.1007/bf01983053
3. Ingberg, S.H.: Tests of the severity of building fires. NFPA Q. **22**(1), 43–61 (1928)
4. Brandon, D., Just, A., Tiso, M.: Parametric fire design–zero-strength layers and charring rates. In: INTER International Network on Timber Engineering Research Proceedings. August 2017, Kyoto, Japan (2017)
5. Wiedenhoeft, A., Eberhardt, T.: Chapter 3: Structure and function of wood. In: Wood Handbook—Wood as an Engineering Material. U.S. Department of Agriculture, Forest Service, Forest Products Laboratory, Madison, WI (2021)
6. Browne, F.L.: Theories of the combustion of wood and its control. US Dept. of Agriculture, Forest Service, Forest Products Laboratory, Madison, WI (1958)
7. Brandon, D., Hopkin, D., Emberley, R., Wade, C.: Timber structures. In: LaMalva, K., Hopkin, D. (eds.) International Handbook of Structural Fire Engineering, pp. 235–322. Springer International Publishing, Cham (2021). https://doi.org/10.1007/978-3-030-77123-2_8
8. CEN: Eurocode 5. Design of Timber Structures. Part 1-2: General. Structural Fire Design (2009)
9. Schaffer, E.L.: An approach to the mathematical prediction of temperature rise within a semi-infinite wood slab subjected to high-temperature conditions (Mathematical prediction of temperature rise within semi-infinite wood slab subjected to high temperature conditions) (1964)
10. Schaffer, E.L., Marx, C., Bender, D., Woeste, F.: Strength Validation and Fire Endurance of Glued-Laminated Timber Beams. United States Department of Agriculture, Forest Products Laboratory, Madison, WI (1986)
11. König, J., Walleij, L.: Timber frame assemblies exposed to standard and parametric fires: part 2: a design model for standard fire exposure. Inst. Träteknisk Forskning. **0001001**, 1–76 (2000)
12. Schaffer, E.L.: Effect of pyrolytic temperatures on the longitudinal strength of dry Douglas-fir. J. Test. Eval. **1**(4), 319–329 (1973)
13. Östman, B.A.L.: Wood tensile strength at temperatures and moisture contents simulating fire conditions. Wood Sci. Technol. **19**(2), 103–116 (1985). https://doi.org/10.1007/bf00353071
14. Wiesner, F., Bisby, L.: The structural capacity of laminated timber compression elements in fire: a meta-analysis. Fire Saf. J. **107**, 114–125 (2019). https://doi.org/10.1016/j.firesaf.2018.04.009
15. Palma, P.: Fire Behaviour of Timber Connections. PhD Thesis, ETH Zurich (2016)
16. Petrycki, A.R., Salem, O.: Structural fire performance of wood-steel-wood bolted connections with and without perpendicular-to-wood grain reinforcement. J. Struct. Fire Eng. **14**(4), 441–460 (2023). https://doi.org/10.1108/JSFE-02-2019-0016
17. Barber, D.: Glulam Connection Fire Test - Summary Report. 12 + Appendices. Arup USA, Washington, DC (2017) https://www.thinkwood.com/wp-content/uploads/2018/01/reThink-Wood-Arup-SLB-Connection-Fire-Testing-Summary-web.pdf
18. Palma, P., Frangi, A., Hugi, E., Cachim, P., Cruz, H.: Fire resistance tests on timber beam-to-column shear connections. J. Struct. Fire Eng. **7**(1), 41–57 (2016)
19. Peng, L.: Performance of Heavy Timber Connections in Fire. PhD Thesis, Carleton University (2010). https://doi.org/10.22215/etd/2010-09680
20. Frangi, A., Erchinger, C., Fontana, M.: Experimental fire analysis of steel-to-timber connections using dowels and nails. Fire Mater. **34**(1), 1–19 (2010)
21. Kruppa, J., Lamadon, T., Rachet, P.: Fire resistance test of timber connections. In: CTICM Ref. INC-00/187-JK/NB (2000)
22. Dhima, D.: Vérification expérimentale de la résistance au feu des assemblages d'éléments en bois. In: CTICM Ref. INC-99/399-DD/NB. CTICM, Paris (1999)

23. Lucherini, A., Torero, J.L.: Defining the fire decay and the cooling phase of post-flashover compartment fires. Fire Saf. J. **141**, 103965 (2023). https://doi.org/10.1016/j.firesaf.2023.103965
24. Gernay, T., Franssen, J.-M.: A performance indicator for structures under natural fire. Eng. Struct. **100**, 94–103 (2015). https://doi.org/10.1016/j.engstruct.2015.06.005
25. Wiesner, F., et al.: Structural capacity in fire of laminated timber elements in compartments with exposed timber surfaces. Eng. Struct. **179**, 284–295 (2019). https://doi.org/10.1016/j.engstruct.2018.10.084
26. Huč, S., Pečenko, R., Hozjan, T.: Predicting the thickness of zero-strength layer in timber beam exposed to parametric fires. Eng. Struct. **229**, 111608 (2021). https://doi.org/10.1016/j.engstruct.2020.111608
27. Gernay, T.: Fire resistance and burnout resistance of timber columns. Fire Saf. J. **122**, 103350 (2021). https://doi.org/10.1016/j.firesaf.2021.103350
28. Wiesner, F., Hadden, R., Deeny, S., Bisby, L.: Structural fire engineering considerations for cross-laminated timber walls. Construct. Build Mater. **323**, 126605 (2022). https://doi.org/10.1016/j.conbuildmat.2022.126605
29. Gernay, T., et al.: Experimental investigation of structural failure during the cooling phase of a fire: timber columns. Fire Mater. **47**(4), 445–460 (2023). https://doi.org/10.1002/fam.3110
30. Wiesner, F., et al.: Structural capacity of one-way spanning large-scale cross-laminated timber slabs in standard and natural fires. Fire Technol. (2020). https://doi.org/10.1007/s10694-020-01003-y
31. Lucherini, A., Pitelková, D.Š., Mózer, V.: Estimating the effective char depth in structural timber elements exposed to natural fires, considering the heating and cooling phase. Eng. Struct. **323**, 119255 (2025). https://doi.org/10.1016/j.engstruct.2024.119255
32. Schmid, J., Just, A., Klippel, M., Fragiacomo, M.: The reduced cross-section method for evaluation of the fire resistance of timber members: discussion and determination of the zero-strength layer. Fire Technol. **51**(6), 1285–1309 (2015). https://doi.org/10.1007/s10694-014-0421-6
33. Richter, K., Steiger, R.: Thermal stability of wood-wood and wood-FRP bonding with polyurethane and epoxy adhesives. Adv. Eng. Mater. **7**(5), 419–426 (2005). https://doi.org/10.1002/adem.200500062
34. Kläusler, O., Clauß, S., Lübke, L., Trachsel, J., Niemz, P.: Influence of moisture on stress–strain behaviour of adhesives used for structural bonding of wood. Int. J. Adhes. Adhes. **44**, 57–65 (2013). https://doi.org/10.1016/j.ijadhadh.2013.01.015
35. Čolić, A., Wiesner, F., Hopkin, D., Spearpoint, M., Wu, W., Bisby, L.: Application of microscale methods to study the heat-induced delamination in engineered wood products bonded with one-component polyurethane adhesives. Int. J. Adhes. Adhes. **135**, 103834 (2024). https://doi.org/10.1016/j.ijadhadh.2024.103834
36. Wiesner, F., Deeny, S., Bisby, L.A.: Influence of ply configuration and adhesive type on cross-laminated timber in flexure at elevated temperatures. Fire Saf. J. **120**, 103073 (2021). https://doi.org/10.1016/j.firesaf.2020.103073
37. Malhotra, H.L., Rogowski, B.F.: Fire resistance of laminated timber columns. Fire Saf. Sci. **671** (1967)
38. Klippel, M.: Fire safety of bonded structural timber elements. In: Eidgenössische Technische Hochschule ETH Zürich, Nr. 21843, Zurich (2014)
39. APA: PRG 320 Standard for Performance-Rated Cross-Laminated Timber. APA-The Engineered Wood Association, Tacoma, WA (2018)

Design of Timber Structures Subjected to Blast Loads

Christian Viau[1(✉)], Abla Krouma[2], and Ghasan Doudak[2]

[1] Department of Civil and Environmental Engineering, Carleton University, Ottawa, ON, Canada
christian.viau@carleton.ca
[2] Department of Civil Engineering, University of Ottawa, Ottawa, ON, Canada

Abstract. The increasing global interest in sustainable and renewable construction materials has led to a rise in the number of mid- to high-rise timber buildings. As interest in these structures increases, understanding their behaviour under blast loads becomes crucial due to the heightened potential risk. The possibility of collapse of tall mass timber structures due to blast loads not only poses risks to occupants but can also cause significant damage to infrastructure. An overview of the key blast parameters and dynamic analysis approach used in blast design are discussed. Material characteristics of lumber, timber, cross-laminated timber, glulam and timber connections under blast loads as well as retrofitting options are reported from research findings and blast design standards. This chapter provides designers and engineers with a roadmap and design framework for blast analysis and design of a new structure or evaluation of blast vulnerability of an existing structure that might require retrofitting.

Keywords: Blast · Design · Glulam · Cross-laminated timber · Light-frame · Connections

1 Introduction and Background

Blast loads, characterized by a sudden release of energy which generates pressurized shock fronts, have shown the potential to cause severe damage to infrastructure. The recent use of wood as a construction material for mid- and high-rise structures has increased considerably in Europe and North America, increasing the risk of these novel structures being subjected to accidental or intentional blast explosions. While timber elements and connections usually have relatively low mass and tend to generally exhibit brittle behaviour (e.g., in flexure or tension perpendicular-to-grain), instances of structures requiring design against blast threats, whether a high-profile structure or simply being in the vicinity of a potential target, may occur. As mass-timber construction (e.g., glue-laminated timber and cross-laminated timber) is becoming more common worldwide, blast design guidance is needed to limit structural damage and maintain an acceptable level of occupant safety.

Design codes and standards, such as the Eurocode 1 [1], aim to guide designers in quantifying and minimizing the risk of progressive collapse as a consequence of

G. Fink et al. (Eds.): *Holistic Design of Taller Timber Buildings*, SPRTRCIENG, pp. 300–312, 2025.
https://doi.org/10.1007/978-3-032-02098-7_23

accidental load situations. One of the principal design philosophies is that of key element design, which considers various possible, albeit rare load scenarios, to which critical load-bearing elements may be exposed [2]. Key element design is typically the method of last resort, utilized when alternative load path design does not generate a safe overall design of the structure [3]. These elements are designed using a static-equivalent pressure (34 kPa) applied on either side of the element; a prescriptive requirement that was derived based on the expected peak loads generated during the event of Ronan Point [4]. This approach has been critiqued for being simplistic and not appropriate for most blast design scenarios [5, 6].

To address the design of structural elements and assemblies against the effects of blast explosions, blast design provisions have been enacted in various countries [7–10] to reduce occupant injuries and fatalities and limit the risk of progressive collapse. The provisions for wood element design found in the current editions of American and Canadian blast standards (i.e., ASCE/SEI 59 and CSA S850) were originally based on a testing program on light-frame wood structures [11]. The study focused on the overall behaviour of the structures and was limited to qualitative experimental results. Since no in-situ properties of the structural members were measured during testing, published data was used to obtain the strength and stiffness of the specimens to conduct a damage assessment and develop design provisions and response limits for wood structures [12]. Significant work towards establishing assessment and design procedures for structural timber elements has been undertaken, allowing for the development of this chapter. The aim is to provide a general overview of the aspects involved in conducting blast analysis and design of timber elements and connections, with a particular focus on tall mass timber buildings.

2 Blast Load Phenomenon and Analysis

2.1 General

Blast explosions are highly dynamic and complex events, producing intense pressures that act over very short durations. When a high explosive detonates, it rapidly releases energy, creating a high-pressure, supersonic shockwave. The expansion of gases from the explosion generates a highly compressed zone of air, which moves outward at speeds exceeding the speed of sound. The shockwave is followed by a negative phase, characterized by a drop in pressure as the expanding gases create a vacuum-like effect. The magnitude of blast load on structural members varies with distance from the center of detonation.

In general, explosions are categorized as either confined or unconfined, depending on whether they occur within an enclosure or in free air. Unconfined explosions generally occur in the atmosphere and can be further sub-classed as free airbursts, airbursts, or surface bursts. The most common explosions affecting infrastructure and their occupants are hemispherical surface explosions. Depending on the location of the explosion relative to the element of interest, surface blast loads can be further divided into three categories: contact, near-field, and far-field loading. The distinction between these categories is done through the use of a fundamental parameter referred to as the scaled distance, Z, based on the Hopkinson-Cranz Scaling Law [13, 14]. Z is defined as:

$$Z = R/W^{1/3} \tag{1}$$

Where R is the standoff distance between the explosion and the target and W is the weight of the explosive charge, usually expressed as the equivalent weight of trinitrotoluene (TNT). Contact and near-contact blast loads occur when the explosive charge is detonated in contact with or very close to a structure (i.e. $Z \approx 0$ m/kg$^{1/3}$). This type of event tends to generate localized high-intensity non-uniform loads, that are difficult to quantify analytically. The high-intensity blast loads create high compressive stress waves that propagate within the structure which can lead to localized material failures, such as perforation through the wood. Close-in detonations produce spherical/hemispherical blast waves and create non-uniform and non-simultaneous loading. Near-field detonations occur when $Z < 1.2$ m/kg$^{1/3}$ [9]. Blast loads from such scenarios are best determined using computational fluid dynamics (CFD) in tandem with structural analysis being conducted with, for example, finite element methods (FEM). If used, equivalent single-degree-of-freedom models should consider load propagation and duration and the sensitivity of the structural elements to impulse (pressure-time history). Finally, far-field blast loads occur for $Z \geq 1.2$ m/kg$^{1/3}$ [9]. When designing against far-field explosions, it is common to assume shock wave planarity at the structure (i.e. uniform and simultaneous loads). The majority of blast design standards and guidelines have been enacted to address far-field blast loads, due to the assumptions and idealizations used to develop general blast design provisions. Due to the complexities involved with near-field and contact blast loads, generalizations are often not appropriate, thus requiring designers to carry out advanced analyses and designs. Such load types are considered outside the scope of this chapter.

2.2 Blast Load Parameters

The effects of blast loads on structures are usually expressed as a function of the amount of pressure above atmospheric pressure caused by the explosion as well as the extent of time during which the pressure wave interacts with the structure. The former is termed *overpressure*, while the latter is quantified as an *impulse*, which is the integral of the applied pressure over the time duration the pressure is applied to the structure. Before any interaction with a structural element, pressure and impulse are termed *incident*. Once the pressure is in contact with the structural element, it reflects from the element and amplifies, hence referred to as *reflected pressure* and the corresponding impulse is the *reflected impulse*. These are key parameters needed in blast analysis and design. Fig. 1 illustrates the incident pressure-time and reflected pressure-time histories from the hypothetical scenario. An incident pressure may be measured using a pressure sensor (denoted P.S. in Fig. 1) suspended in the air, while reflected pressure can be obtained from a P.S. mounted to the front face of the building. The most utilized method to model the pressure-time history is the modified Friedlander equation [15], which is characterized by an instantaneous rise in pressure above atmospheric pressure (P_{atm}) followed by an exponential decay (see Fig. 1). While the negative phase of the curve can be critical for the design of certain components (e.g., window anchorages), the positive phase tends to govern the design of primary structural elements (e.g., columns).

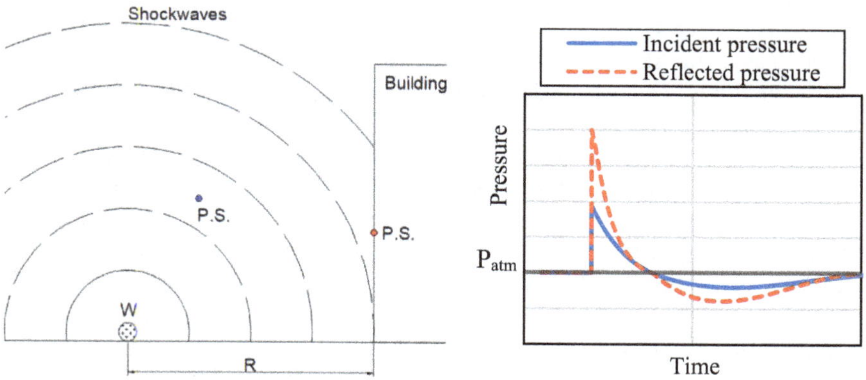

Fig. 1. Hypothetical surface burst scenario.

Empirical test data is often used to determine key blast design parameters as a function of Z. The most widely used sets of test data are the Kingery and Bulmash (K&B) curves [16]. These curves describe the properties of blast waves, such as peak overpressure, impulse, and arrival time, based on the distance from the explosion and the weight of the explosive in TNT-equivalent. The K&B curves form the foundation of widely used blast guidance manuals (e.g., [8, 17]), software (e.g., [18, 19]), and blast design standards (e.g., [9, 10]) in terms of obtaining blast load demands on structures. These curves have been developed as a function of the scaled distance Z and can therefore be adjusted based on various types of explosions.

2.3 Dynamic Analysis Methods

Blast events involve loading durations significantly smaller than those observed during earthquakes and wind loading. The interaction between the shockwave and the structure lasts only a fraction of a second and generates inertial forces and kinetic energies that must be accounted for in analyses. Minimizing the computational costs while obtaining accurate results presents one of the main challenges in modelling blast loads on structures. High-fidelity modelling techniques, such as finite element analysis (FEA), have been successfully used by researchers to describe and predict the behaviour of various structural timber elements subjected to blast loading [20]. Continuum-based FE models have shown to be capable of capturing the overall failure mode, and damage propagation, as well as predicting the resistance curves with good accuracy, however, the accuracy of such models is heavily reliant on user input and adequate failure criteria that capture dependent and independent stress interactions within the wood material [20, 21]. While developing FEA models using these tools allows for great flexibility in the material and loading inputs, such information may not be readily available in the case of blast analysis for both the loads and material resistance. Due to the uncertainties associated with blast events, especially when characterizing the blast loading, designers typically resort to much simpler analysis tools, such as the equivalent single-degree-of-freedom (SDOF) method. As shown in Fig. 2a, the method consists of transforming an actual structural element with distributed properties into an equivalent system consisting of a spring, to describe the stiffness of the wood element, (k_{wood}) and a lumped mass (m_{wood}). This is done using transformation factors, which consider the actual distribution of load, boundary conditions, and assumed deflected shape [22]. Since its implementation, several official documents have adopted this methodology (e.g. [8, 23]). One main limitation of SDOF analysis is that it can only capture a single deflected shape, which is usually approximated by a quadratic or sinusoidal equation. Higher-order vibrational modes are therefore not captured, which can lead to unconservative results, depending on the structure. Another limitation is that boundary conditions are typically idealized as being a combination of pinned, fixed, or free. Methods for including the translational and rotational stiffness and behaviour within the connections have been published, including two-degree-of-freedom (TDOF) modelling [24–26]. This is shown in Fig. 2b, where the behaviour of the connection within the system is captured by lumping the connection behaviour into an equivalent spring, similar to SDOF methodology for structural elements.

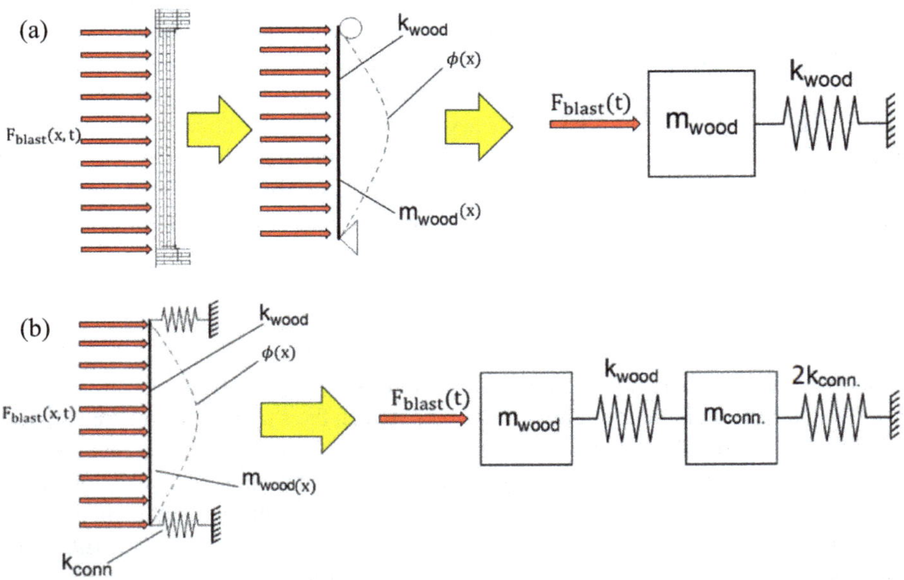

Fig. 2. Idealization of a blast-loaded CLT wall using (a) SDOF analysis; (b) TDOF analysis.

3 Material Properties of Wood Under Blast Loads

When structural elements are subjected to blast loads, their behaviour is affected by high strain-rates (in the range of milliseconds), and they tend to exhibit increases in material properties such as strength and stiffness. Comprehensive experimental research programs, usually accomplished through live explosion or shock tube testing, are necessary to characterize material behaviour under high strain-rates. The benefit of laboratory experimentation over live arena testing is the richness of information generated, including pressure-, displacement-, reaction, and strain-time histories. An example of a shock tube facility is that found at the University of Ottawa's structural lab, which has produced extensive data for light-frame wood stud walls, CLT panels, glulam beams, GLT panels, and connections.

Due to the rarity of blast events and the cost limitations associated with keeping elements elastic, blast designers rely on average material properties. The Canadian blast design standard, CSA S850 [9], is used here as an example due to its meticulous and relatively complete presentation of the design procedure. In the standard, the dynamic strength, S_D, is defined as:

$$S_D = S_S \times SIF \times DIF \quad (2)$$

Where S_S is the specified static strength, *SIF* is the strength increase factor, and *DIF* is the dynamic increase factor. The *SIF* accounts for the material variability and transforms specified design resistance to reflect near-average values. Within the Canadian design framework for wood structures, the SIF represents the ratio taken between the mean and fifth percentile strength properties. The values for visually-graded and machine stress-rated (MSR) lumber are based on an extensive in-grade testing campaign [27], where SIF values in the ranges of 1.9 to 3.9 and 1.5 to 1.8, respectively, were obtained. Published studies on glulam elements under flexure test reported *SIF* between 1.2 to 1.3 [28]. As shown in Table 1, the CSA S850 specifies lower bound *SIF* values for these wooden elements. As observed, *SIF* values for engineered wood products are lower than solid-sawn lumber due to reduced variability in material properties.

Table 1. Increase Factors for Blast Design of Wood per CSA S850 [9].

Component	SIF	DIF
Visually-graded lumber & sawn timber	1.9	1.4
Machine stress-rated lumber	1.5	1.4
Cross-laminated timber	1.2	1.2
Glulam	1.2	1.1
Connections	1.0	1.0

DIF values are obtained by relating dynamic resistance, whereby failure occurs within milliseconds, to those obtained from static tests, where failure is typically achieved around one to ten minutes. Current values in CSA S850 have been obtained through extensive experimental test campaigns on studs and stud walls [29–31], glulam beams and columns [32–36], and cross-laminated timber (CLT) panels [37, 38]. It can be observed that engineered wood products tend to have lower *DIF* values, due to the presence of finger joints, adhesives, and the use of higher-quality lumber in their manufacturing [39].

While work on timber connections under blast loads has been undertaken and *DIF* values for connections have been reported [32, 37, 40, 41], it is still premature to establish clear and simplified design requirements for connections. The magnitude of *DIF* in connections has been observed to correlate with the governing failure mode, whereby failure predominantly in wood crushing tends to exhibit higher increases compared to those failing in fastener yielding and rupturing. A connection design procedure based on capacity-based design has been proposed [32, 42], whereby connections are designed to protect against brittle failures while allowing for some energy dissipation through connection yielding. The former point ensures that ultimate failure occurs in the load-bearing element and not the connections, while the latter attempts to optimize possible sources of inelastic energy dissipation. Details on this design philosophy will be discussed in greater detail in the next section.

The design requirement outlined in Eq. (2) can also be utilized when considering other timber design codes in other countries. For example, the design strength property for timber design under blast loading using Eurocode 5 can be expressed as:

$$X_{d,blast} = (k_{mod} \times X_k/\gamma_M) \times SIF \times DIF \tag{3}$$

Where k_{mod} is the modification factor for duration of load and moisture content, γ_M is the partial factor for a material property, and X_k is the characteristic strength value. For a blast event, γ_M is taken as unity for accidental loading. Maintaining parallels with the CSA O86's load duration factor, the value of k_{mod} obtained from Eurocode 5 should correspond to an approximate load duration of 10 minutes (i.e., quasi-static testing). While Eurocode 5 specifies a k_{mod} of 1.1 for "instantaneous" loads, such a load-duration class includes accidental loads [43]. As impact loads may fall under the category of "accidental loads", k_{mod} may be conservatively taken as 1.0 when applying Eq. (3).

Care must be taken when deciding on a value for *SIF* and *DIF*, as those presented in Table 1 may need to be verified for lumber sourced in the country where they are designed. Statistical analyses conducted on the grading classes found within Eurocode 5 are expected to yield different *SIF* values. For example, based on probabilistic analyses conducted on published quasi-static bending test results for glulam beams [44] and CLT panels from Central Europe [45], *SIF* values of 1.3 and 1.1 are obtained, respectively, which are similar to the values given in CSA S850 [9].

4 Design Approaches

Whether conducting the design and analysis of a new structure or assessing the blast vulnerability of an existing structure, the performance of critical load-bearing elements must be assessed against Design Basis Threats (DBT). Determined by the designer alongside the building owner and other parties (e.g. government officials, police, etc.), these consist of likely blast scenarios that the structure may be subjected to during its design life. A DBT outlines the location of the detonation, the type of charge, and the weight of the charge. Based on this info, designers can utilize established empirical data (e.g. K&B curves) or software to determine the blast parameters, including peak reflected pressure and impulse, positive phase duration, and secondary gases as well as explosive effects (e.g. fireballs).

Using established dynamic analysis methods, such as SDOF analysis, the behaviour of a structural timber element against a given DBT is then assessed. To accomplish this, the behaviour of the element is described through a resistance curve, relating the internal force acting against the blast load versus the mid-span deflection [22]. Guidance on developing resistance curves for light-frame wood stud walls, glulam elements, and CLT panels can be found in the literature [6, 22, 33, 38, 46]. In general, a bilinear resistance curve was proposed to capture the behaviour of light-frame wood stud walls, particularly the interaction between the sheathing and the studs, which provides some form of inelastic energy dissipation. Comparatively for glulam and studs in isolation, a linear elastic resistance curve best describes the overall behaviour of these elements. Finally, CLT panels have been shown to have some post-peak resistance due to the presence of transversal laminates, which act as reinforcement for the longitudinal laminates. Therefore, a staircase-shaped resistance curve can be used to describe the flexural behaviour of CLT, with the number of total drops being a function of the number of plies. Generalized resistance curves for these structural elements, which have been normalized regarding the peak dynamic resistance of the element in question (R_{peak}) and the mid-span deflection at the elastic limit (Δ_e), are shown in Fig. 3. The values for relative displacements between ultimate failure and elastic limit (i.e., 1.0, 2.0 and 2.5 for studs/glulam, stud walls, and CLT, respectively) can be obtained from appropriate blast design standards [9]. A summary of the approaches for how to obtain the resistance curves is provided in [39].

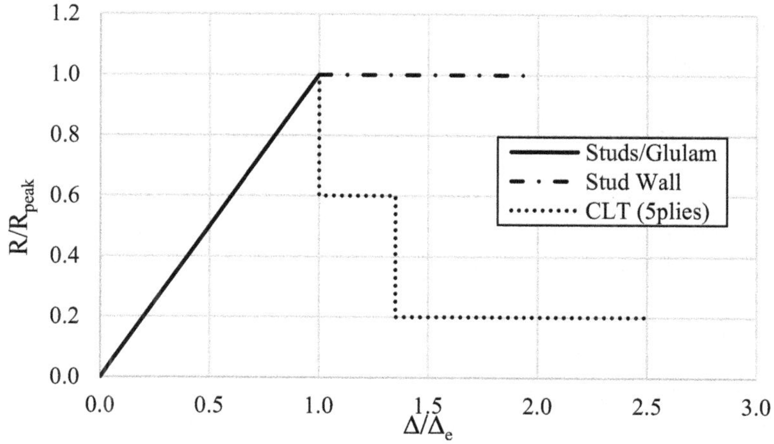

Fig. 3. Generalized Resistance Curves for Timber Flexural Elements.

The design of connections for timber assemblies is critical to ensuring the life safety of building occupants and preventing progressive collapse by maintaining overall structural integrity. Current blast design codes, whether Canadian or American, provide very little guidance in terms of how to address connection-specific designs for timber assemblies, however, research on the topic has been conducted on connections for light-frame wood stud wall connections [47], bolted connections [32, 40], self-tapping screws [48], steel angles [37, 48], as well as innovative connections to resist blast loads [42, 49] have been proposed. In instances where connections are overdesigned relative to the structural element and are significantly stiffer such that they do not allow for significant translation in the connections (i.e., pin), SDOF can be used [24, 25]. Alternatively, TDOF or other methods (FEA) may be required.

Current blast design provisions for timber structures follow the principle of capacity-based design, whereby timber connections are to be overdesigned to enforce the requirement that failure is reached within the load-bearing element before any damage in the connections [9]. When establishing the resistance curve of an element of interest (see Fig. 3), the peak dynamic resistance (R_{peak}) can be used to determine the design force at the connection, namely half of R_{peak}. Therefore, to ensure member failure before connection failure, the following design objective is required to be met, for the case of two equally-loaded end connections:

$$R_{conn} \geq \alpha \times R_{peak} \times 0.5 \tag{4}$$

where R_{conn} is the mean capacity of the connection under blast and α is the overstrength factor. CSA S850 [9] stipulates that α should be taken as 1.2, however, research has shown that this may not be sufficiently conservative depending on the connection type [32].

Connections designed and detailed to yield and deform in a controlled manner have been shown to dissipate blast energies and ultimately delay failure in the wood element. The capacity-based design methodology described above can be extended to take this into account by ensuring that the force level in the connection associated with ductile failure, such as the yielding of steel, is designed relative to the peak resistance R_{peak} such that the energy dissipating mechanism always occurs before the peak resistance. This can be achieved by the following design objective:

$$R_{conn,ductile} \leq \gamma \times R_{peak} \times 0.5 \tag{5}$$

where $R_{conn,ductile}$ is the mean capacity of the connection for its ductile failure mode and γ is the yield factor. Values of γ can be obtained by conducting a probabilistic analysis of the distribution of $R_{conn,ductile}$ and R_{peak} to ensure that an appropriate value is selected thus guaranteeing that yielding will occur in the connection before wood failure [42].

Retrofit options applied to timber elements to resist blast loads have been explored through experimental and analytical studies. Wood-based sheathing panels found in light-frame wood assemblies, such as oriented-strand boards (OSB), were observed to fail before the load-bearing wall elements reaching their flexural failure, generating hazardous flying debris and allowing for ingress of blast pressures [29]. To mitigate this, sheathing retrofitting options such as welded-wire-mesh reinforcement, catcher systems, other wood-based sheathing, and corrugated steel panels were found to mitigate these hazards in a cost-effective manner [29, 50].

The application of fibre-reinforced polymers (FRP) in different configurations has also been studied for glulam elements [34, 51], light-frame stud walls [52], and CLT panels [53]. In general, the addition of FRP reinforcement on the tension-side fibres led to significant increases in stiffness and strength, however, brittle failure via debonding of the FRP layer from the wood was also observed [34, 51, 52]. Combined with partial or full confinement, this effectively delayed brittle failure of the glulam and CLT elements and enhanced post-peak behaviour in terms of post-peak capacity and ductility. Overall, the addition of FRP confinement was found to delay crack development and bridge defects, as well as confining rolling-shear deformations to delay the development of this failure mode [53]. Near-surface mounted rod/bar elements have also been implemented within glulam beams to improve performance in flexure (e.g., [54, 55]). The performance of such retrofits was found to heavily depend on the quality of the workmanship and the quality of the bond between wood and rod material, as well as whether the implementation of the retrofit allowed for crushing in the wood compression fibres to be engaged.

5 Concluding Remarks and Outlook

With the advancement of mass-timber technologies as well as the desire to use more wood in buildings due to sustainability and carbon emissions targets, an ever-increasing stock of mass-timber buildings has become a reality. The current state of knowledge on how mass-timber elements and connection behave under blast loads allows designers to properly consider the behaviour of mass-timber to resist blast loads, whether for the intentional use of mass-timber in a structure requiring blast design considerations or whether simply an existing structure that could be in the vicinity of such an event (e.g., near a high-profile target). While significant progress has been made in understanding the response of individual elements and connections under blast loading, it is crucial to study how these structures behave at the system and building levels. Future research must prioritize system-level investigations to comprehensively assess the susceptibility of mass-timber structures to disproportionate and progressive collapse. These studies are essential to ensure that failures do not propagate in a manner that compromises the entire structure. Addressing these vulnerabilities will involve developing robust mitigation strategies that enhance structural redundancy, improve load redistribution, and prevent cascading failures. A key focus must also be placed on reinforcing mass-timber systems against undesirable brittle failure modes, particularly in the connections, to enhance ductility and ensure energy dissipation. Ultimately, future studies in this field will require a holistic design philosophy that integrates sound design methods for mass-timber elements and connections, energy-dissipating technologies, as well as advanced computational modelling. Experimental testing at scales presents the biggest challenge to the continuance of this field of study, however, these are required to ensure that the proposed methodologies possess sufficient reliability and effectiveness.

References

1. Eurocode 1: Eurocode 1 - Actions on structures - Part 1-7: General actions - Accidental actions, in EN 1991-1-7, CEN. (2006)
2. Ellingwood, B., Smilowitz, R., Dusenberry, D., Duthinh, D., Lew, H., Carino, N.: Best practices for reducing the potential for progressive collapse in buildings. In: NIST Interagency/Internal Report (NISTIR). National Institute of Standards and Technology, Gaithersburg, MD (2007)
3. Cormie, D., Mays, G., Smith, P. (eds.): Blast Effects on Buildings, 2nd edn, p. 356, Reston, VA, Thomas Telford Ltd (2009)

4. ISE: Practical Guide to Structural Robustness and Disproportionate Collapse in Buildings. Institution of Structural Engineers, London (2010)
5. Arup: Review of International Research on Structural Robustness and Disproportionate Collapse. Department for Communities and Local Government, London (2011)
6. Doudak, G., Viau, C., Lacroix, D.: Proposed design methodologies for timber assemblies subjected to blast loading. In: Görlacher, R. (ed.) International Network on Timber Engineering Research Meeting 55 (INTER 2022), August 21-25, pp. 1–15. Timber Scientific Publishing, KIT Holzbau und Baukonstruktionen, Bad Aibling, Germany (2022)
7. Unified Facilities Criteria Program: Design of Buildings to Resist Progressive Collapse (UFC 04-023-03). United States of America Department of Defense, Washington, DC (2009)
8. Unified Facilities Criteria Program: Structures to Resist the Effects of Accidental Explosions (UFC 03-340-02). United States of America Department of Defense, Washington, DC (2008)
9. CSA: Design and assessment of buildings subjected to blast loads. In: CSA S850. CSA Group, Mississauga, ON (2023)
10. ASCE: Blast Protection of Buildings. In: ASCE/SEI, pp. 59–22. American Society of Civil Engineers, Reston, VA (2022)
11. Marchand, K.A.: BAIT, BASS & RODS Testing Results. Applied Research Associates: Prepared for the USAF Force Protection Battlelab (2002)
12. Oswald, C.J.: Component Explosive Damage Assessment Workbook (CEDAW) Methodology Manual V1.0. Prepared for the Protective Design Center, US Army Corps of Engineers, San Antonio, TX (2005)
13. Cranz, C.: Lehrbuch der Ballistik. Berlin (1926)
14. Hopkinson, B.: British Ordnance Board Minutes, in Report 13565 (1915)
15. Friedlander, F.G.: The diffraction of sound pulses I. Diffraction by a semi-infinite plane. Proc. R. Soc. Lond. Ser. A Math. Phys. Sci. **186**(1006), 322–344 (1946)
16. Kingery, C.N., Bulmash, G.: Air blast parameters from TNT spherical air burst and hemispherical surface burst. In: Technical report BRL; 02555; ARBRL-TR-02555. U. S. Army Ballistic Research Laboratory, Aberdeen Proving Ground, MD (1984)
17. Unified Facilities Criteria Program: Design and Analysis of Hardened Structures to Conventional Weapons Effects (UFC 03-340-01). United States of America Department of Defense, Washington, D.C. (2002)
18. Jacques, E.: RCBlast. Ottawa, ON (2014). http://www.rcblast.ca
19. Viau, C.: BlasTDOF. Ottawa, ON (2019). www.blastdof.com
20. Oliveira, D., Viau, C., Doudak, G.: Modelling the behaviour of heavy and mass timber members subjected to blast loads. Eng. Struct. **291**, 116397 (2023)
21. Oliveira, D., Viau, C., Doudak, G.: Modelling of heavy timber members subjected to sequential shocktube-simulated blast loads. In: CSCE 2022 Annual Conference. Canadian Society for Civil Engineering, Whistler, BC (2022)
22. Biggs, J.M.: Introduction to Structural Dynamics, p. 341. McGraw-Hill, New York, NY (1964)
23. PDC: Methodology manual for the single-degree-of-freedom blast effects design spreadsheets (SBEDS). In: Technical Report 06-01 Rev 1. U.S. Army Corps of Engineers, Protective Design Center (PDC), Omaha, NE (2008)
24. Viau, C., Doudak, G.: Dynamic analysis methods for modelling timber assemblies subjected to blast loading. Eng. Struct. **233**, 111945 (2021)
25. Viau, C., Doudak, G.: Analytical modelling of heavy timber assemblies with realistic boundary conditions subjected to blast loading. In: CSCE 2019 Annual Conference. Canadian Society for Civil Engineering, Laval, QC (2019)
26. Hadianfard, M.A., Shekari, M.: An equivalent single-degree-of-freedom system to estimate nonlinear response of semi-fixed flexural members under impact load. Iran. J. Sci. Technol. Trans. Civ. Eng. (2018)

27. Barrett, J.D., Lau, W.: Canadian Lumber Properties. Canadian Wood Council, Ottawa, ON (1994)
28. Foschi, R.O., Folz, B., Yao, F.: Realibility-based design of wood structures. In: Structural Research Series, Report No. 34. University of British Columbia, Vancouver, BC (1989)
29. Viau, C., Doudak, G.: Investigating the behavior of light-frame wood stud walls subjected to severe blast loading. J. Struct. Eng. **142**(12), 04016138 (2016)
30. Lacroix, D.N., Doudak, G.: Investigation of dynamic increase factors in light-frame wood stud walls subjected to out-of-plane blast loading. J. Struct. Eng. **141**(6), 04014159 (2015)
31. Jacques, E., Lloyd, A., Braimah, A., Saatcioglu, M., Doudak, G., Abdelalim, O.: Influence of high strain-rates on the dynamic flexural material properties of spruce–pine–fir wood studs. Can. J. Civ. Eng. **41**(1), 56–64 (2014)
32. Viau, C., Doudak, G.: Behavior and modeling of glulam beams with bolted connections subjected to shock tube-simulated blast loads. J. Struct. Eng. **147**(1), 04020305 (2021)
33. Lacroix, D.N., Doudak, G.: Effects of high strain rates on the response of glulam beams and columns. J. Struct. Eng. **144**(5), 04018029 (2018)
34. Lacroix, D.N., Doudak, G.: Experimental and analytical investigation of FRP retrofitted glued-laminated beams subjected to simulated blast loading. J. Struct. Eng. **144**(7), 04018089 (2018)
35. Lacroix, D.N., Doudak, G.: Determining the dynamic increase factor for glued-laminated timber beams. J. Struct. Eng. **144**(9), 04018160 (2018)
36. Wight, N., Viau, C., Heffernan, P.: Behaviour of glued-laminated timber beams under impact loading. Can. J. Civ. Eng. (2024)
37. Viau, C., Doudak, G.: Behaviour and modelling of cross-laminated timber panels with boundary connections subjected to blast loads. Eng. Struct. **197**, 109404 (2019)
38. Poulin, M., Viau, C., Lacroix, D.N., Doudak, G.: Experimental and analytical investigation of cross-laminated timber panels subjected to out-of-plane blast loads. J. Struct. Eng. **144**(2), 04017197 (2018)
39. Doudak, G., Viau, C., Lacroix, D.N.: Proposed design methods for timber members subjected to blast loads. J. Perform. Constr. Facil. **36**(3), 04022020 (2022)
40. McGrath, A., Doudak, G.: Investigating the response of bolted timber connections subjected to blast loads. Eng. Struct. **236**, 112112 (2021)
41. Viau, C., Doudak, G.: Effect of high strain-rates on heavy timber connections. In: CSCE 2019 Annual Conference. Canadian Society for Civil Engineering, Laval, QC (2019)
42. Viau, C., Doudak, G.: Energy-absorbing connection for heavy-timber assemblies subjected to blast loads–concept development and application. J. Struct. Eng. **147**(4), 04021027 (2021)
43. CEN, Eurocode 5: Design of timber structures. In: ENV 1995-1-1. Comité Européen de Normalisation, Brussels (2004)
44. Schilling, S., Palma, P., Steiger, R., Frangi, A.: Probabilistic description of the mechanical properties of glued laminated timber made from softwood. In: Görlacher, R. (ed.) International Network on Timber Engineering Research Meeting 54 (INTER 2021), Online, August 16-19, 2021, p. 349. Timber Scientific Publishing (2021)
45. Fink, G., Kohler, J., Brandner, R.: Application of European design principles to cross laminated timber. Eng. Struct. **171**, 934–943 (2018)
46. Viau, C., Lacroix, D.N., Doudak, G.: Damage level assessment of response limits in light-frame wood stud walls subjected to blast loading. Can. J. Civ. Eng. **44**(2), 106–116 (2016)
47. Viau, C., Doudak, G.: Investigating the behaviour of typical and designed wall-to-floor connections in light-frame wood stud wall structures subjected to blast loading. Can. J. Civ. Eng. **43**(6), 562–572 (2016)
48. Côté, D., Doudak, G.: Experimental investigation of cross-laminated timber panels with realistic boundary conditions subjected to simulated blast loads. Eng. Struct. **187**, 444–456 (2019)

49. Bérubé, A., Doudak, G.: Investigation of energy-absorbing connections for application in mass timber structures. Eng. Struct. **297**, 117018 (2023)
50. Lacroix, D.N., Doudak, G., El-Domiaty, K.: Retrofit options for light-frame wood stud walls subjected to blast loading. J. Struct. Eng. **140**(4), 04013104 (2014)
51. Lacroix, D., Doudak, G.: Towards enhancing the post-peak performance of glued-laminated timber beams using multi-directional fibre reinforced polymers. Eng. Struct. **215**, 110680 (2020)
52. Lacroix, D., Viau, C., Battelli, E., Doudak, G.: Enhancing the performance of light-frame wood studs using glass fibre-reinforced polymers. Eng. Struct. **245**, 112973 (2021)
53. Lopez-Molina, A., Doudak, G.: Retrofit techniques for Cross-Laminated Timber (CLT) elements subjected to blast loads. Eng. Struct. **197**, 109450 (2019)
54. Gentile, C., Svecova, D., Rizkalla, S.H.: Timber beams strengthened with GFRP bars: development and applications. J. Compos. Constr. **6**(1), 11–20 (2002)
55. Yang, H., Liu, W., Lu, W., Zhu, S., Geng, Q.: Flexural behavior of FRP and steel reinforced glulam beams: experimental and theoretical evaluation. Construct. Build Mater. **106**, 550–563 (2016)

Towards a Multi-hazard Framework for the Design of Taller Timber Structures

Christian Viau[1(✉)], Onur Kaplan[2], Abla Krouma[3], Gabriele Tamagnone[4], Daniel Brandon[5], Felix Wiesner[6], Tuba Tatar[7], and Daniele Casagrande[8]

[1] Department of Civil and Environmental Engineering, Carleton University, Ottawa, ON, Canada
christian.viau@carleton.ca

[2] Institute of Earth and Space Sciences, Eskisehir Technical University, Eskisehir, Turkey

[3] Department of Civil Engineering, University of Ottawa, Ottawa, ON, Canada

[4] Mechanics and Civil Engineering Laboratory, University of Montpellier, Montpellier, France

[5] Division of Fire Safety Engineering, Lund University, Lund, Sweden

[6] Department of Wood Science, University of British Columbia, Vancouver, BC, Canada

[7] Department of Civil Engineering, Sakarya University, Sakarya, Turkey

[8] Department of Civil, Environmental and Mechanical Engineering, University of Trento, Trento, Italy

Abstract. Multi-hazard events, though infrequent and rare, are responsible for a substantial portion of global economic losses. This book chapter investigates the complexities of multi-hazard events and explores the design challenges and methodologies in multi-hazard risk assessment and structural resilience for wooden structures. Special attention is given to the interaction between earthquakes, fires, and blast loads on timber structures, highlighting the state of the art, the gaps in current research, and the potential for probabilistic design and analysis approaches to enhance understanding and improve structural resilience and robustness. The limitations of existing design codes and standards are reviewed and frameworks for assessing cumulative damage and risk under multi-hazard scenarios are discussed. The book chapter aims to provide the state of the art on the topic and guide future research and practical applications in designing timber structures capable of enduring multiple hazards, thereby enhancing the safety and sustainability of the built environment.

Keywords: Timber · Seismic · Fire · Blast · Multi-hazard · Design

1 Introduction and Background

Multi-hazard events, which involve the simultaneous, cascading, or cumulative occurrence of multiple natural hazards, pose a serious threat to human lives and property. This danger arises mainly from the cumulative and cascading impacts caused by the interaction of different natural hazards across space and time. However, identifying these events is difficult due to the complex interactions between hazards and the limited availability of multi-hazard data. Globally, about 19% of the 16,535 disasters recorded

G. Fink et al. (Eds.): *Holistic Design of Taller Timber Buildings*, SPRTRCIENG, pp. 313–324, 2025.
https://doi.org/10.1007/978-3-032-02098-7_24

in the EM-DAT Database in the last decade are classified as multi-hazard events [1]. The identified multi-hazard events were reclassified into four distinct categories: (1) preconditioned/triggering events, (2) multivariate events, (3) temporally compounding events, and (4) spatially compounding events. These multi-hazard categories were primarily based on the typology that classified compound events into preconditioned, multivariate, temporally compounding, and spatially compounding categories [2]. Despite their lower occurrence, these events are responsible for nearly 59% of the estimated global economic losses, while single hazard events tend to result in higher fatalities [1]. Most multi-hazard events are associated with floods, storms, fires, and earthquakes, with landslides frequently emerging as secondary hazards, often triggered by these primary events. The majority of these multi-hazard events display preconditioned/triggering and multivariate characteristics. Multi-hazard events are more common in Asia and North America, whereas Europe sees a higher occurrence of temporal overlaps of multiple hazards. The Great Lisbon Earthquake of 1755, the Typhoon Nina-Banqiao disaster in 1975, and the 2010 Haiti earthquake are key examples of multi-hazard events. The Lisbon earthquake led to widespread structural destruction from ground acceleration, followed by a tsunami, fires, and aftershocks. Typhoon Nina in China caused heavy rainfall and winds, leading to the failure of 61 dams and severe flooding. Similarly, the Haiti earthquake triggered cascading hazards, including landslides, a tsunami, and aftershocks, amplifying its impact.

The growing interest in multi-hazard risk assessment and structural resilience has led to diverse methodologies and frameworks. Building taxonomy for multi-hazard risk assessment (GEM4ALL) [3], methodologies for assessing cumulative damage from sequential multi-hazard events [4], risk analysis methods for multi-hazard scenarios [5] are just some of the directions pursued by researchers around the globe to investigate the subject. However, researchers have also highlighted the infancy of multi-hazard engineering, noting the need for integrated resilience strategies [6], the shortcomings of current design codes [7], and the necessity of tailored multi-hazard criteria for enhancing structural stability against combined threats [8]. Lou et al. [9] and Meacham [10] developed material-agnostic performance-based frameworks for Post-Earthquake Fires (PEF) assessment, emphasizing the importance of a more extensive database, fragility curves for multi-hazard resilience, and collaboration between fire and seismic engineering. Studies on fire-induced material degradation indices [11], fragility curves [12], and full-scale testing under combined earthquake and fire loading [13] offer valuable knowledge. Other key factors to consider are damage to passive fire protection systems from seismic events impact fire resistance [14] as well as the very factors leading to PEF [15]. In an attempt to implement the performance-based design approach in the case of post-earthquake fire, Meacham [16] conducted shake table tests on a full-scale, five-story reinforced concrete building and proposed a model to performance-based seismic and fire design.

The dynamic behaviour of reinforced concrete (RC) slabs was investigated under simultaneous fire and explosion, and it was found that fire reduces the RC slab's ability to withstand blast loading [17]. Xue et al. [18] proposed a novel damage assessment methodology for shear wall structures facing earthquake–explosion disaster chains, revealing that initial seismic damage exacerbates vulnerabilities to subsequent explosions, often elevating damage levels to severe states. Verma et al. [19] examined reinforced concrete structures under combined earthquake and blast scenarios, demonstrating that such multi-hazard conditions result in the most severe structural responses due to compounded localized and widespread impacts. Finally, Quiel and Marjanishvili [20] examined progressive collapse in steel-framed buildings following blast and fire, underscoring the necessity of multi-hazard approaches for enhancing structural resilience.

While extensive research has been conducted on timber structures concerning individual hazards such as earthquakes, blast loads, or fires, the complex interactions between multiple hazards remain underexplored and are mostly limited to other structural materials and forms. The exception to this is work by Tesfamariam [21], in which the design of taller timber buildings is discussed for earthquake and wind loads within a multi-hazard performance-based design framework. This gap in the literature limits understanding of how tall timber structures respond when exposed to the cumulative or triggered effects of more than one hazard. This book chapter aims to guide and provide a succinct overview of pertinent literature and design considerations on the subject through the existing knowledge.

2 Overview of Accidental Load Situations

Accidental load situations, such as earthquakes, fires, and blast loads, represent extreme events that can pose significant challenges to the integrity and safety of structures and occupants. These loads are rare but have the potential to cause catastrophic damage if not properly accounted for in the design and evaluation process. Designers and building officials consider these load scenarios individually, as each has unique characteristics, mechanisms, and impacts on structural behaviour. Understanding the specific demands and responses associated with each accidental load is critical before exploring their potential interactions or cascading effects. This foundational knowledge allows engineers to implement appropriate design strategies and ensure compliance with safety standards, ultimately enhancing the resilience of built environments.

2.1 Earthquakes

Timber lateral load-resisting systems are recognized to be a viable alternative to other structural types in seismic-prone areas, even those of moderate to high seismicity. Several studies have shown that timber buildings perform well under strong ground motions primarily because of their high strength-to-mass ratio [22] and ability to dissipate energy in ductile metal connections. However, throughout history, timber structures have exhibited varying degrees of resilience to seismic activity shaped by structural types, material properties, and design regulations.

Following the 1994 Northridge Earthquake, light-frame timber buildings attracted the attention of structural engineers and researchers due to the significant damage observed especially in two to three-storey apartment buildings [23]. Most wood-frame buildings in the area were poorly engineered and lacked a seismic-force-resisting system, triggering significant losses in wood-frame constructions. Consequently, in order to minimize earthquake losses to wood-frame construction and develop specific performance-based seismic design (PBSD) approaches, in 1998 the CUREE project [24] was launched as a series of coordinated and comprehensive scientific investigations in North America. In 2002, Filiatrault and Folz [25] conducted a comprehensive literature review on force-based seismic design, showing a strong correlation between inter-storey drift during an earthquake and the subsequent damage to wood-frame structures. Rosowsky & Ellingwood [26] and van de Lindt & Walz [27] established the performance levels in terms of inter-storey drift limits to control non-structural and structural damage in wood-frame timber buildings and developed specific design procedures accordingly. In 2005, the NEESWOOD research project [28] aimed at developing a PBSD approach for mid-rise wood-frame constructions. Shake table tests on a two- and a six-storey light-frame building were conducted [29, 30] (including tests on some damping systems and half-scaled base-isolation tests [23]) and a specific non-linear time history analysis software package, was developed. Moreover, a proposal for the application of a direct displacement design (DDD) approach to wood-frame buildings was presented. In Italy, Tomasi et al. [31] investigated the seismic performance of multi-storey light-frame timber buildings through a shake table test conducted on a three-storey timber light-frame building. A highly dissipative behaviour, with most plastic deformations concentrated in the sheathing-to-framing joints and the hold-downs and angle brackets, was observed.

The studies on the seismic performance of mass timber (e.g. Cross-Laminated Timber - CLT) structural systems started in the early 2000s, showing significant differences with those conducted on wood-frame buildings. The energy dissipation in CLT buildings is in fact concentrated in the mechanical anchors (i.e. hold-down, angle brackets) when single-panel shearwalls are used whereas vertical joints are designed to dissipate energy in segmented shearwalls [32]. Within the SOFIE [33] and SERIES [34] projects, shake-table tests were conducted on a seven- and three-storey building made with multi-panel and single-panel shearwalls, respectively. Gavric et al. [35] and Flatscher et al. [36] conducted experimental tests on hold-down and angle brackets showing a ductile failure mechanism when hold-downs were subjected to tensile loads and angle brackets to shear load.

Despite most of the current design standards for the seismic design of timber buildings (e.g. Eurocode 8 [37]) primarily follow a deterministic force-based procedure, in the last decade, several studies have applied risk-based approaches to investigate the seismic performance of timber buildings and determine the structural behaviour q-factors (i.e. *R*-factors in North America) of different structural types. For CLT structures, for example, Van de Lindt et al. [38] calculated the *R*-factor for CLT platform-type buildings using the FEMA P695 [39] procedure while Rinaldi et al. [40] used a risk-consistent approach to verify the values of *q*-factors reported in the second generation of Eurocode 8 according to the *q*-factor estimation methodology [41]. Tesfamariam et al. [42] assessed the *R*-factor for a 10-storey hybrid timber-reinforced concrete building while non-linear incremental dynamic analyses (IDA) were conducted by Morshedi et al. [43] to assess the *R*-factors of light-timber frame shearwalls.

2.2 Fires

Timber buildings have burned down in the past mainly because they lacked proper fire safety features. These features include adequate fire compartments, detection systems, and suppression methods. Historical disasters like The Great Fire of London in 1666 and The Boston Fire in 1872 happened because the fire safety technology at the time could not stop small fires from propagating. Many building regulations today, which limit the height of timber buildings to about four stories, were created in response to these massive urban fires from the past [44]. Modern structural systems and technologies for detecting and suppressing building fires suggest no continuing reason to limit the permissible heights of timber buildings prescriptively [22].

In most jurisdictions, fire safety regulations follow a deterministic approach, designing buildings to withstand specific, standardized fire exposures, typically a fire resistance test (e.g., EN1363-1) for a set duration. However, this method does not account for variations in fire scenarios: many buildings may not experience significant fires, while some may encounter more severe events than anticipated within the deterministic approach. In a deterministic approach, it is not straightforward to quantify the effect of risk-reducing measures, such as sprinklers, fire-stopping, and/or improved firefighting access. Instead, deterministic regulations are adjusted over time based on experience [45], such as by high-consequence fire incidents or lack thereof.

A probabilistic approach may enable societal cost optimization concerning acceptable losses, as a deterministic approach applied to all buildings can be overly conservative and costly. Probabilistic fire safety design aims primarily to ensure life safety but can also address property protection. A possible approach for fire safety engineering is the use of Farmer's diagrams, better known as F-N curves, which set limits on the frequency (F) of incidents with a specific number of casualties (N). These curves emphasize that the likelihood of multiple fatalities should be extremely low, with incidents causing more than 10 fatalities considered nearly unacceptable, as shown in Fig. 1. The ALARP principle (As Low as Reasonably Practicable) allows designs to bypass strict F-N curve limits if compliance is infeasible.

Property protection, a secondary goal of fire safety engineering, accepts certain probabilities of failure. However, regulations do not explicitly specify limits for failure probabilities due to fires. Some regulations indirectly offer property protection, such as Eurocode 0 [46], which specifies a reliability index of $\beta = 3.8$, equating to a probability of approximately $P = 7.23 \times 10^{-5}$ over 50 years for moderate structural failure. Notably, no reliability index is provided for more substantial failures such as progressive collapse.

Implementing performance-based design faces significant challenges due to a lack of data, the complexity of required analyses, and inherent uncertainties. Probabilistic design depends on statistical data, such as fire load densities, and fire growth rates. However, fires are relatively rare and poorly documented events and outliers are poorly represented in statistical distributions, limiting the ability of a probabilistic design approach to foresee high-consequence events. The ALARP principle adds complexity, requiring a balance between financial feasibility and life safety, which is hard to quantify. Additionally, many stakeholders, including fire brigades, local authorities, and designers, often trust deterministic regulations more than probabilistic approaches due to a limited understanding of risk-based methods.

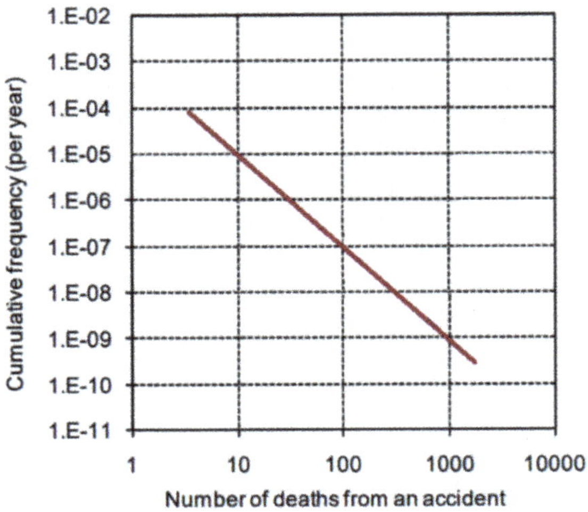

Fig. 1. F-N curve for fire events in the Netherlands.

2.3 Blast Loads

Blast loads tend to occur nearly instantly and release enormous amounts of energy, leading to a dynamic response, localized damage, and a high strain rate effect of the wood material. Examples of an explosion followed by a fire are the Oklahoma City bombing (USA, 1995) and UNBC's Wood Innovation Research Laboratory Explosion and Fire (Canada, 2023), while the case of a fire followed by an explosion is the Port of Beirut Explosion (Lebanon, 2020). The behaviour of timber elements under blast loading has been investigated in recent years, and the design of timber structures subjected to blast loads is explained thoroughly in another Book Chapter. It is important to note that design against blast loads is not required in the standards, but it is upon request of the owner or government authorities.

3 Multi-hazard Scenarios

When combined load effects, such as those from fire, earthquakes, and blast loads are considered in the multi-hazard design framework, performance-based design becomes a necessity. Performance-based design liberates the designer and provides opportunities to communicate with the public (e.g., owners, insurers, occupants) to understand the needs and expectations better and allows the designers to provide solutions not being limited to prescriptive regulations that are not capable of delivering required or expected performance, especially in the multi-hazard scenarios. Multi-hazard events often have more severe impacts than single-hazard events and present challenges for traditional design. In addressing multi-hazard events, adopting a probabilistic approach, despite its inherent challenges, becomes increasingly necessary. Rare, high-impact events—where the primary goal is to prevent mass casualties—present unique scenarios where lower-probability outcomes, such as moderate structural damage or minimal casualties, may already align with acceptable thresholds. This stems from the rarity of such events inherently limiting their overall risk. By focusing on mitigating catastrophic outcomes, probabilistic methods provide a rational framework for balancing safety and resource allocation in the face of complex, multi-hazard scenarios.

3.1 Earthquakes and Fires

Both earthquakes and large building fires are relatively rare events, although they are not statistically independent with fires often starting in the aftermath of earthquakes. A joint deterministic approach for combined earthquakes and fires would be prohibitively costly, both in terms of test requirements and design implementation. This also highlights the opportunities for a probabilistic approach to combine these events.

While conventional fire design is heavily focused on deterministic solutions, earthquake design usually involves probabilistic events and design against expected earthquakes (rather than a 'standard earthquake'). There are also design aspects between earthquakes and timber fire design that are fundamentally opposed. For example, earthquake design has a clear sacrificial order in which limited damage from major earthquakes is part of the design in order to prevent overall collapse. Paired with repair considerations this favors exposed, easily accessible connections that absorb damage from deformations but can rapidly be accessed and repaired. In contrast for timber design in fire, connections are often placed within timber elements to avoid heating and the resultant weakening of steel components and the timber adjacent to this steel.

Wood is combustible, which distinguishes it from other materials that are commonly used for tall buildings, such as steel or concrete. In most current design approaches the amount of exposed timber is limited by encapsulation with fire-rated plasterboard. This can help to ensure that self-extinction of timber can be achieved and limits the overall fire size [47, 48]. Therefore, beyond the direct effects of combined earthquakes and fires on the structure, the potential contribution of timber due to damage to encapsulation must be considered in probabilistic analysis. This can also include damage to other fire safety measures and the lack of fire service intervention.

3.2 Blast Loads and Fires

The design and analysis of structural elements subjected to blast loading in CSA S850 [49] defines the dynamic strength, S_D, provided as:

$$S_D = S_S \times SIF \times DIF \tag{1}$$

Where *SIF* is the strength increase factor, *DIF* is the dynamic increase factor, and S_S is the specified static strength. While the *DIF* incorporates the effects of high strain rates onto the material properties, which tend to be beneficial (i.e., greater or equal to unity), the SIF acts to transform design-level specified strengths to near-average strength values. The use and magnitudes of the *SIF* in blast design resemble the current Canadian [50] and European [51] wood design provisions for fire scenarios.

There are currently no design standards or guidelines for cases of succeeding or simultaneous fire and blast scenarios. In addition, due to the extreme rarity of both loads acting concurrently, no documented case study nor literature exists on the topic of timber structures. With that being said, engineering judgement can be applied to develop rational design approaches for determining potential interaction effects for pre- and post-blast fires. Fires may occur as a direct or indirect result of an explosion or may be the source of an explosion (e.g., fire causing a gas line to ignite and explode). Concerning the former, localized deformations in structural and non-structural elements brought on by the blast event may jeopardize the integrity of fire compartments and otherwise safe evacuation routes, particularly in light-frame wood assemblies. While structural elements may only be subjected to superficial damage levels following blast loading, such cracks may present a potential path for fire development to penetrate through walls and other compartments. In such cases, one may conservatively assume failure of compartmentalization. If consideration of potential cases of blast events is required, for example, for high-profile assets or proximity to high-profile assets, one may plan evacuation routes to be outside the perimeters of the building, to move them as far as possible to the effects of the shockwaves.

For the latter (i.e., blasts occurring due to a fire event), designers should consider the potential loss of cross-sectional area and mass, both of which would contribute to an element's response to a blast load. Estimating the effective cross-section depends on the zero-strength layer and char depth, which can be obtained from the charring rate and fire exposure time.

4 Outlook

The increasing number of taller timber structures and the need for more resilient and safe tall structures require addressing the complex challenges posed by multi-hazard scenarios. Experimental testing for scenarios of fire and explosions is highly complex due to cost and testing conditions. The ability to predict materials' behaviour and damage level under explosions and fire is uncertain, and many factors play a role in this, such as the sequence of events and type of Engineered Wood Products (EWP) being used in the design. The current knowledge available in case of fire and explosion events was mainly developed using advanced analysis using FE packages. Additionally, the behaviour of connections under such combined or successive events is not investigated. With the current lack of comprehensive design guidelines for multi-hazard design or studies on the response of the building under such events, redundancy and robustness of taller timber structures are paramount, and the design in such cases should be towards including alternative load paths to prevent progressive collapse.

References

1. Lee, R., White, C.J., Adnan, M.S.G., Douglas, J., Mahecha, M.D., O'Loughlin, F.E., Patelli, E., Ramos, A.M., Roberts, M.J., Martius, O., Tubaldi, E., Van Den Hurk, B., Ward, P.J., Zscheischler, J.: Reclassifying historical disasters: from single to multi-hazards. Sci. Total Environ. **912**, 169120 (2024). https://doi.org/10.1016/j.scitotenv.2023.169120
2. Zscheischler, J., Martius, O., Westra, S., Bevacqua, E., Raymond, C., Horton, R.M., van den Hurk, B., AghaKouchak, A., Jézéquel, A., Mahecha, M.D., Maraun, D., Ramos, A.M., Ridder, N.N., Thiery, W., Vignotto, E.: A typology of compound weather and climate events. Nat. Rev. Earth Environ. **1**, 333–347 (2020). https://doi.org/10.1038/s43017-020-0060-z

3. Silva, V., Brzev, S., Scawthorn, C., Yepes, C., Dabbeek, J., Crowley, H.: A building classification system for multi-hazard risk assessment. Int. J. Disaster Risk Sci. **13**, 161–177 (2022). https://doi.org/10.1007/s13753-022-00400-x
4. Marasco S, Noori AZ, Kammouh O, Cardoni A, Cimellaro P A New Approach to Multi-Hazard Analysis (2018)
5. Wang, J., He, Z., Weng, W.: A review of the research into the relations between hazards in multi-hazard risk analysis. Nat. Hazards. **104**, 2003–2026 (2020). https://doi.org/10.1007/s11069-020-04259-3
6. Bruneau, M., Barbato, M., Padgett, J.E., Zaghi, A.E., Mitrani-Reiser, J., Li, Y.: State of the art of multihazard design. J. Struct. Eng. **143**, 03117002 (2017). https://doi.org/10.1061/(ASCE)ST.1943-541X.0001893
7. Roy, T., Matsagar, V.: Multi-hazard analysis and design of structures: status and research trends. Struct. Infrastruct. Eng. **19**, 845–874 (2023). https://doi.org/10.1080/15732479.2021.1987481
8. Carlson, N., Saadeghvaziri, M.A.: On multi-hazard considerations in design of structures. In: 9th US National and 10th Canadian Conference on Earthquake Engineering 2010, Including Papers from the 4th International Tsunami Symposium, pp. 4829–4838 (2010)
9. Lou, T., Wang, W., Izzuddin, B.A.: A framework for performance-based assessment in post-earthquake fire: Methodology and case study. Eng. Struct. **294**, 116766 (2023). https://doi.org/10.1016/j.engstruct.2023.116766
10. Meacham, B.J.: Toward a risk-informed performance-based approach for post-earthquake fire protection design of buildings. In: Harada, K., Matsuyama, K., Himoto, K., Nakamura, Y., Wakatsuki, K. (eds.) Fire Science and Technology 2015, pp. 73–85. Springer Singapore, Singapore (2017)
11. Ni, S., Birely, A.C.: A simplified model for the post-fire earthquake flexural response of reinforced concrete walls with boundary elements. Eng. Struct. **175**, 721–730 (2018). https://doi.org/10.1016/j.engstruct.2018.08.044
12. Covi, P., Tondini, N., Sarreshtehdari, A., Elhami-Khorasani, N.: Development of a novel fire following earthquake probabilistic framework applied to a steel braced frame. Struct. Saf. **105**, 102377 (2023). https://doi.org/10.1016/j.strusafe.2023.102377
13. Shah, A.H., Sharma, U.K., Kamath, P., Bhargava, P., Reddy, G.R., Singh, T.: Fire performance of earthquake-damaged reinforced-concrete structures. Mater. Struct. **49**, 2971–2989 (2016). https://doi.org/10.1617/s11527-015-0699-y
14. Risco, G.V., Zania, V., Giuliani, L.: Numerical assessment of post-earthquake fire response of steel buildings. Saf. Sci. **157**, 105921 (2023). https://doi.org/10.1016/j.ssci.2022.105921
15. Gökşen, F., Takva, Ç., Takva, Y., İlerisoy, Z.Y.: Post-earthquake fires: risk assessment and precautions. Mehran Univ. Res. J. Eng. Technol. **43**(1) (2024). https://doi.org/10.22581/muet198224012710
16. Meacham, B.J.: Post-earthquake fire performance of buildings: summary of a large-scale experiment and conceptual framework for integrated performance-based seismic and fire design. Fire Technol. **52**, 1133–1157 (2016). https://doi.org/10.1007/s10694-015-0523-9
17. Li, H., Chen, W., Hao, H.: Performance of reinforced concrete slabs subjected to simultaneous fire and blast loads. Eng. Struct. **311**, 118133 (2024). https://doi.org/10.1016/j.engstruct.2024.118133
18. Xue, Z., Huo, L., Ying, P., Li, H.: Damage assessment of shear wall structures in an earthquake–blast disaster chain. Appl. Sci. **12**, 11781 (2022). https://doi.org/10.3390/app122211781
19. Verma, S., Goel, M.D., Sirdesai, N.N.: Vulnerability analysis of earthquake-driven blast load on a reinforced concrete structure. Proc. Inst. Civil Eng. Struct. Build. **177**, 159–176 (2024). https://doi.org/10.1680/jstbu.22.00165

20. Quiel, S.E., Marjanishvili, S.M.: Fire resistance of a damaged steel building frame designed to resist progressive collapse. J. Perform. Constr. Facil. **26**, 402–409 (2012). https://doi.org/10.1061/(ASCE)CF.1943-5509.0000248
21. Tesfamariam, S.: Performance-based design of tall timber buildings under earthquake and wind multi-hazard loads: past, present, and future. Front. Built. Environ. **8**, 848698 (2022). https://doi.org/10.3389/fbuil.2022.848698
22. Smith, I., Frangi, A.: Overview of design issues for tall timber buildings. Struct. Eng. Int. **18**, 141–147 (2008). https://doi.org/10.2749/101686608784218833
23. van de Lindt, J.W., Liu, H., Symans, M.D., Shinde, J.K.: Seismic performance and modeling of a half-scale base-isolated wood frame building. J. Earthq. Eng. **15**, 469–490 (2011). https://doi.org/10.1080/13632469.2010.498561
24. Seible F, Filiatrault A, Uang C-M: Proceedings of the Invitational Workshop on Seismic Testing, Analysis and Design of Woodframe Testing, CUREE - Caltech Woodframe Project Report W-01 (1999)
25. Andre, F., Bryan, F.: Performance-based seismic design of wood framed buildings. J. Struct. Eng. **128**, 39–47 (2002). https://doi.org/10.1061/(ASCE)0733-9445(2002)128:1(39)
26. Rosowsky, D.V., Ellingwood, B.R.: Performance-based engineering of wood frame housing: fragility analysis methodology. J. Struct. Eng. **128**, 32–38 (2002). https://doi.org/10.1061/(ASCE)0733-9445(2002)128:1(32)
27. van de Lindt John, W., Walz, M.A.: Development and application of wood shear wall reliability model. J. Struct. Eng. **129**, 405–413 (2003). https://doi.org/10.1061/(ASCE)0733-9445(2003)129:3(405)
28. van de Lindt, J.W., Rosowsky, D.V., Filiatrault, A., Symans, M.D., Davidson, R.A.: The NEESWood Project in Review. In: Proceedings of the 9th U.S. National and 10th Canadian Conference on Earthquake Engineering (2010)
29. Filiatrault, A., Christovasilis, I.P., Wanitkorkul, A., van de Lindt, J.W.: Experimental seismic response of a full-scale light-frame wood building. J. Struct. Eng. **136**, 246–254 (2010). https://doi.org/10.1061/(ASCE)ST.1943-541X.0000112
30. van de Lindt John, W., Shiling, P., Pryor, S.E., Shimizu, H., Isoda, H.: Experimental seismic response of a full-scale six-story light-frame wood building. J. Struct. Eng. **136**, 1262–1272 (2010). https://doi.org/10.1061/(ASCE)ST.1943-541X.0000222
31. Tomasi, R., Sartori, T., Casagrande, D., Piazza, M.: Shaking table testing of a full-scale prefabricated three-story timber-frame building. J. Earthq. Eng. **19**, 505–534 (2015). https://doi.org/10.1080/13632469.2014.974291
32. Izzi, M., Casagrande, D., Bezzi, S., Pasca, D., Follesa, M., Tomasi, R.: Seismic behaviour of Cross-Laminated Timber structures: a state-of-the-art review. Eng. Struct. **170**, 42–52 (2018). https://doi.org/10.1016/j.engstruct.2018.05.060
33. Ceccotti, A., Sandhaas, C., Okabe, M., Yasumura, M., Minowa, C., Kawai, N.: SOFIE project – 3D shaking table test on a seven-storey full-scale cross-laminated timber building. Earthq. Eng. Struct. Dynam. **42**, 2003–2021 (2013). https://doi.org/10.1002/eqe.2309
34. Flatscher, G., Schickhofer, G.: Shaking-table test of a cross-laminated timber structure. Proc. Inst. Civil Eng. Struct. Build. **168**, 878–888 (2015)
35. Gavric, I., Fragiacomo, M., Ceccotti, A.: Cyclic behaviour of typical metal connectors for cross-laminated (CLT) structures. Mater. Struct. **48**, 1841–1857 (2015). https://doi.org/10.1617/s11527-014-0278-7
36. Flatscher, G., Bratulic, K., Schickhofer, G.: Experimental tests on cross-laminated timber joints and walls. Proc. Inst. Civil Eng. Struct. Build. **168**, 868–877 (2015)
37. EN 1998: EN 1998 Eurocode 8: Design of structures for earthquake resistance - Part 1: General rules, seismic actions and rules for buildings (2002)

38. van de Lindt John, W., Omar, A.M., Douglas, R., Philip, L., Shiling, P., Marjan, P.: Seismic performance factors for cross-laminated timber shear wall systems in the United States. J. Struct. Eng. **146**, 04020172 (2020). https://doi.org/10.1061/(ASCE)ST.1943-541X.0002718
39. ATC FEMA P695: Quantification of Building Seismic Performance Factors. Applied Technology Council, US Department of Homeland Security, FEMA (2009)
40. Rinaldi, V., Casagrande, D., Fragiacomo, M.: Verification of the behaviour factors proposed in the second generation of Eurocode 8 for cross-laminated timber buildings. Earthq. Eng. Struct. Dynam. **52**, 910–931 (2023). https://doi.org/10.1002/eqe.3792
41. Vamvatsikos, D., Castiglioni, C., Bakalis, K., Calado, L., D'Aniello, M., Degee, H., Hoffmeister, B., Pinkawa, M., Proenca, J.M., Kanyilmaz, A., Morelli, F., Stratan, A., Vayas, I.: I.11.47: A risk-consistent approach to determine behavior factors for innovative steel lateral load resisting systems. In: ce/papers, vol. 1, pp. 3434–3443 (2017). https://doi.org/10.1002/cepa.398
42. Tesfamariam, S., Skandalos, K., Goda, K., Bezabeh, M.A., Bitsuamlak, G., Popovski, M.: Quantifying the ductility-related force modification factor for 10-Story Timber–RC hybrid building Using FEMA P695 procedure and considering the 2015 NBC seismic hazard. J. Struct. Eng. **147**, 04021052 (2021). https://doi.org/10.1061/(ASCE)ST.1943-541X.0003007
43. Morshedi, E., Doudak, G., Fazileh, F., Fathi-Fazl, R.: Seismic performance assessment of wood light-frame shearwalls using the performance-based unified procedure. J. Struct. Eng. **150**, 04024172 (2024). https://doi.org/10.1061/JSENDH.STENG-13742
44. Canadian Wood Council: Fire Safety Design in Buildings. Canadian Wood Council, Ottawa, ON (1996)
45. Buchanan, A.H., Abu, A.K.: Structural Design for Fire Safety. John Wiley & Sons, Ltd (2017)
46. EN 1990: EN 1990 Eurocode: Basis of Structural Design (2002)
47. Xu, H., Pope, I., Gupta, V., Cadena, J., Carrascal, J., Lange, D., McLaggan, M.S., Mendez, J., Osorio, A., Solarte, A., Soriguer, D., Torero, J.L., Wiesner, F., Zaben, A., Hidalgo, J.P.: Large-scale compartment fires to develop a self-extinction design framework for mass timber—Part 1: Literature review and methodology. Fire Saf. J. **128**, 103523 (2022). https://doi.org/10.1016/j.firesaf.2022.103523
48. Hadden, R.M., Bartlett, A.I., Hidalgo, J.P., Santamaria, S., Wiesner, F., Bisby, L.A., Deeny, S., Lane, B.: Effects of exposed cross laminated timber on compartment fire dynamics. Fire Saf. J. **91**, 480–489 (2017). https://doi.org/10.1016/j.firesaf.2017.03.074
49. CSA: CSA S850: Design and Assessment of Buildings Subjected to Blast Loads (2023)
50. CSA: CSA O86: Engineering Design in Wood (2024)
51. EN 1995-1-2: Eurocode 5: Design of Timber Structures - Part 1-2: General - Structural Fire Design (2004)

Circularity of Taller Timber Buildings

Adaptability in Wood Construction: Barriers to Flexible Multi-Storey Timber Buildings

E. A. K. J. Vandamme[1(✉)], M. Rinke[1], and R. N. Passarelli[2]

[1] University of Antwerp, Antwerp, Belgium
esther.vandamme@uantwerpen.be
[2] Hasselt University, Hasselt, Belgium

Abstract. In an era marked by rapid and constant change, designers are challenged to create solutions that are not only sustainable but also adaptable to an ever-evolving environment. Wood, as a renewable and versatile resource, plays a pivotal role in this context. Also, timber buildings and structures are in a perpetual state of transition, and enabling these transformations is crucial for the sustainable development of our built environment.

This chapter discusses adaptability for multi-storey timber buildings, addressing the current lack of empirical data and lived experience in this area. The research aims to identify the technical barriers to adaptation of multi-storey timber buildings and which research currently addresses them.

The methodology comprises an in-depth literature review on adaptability in timber engineering and architecture and a comparative analysis of adaptability research.

The study highlights the importance of designing timber buildings with adaptability in mind, considering aspects such as acoustics, prefabrication, standardisation, span, fire safety, services and moisture planning. The findings suggest that while timber structures are designed for longevity, their potential for adaptation is often insufficient, emphasizing the need for a more holistic approach to design that includes whole-building scales, while adaptability research does not yet incorporate these timber-specific aspects.

Keywords: Multi-Storey Timber Building · Circular Economy · Wood Construction · Adaptability · Flexibility · Building Adaptation

1 Introduction

Time is a crucial factor when discussing the effectiveness of biobased materials, particularly in timber construction. To match a benchmark forest's carbon, the carbon stored in wood products plus the remaining forest carbon requires over 150 years of timber use [1]. This carbon balance exceeds by far the average lifespan of buildings. Despite being designed for durability, more than half of all buildings are demolished due to vacancy rather than technical deficiencies, indicating insufficient adaptability [2]. Current discussions about new timber buildings often focus on technological aspects, neglecting spatial-architectural and organizational-functional considerations [3]. We can close this

G. Fink et al. (Eds.): *Holistic Design of Taller Timber Buildings*, SPRTRCIENG, pp. 327–334, 2025.
https://doi.org/10.1007/978-3-032-02098-7_25

gap between the carbon balance of wood and a building's obsolescence by applying circular design strategies.

MTBs offer a promising solution by fitting into both the technical and biological cycle of the Circular Economy (CE), defined as an economy that is restorative and regenerative by design [4]. While engineered wood products (EWPs) are often downcycled for energy recovery, maximizing their potential requires integration into the technical cycle through circularity design. Efforts to translate CE concepts into specific frameworks for the building industry are ongoing. Cheshire [5] outlines five design principles for applying circular economics to construction, including building in layers, designing out waste, design for disassembly (DfD), selecting materials and designing for adaptability (DfA) [6].

Although referred to as 'flexibility' in several publications [7], adaptability is defined as the capacity of a building to accommodate the evolving demands of its users and environment effectively, thus maximizing value through life [8]. An adaptable building facilitates adaptation: any work to a building over and above maintenance to change its capacity, function or performance [9]. Scenarios of change include climate change (e.g. extreme weather events or migration of pests), changes in use (e.g. office to housing) and extensions (e.g. top-up construction).

While discussions on circular timber construction often mention design for disassembly [10], experts rate it as the least effective design-based enabler for adaptability [2], as adaptability often requires layout changes and systems updates that transcend the DfD scope. Furthermore, design for circularity on a building level keeps the most carbon in place, as 14% of embodied greenhouse gas emissions can be prevented if a structure is reused as a whole instead of recycling its materials due to the associated mixed demolition waste [11].

Timber is perceived as less adaptable and more complex for alteration than other structural materials [12, 13]. However, the environmental benefits of adaptability may outweigh those of using timber alone, highlighting the need for efficient use of this renewable resource [12]. This chapter explores which physical adaptation barriers within the design process for multi-storey timber buildings contribute to this perception.

2 Methods

To investigate adaptability in timber construction, we performed an extensive literature review using Scopus and Web of Science databases in January 2025. We targeted keywords such as "adaptability," "flexibility," "timber," "wood," "design," "building," and "architecture" to uncover key design principles and their applications in multi-storey timber buildings (MTBs), with a focus on environmental objectives and circular strategies.

We examined recent literature on timber architecture (post-2020) and integrated these insights with practical applications discussed by scholars. By synthesizing and summarizing existing research, this review aimed to identify knowledge gaps and general barriers to adaptation in multi-storey timber buildings. It is important to note that this review aimed to highlight the state-of-the-art application of adaptable design strategies and barriers to adaptation in mass timber construction, rather than provide an exhaustive study.

Employing an inductive approach, we analysed the papers based on identified adaptation barriers. Given the research objectives, thematic analysis was chosen to identify and categorize circular economy (CE) design strategies during the initial stage of the review. This approach allowed us to synthesize key findings from the reviewed literature, identify gaps, and outline future research directions.

3 Timber-Specific Adaptation Barriers

3.1 Acoustics

Acoustic challenges in timber construction significantly impact adaptability. The low mass of timber necessitates multi-layered systems to meet sound insulation requirements, complicating future modifications. These systems typically include double-leaf party walls, acoustic decoupling, false ceilings, and additional mass through screeds or infills. Exposed timber elements spanning multiple units are often unfeasible due to acoustic decoupling needs, and solid timber partition walls must be doubled for sound insulation [14].

Changes in apartment layout require adjustments to floor decoupling cuts, involving substantial structural alterations. Walls separating corridors from non-living spaces like hallways are built thinner than adjacent to living rooms due to different acoustic specifications [15]; leaving little room for change of use. While additional layers of gypsum boards, suspended ceiling systems, fillers, intermediate insulation and decoupling elastomers enhance acoustic performance, they compromise structural clarity and future adaptability. The key challenge lies in balancing immediate cost-effectiveness, acoustic and structural performance with long-term flexibility in timber structures.

3.2 Prefabrication, Connections and Standardisation

Prefabrication in modern timber construction offers efficient assembly but can create challenges for adaptability. The integration of structure, facade, and services in compact modules may hinder selective replacement or upgrading, particularly when services are concealed behind structural linings. Custom-designed solid timber products like cross-laminated timber (CLT) may have limited reuse potential in different contexts. Transport size restrictions can impact room dimensions and ceiling heights, with panelised buildings typically having net ceiling heights between 2.4 and 3.5 m [16], whereas in other building systems this is less a constraint.

Connections must withstand multiple assembly-disassembly cycles while resisting friction, creep, and corrosion [17]. Deconstruction effort and cost depend on element size and weight, connection type and number, damage potential, removal direction, health and safety regulations and location within the building. While hooking brackets allow vertical disassembly, current market limitations in CLT wall-to-floor connections restrict horizontal disassembly, creating friction between modular design intentions and technical feasibility [18]. Disassembly may require heavy equipment like cranes and temporary scaffolding, similar to the construction phase, limiting feasibility.

Prefabrication trends favour rectilinear volumes and regular flat extrusions for industrial efficiency. Standardized modular grids, often based on 625 mm multiples due to the size of wood-based boards, can impose constraints on interior layouts and facade design. Deviations from this geometry increase costs, potentially limiting design freedom and adaptability to complex urban contexts or future volume requirements [19]. While the rectangular grid plan layout offers simplicity, legibility and spatial planning, it may overlook architectural expression and spatial optimisation through its structure.

3.3 Structural Depth and Span

Design for Adaptability (DfA) principles, which favour open-plan, post-and-beam structures with large spans and tall floor-to-floor heights, face challenges in timber construction. Serviceability constraints limit floor and beam spans and larger spans and ceiling heights increase material costs and reduce resource efficiency [12]. The adaptability of timber structures is also shaped by their configuration. Vertically offset floor plans are generally unfavourable, while creating large column-free spaces demands significant effort. Conventional timber elements restrict flexibility because of their unidirectional load-bearing nature. While CLT slabs can span up to 6 m, commercial spans of 9–12 m often require composite timber-concrete or hollow box slabs such as Kerto Ripa (up to 8 m) or Lignatur (up to 10 m). Adapting a building with shorter timber spans may necessitate significant structural interventions [20]. Furthermore, timber systems with spans comparable to concrete often result in greater structural depth, with beams and girders impacting partition wall placement and limiting interior flexibility.

Volumetric aspects further limit adaptability. To achieve similar spaciousness as steel structures, timber buildings often require larger volumes, increasing enclosure areas and energy loss. Height restrictions in timber buildings exacerbate this issue, as additional structural depth reduces ceiling heights—one of the key barriers to adaptability [12, 21]. Addressing these limitations in span, volumetry, and structural rigidity is critical to achieving spatial overcapacity for adaptable multi-storey timber buildings.

3.4 Lateral stability

Timber buildings, due to the lower flexural stiffness of wood, depend on specific lateral force-resisting systems with shear walls, cores, and bracing to manage lateral loads such as winds or earthquakes. These systems are critical for structural integrity and include horizontally planar diaphragms or bracing and vertically shear walls, cores, bracing, or moment-resisting frames. Altering these systems can jeopardize stability, necessitating meticulous analysis and often extensive retrofitting.

While popular in mid-rise timber buildings, shear walls limit internal layout flexibility since they must align vertically across storeys. For instance, solid wood shear walls may restrict adding openings, making them unsuitable for flexible spaces like offices. Conversely, bracing can be architecturally integrated to minimize its impact on open spaces. Moment-resisting connections are hard to achieve, making large rigid frames without braced façades infeasible [14]. Bracing elements such as glulam or steel may conflict with façade permeability, refitability or later additions like balconies. In modular construction, bracing is often integrated into edge modules using steel-wood combinations, complicating future modifications [22].

Central cores are another common solution in multi-storey timber buildings (MTBs), offering structural stability while optimizing floor plans by enhancing circulation and natural light access. However, as core are semi-permanent, their materiality and location govern refitability and scalability.

In seismic regions, the ductility of connections allows for deformation without failure. However, permanent deformation of connections (e.g. unidirectional carpentry joints) hinders disassembly and reuse [17].

To ensure adaptability in timber construction, structural components for lateral stability must be carefully coordinated with architectural design to avoid conflicts with programmatic spaces.

3.5 Fire Safety

Timber's predictable charring rates allow for engineered fire resistance, yet building codes often impose stringent fire-resistance requirements on MTBs that complicate adaptation without extensive structural modifications. Effective fire compartmentation is a key constraint, limiting floor plan areas to avoid reliance on immovable interior firewalls [23]. Solutions like fire shutters at each ceiling level and separate shafts per fire compartment restrict spatial reconfiguration.

To achieve "highly fire-retardant" classifications, floor assemblies incorporating materials such as wet screed or gypsum fibre dry screed offer both safety and design flexibility. However, regulatory frameworks often restrict the use of exposed timber, particularly in escape routes. Encapsulation for fire protection also hinders demountability and impairs the simplicity, accessibility, and legibility of the structure, reducing its adaptability.

Finally, repairability is critical for timber structures damaged by fire or water. Yet, initial designs rarely consider the possibility of modifications to load-bearing elements, increasing the likelihood of full demolition after localized damage [13].

3.6 Services

The integration of building services, such as HVAC systems, fire protection, electrical conduits, and plumbing, poses significant adaptability challenges due to differing requirements for office and residential uses. Service layouts, whether linear or dispersed with multiple shafts and structural perforations, enhance flexibility during use but create friction between structure and services during future modifications [16].

Structural grid and beam orientation also influence service integration. While joisted floors without integrated channels facilitate modifications, they may require increased floor heights. Strategic placement of service openings mid-span minimizes structural impact but constrains refitability. Embedding services within floors can conflict with fire safety and acoustic requirements, limiting future changes unless modular designs or access panels are implemented.

Integrated timber elements further constrain adaptability. Pre-planned openings such as airtight cavity wall sockets are essential for maintaining airtightness and fire safety, with post-construction modifications potentially compromising these properties. Also, in milled recesses of CLT walls, repeated rerouting can compromise wall stability, fire rating, and soundproofing due to the difficulty of properly refilling these cavities.

3.7 Moisture, Airtightness and Thermal Insulation

Moisture, airtightness, and thermal insulation are critical considerations in timber construction due to the hygroscopic nature of wood, which demands robust moisture control strategies to preserve the structural integrity and durability of timber elements. While thermal bridging is less significant in timber structures, addressing moisture concerns requires careful implementation of (moisture-adaptive) vapor barriers, particularly at junctions between structural components. However, integrating the structure behind the vapor barrier or airtightness layer in façades reduces refitability [14].

Timber structures are particularly vulnerable to water damage, necessitating meticulous sealing of leak-prone areas. These areas may change during conversions, further complicating adaptation. Unlike mineral buildings where such issues are rarely structurally critical, maintaining moisture control adds complexity to façade modifications or wet room relocations, thereby impacting adaptability in MTBs.

4 Limitations and Opportunities for Timber-Inclusive Adaptability Frameworks

Several material-specific challenges remain largely unaddressed in the timber literature on adaptability. Current strategies for adaptability are often too broad and fail to consider the technical requirements and their timber-specific solutions. For instance, acoustic measures are only mentioned by Birk (2023), while Design for Manufacturing and Assembly is discussed by Ottenhaus et al. (2023), Öberg, Jockwer, and Goto (2024), and Hasani and Riggio (2025). Fire safety measures are covered by Laboy (2022), Birk (2023), Öberg, Jockwer, and Goto (2024), and Hasani and Riggio (2025). The focus of current adaptability research is primarily on structural depth and span, which is recognized as a barrier in timber literature [14].

Moreover, lateral force resistance systems are not considered a barrier to adaptation in the adaptable timber cases studied by Hasani and Riggio (2025), as these elements have been strategically placed. In the project descriptions of these predominantly low-rise buildings, adaptability is highlighted, suggesting that the placement of elements such as shear walls was a key consideration during the design process. This contrasts with regular timber buildings, which may overlook this aspect. Therefore, the importance of lateral stability measures for regular multi-storey timber buildings (MTBs) remains significant.

Additionally, unlike brick architecture, where uni-layer load-bearing walls can serve multiple functions, planar timber elements rarely exhibit such overcapacity. These gaps highlight the need to reassess the existing adaptability strategies for timber construction. Given the numerous adaptability frameworks available, it is essential to verify the incorporation of timber-specific adaptation barriers and identify the most valid framework to serve as the basis for a timber-inclusive adaptability framework.

5 Conclusion and Future Research

This study reveals a significant gap between existing adaptability criteria and the unique characteristics of timber construction in multi-storey timber buildings (MTBs). Current design strategies, while valuable, often lack aspects specific to timber construction, such as acoustics, fire safety and lateral stability. To address this, future research should include an assessment of the current adaptability frameworks with a case-study analysis of timber building systems.

The study is limited by the accessibility of literature and real world-applications. However, it sets the stage for developing specific design strategies for adaptable MTBs and potentially a material-inclusive adaptability framework. By addressing these priorities, future research can create robust tools that enhance MTB adaptability and contribute to broader sustainability goals, unlocking timber's full circular potential.

References

1. Ximenes, F.D.A., George, B.H., Cowie, A., Williams, J., Kelly, G.: Greenhouse gas balance of native forests in New South Wales, Australia. Forests. **3**(3), 653–683 (2012). https://doi.org/10.3390/f3030653
2. Ross, B.E., Chen, D.A., Conejos, S., Khademi, A.: Enabling adaptable buildings: results of a preliminary expert survey. Procedia Eng. **145**, 420–427 (2016). https://doi.org/10.1016/j.proeng.2016.04.009
3. Vandamme, E. Rinke, M.: Adaptability in multi-storey timber buildings - towards differentiated durability layers in architecture. In: *WCTE 2023*, Oslo, Norway: WCTE 2023, pp. 3634–3643 (2023). https://doi.org/10.52202/069179-0473
4. Ellen MacArthur Foundation: Towards the Circular Economy Vol.1: economic and business rationale for an accelerated transition, p. 1. Ellen MacArthur Foundation (2013) https://ellenmacarthurfoundation.org/towards-the-circular-economy-vol-1-an-economic-and-business-rationale-for-an
5. Cheshire, D.: Building Revolutions: Applying the Circular Economy to the Built Environment. Routledge (2019) https://doi.org/10.4324/9780429346712
6. Campbell, A.: Mass timber in the circular economy: paradigm in practice? Proc. Inst. Civil Eng. Eng. Sustain. **172**(3), 141–152 (2019). https://doi.org/10.1680/jensu.17.00069
7. Estaji, H.: A Review of flexibility and adaptability in housing design. Int. J. Contemp. Archit. The New ARCH. **4**(2), 37–49 (2017). https://doi.org/10.14621/tna.20170204
8. Schmidt III, R., Austin, S.: Adaptable Architecture. Routledge (2016). https://doi.org/10.4324/9781315722931
9. Chudley, R.: The maintenance and adaptation of buildings. Longman, London; New York (1981) http://archive.org/details/maintenanceadapt0000chud. Accessed 7 Mar 2025
10. Abad, F., Rameezdeen, R., Chileshe, N.: Circular economy design strategies in mass timber construction: a systematic literature review. Smart Sustain. Built Environ. (2024). https://doi.org/10.1108/SASBE-05-2024-0183
11. Kröhnert, H., Itten, R., Stucki, M.: Comparing flexible and conventional monolithic building design: life cycle environmental impact and potential for material circulation. Build. Environ. **222**, 109409 (2022). https://doi.org/10.1016/j.buildenv.2022.109409
12. Laboy, M.M.: Reimagining low-carbon futures: architectural and ecological tradeoffs of mass timber for durable buildings. Archit. Struct. Constr. **2**(4), 723–741 (2022). https://doi.org/10.1007/s44150-022-00048-7

13. Öberg, V., Jockwer, R., Goto, Y.: Design for structural adaptation in timber buildings: industry perspectives and implementation roadmap for Sweden and Australia. J. Build. Eng. **98**, 111413 (2024). https://doi.org/10.1016/j.jobe.2024.111413
14. Kaufmann, H., Krötsch, S., Winter, S.: Manual of Multi-Storey Timber Construction. Edition Detail, Munich (2018) https://anet.be/record/opacuantwerpen/c:lvd:14741791
15. Tuure, A., Ilgın, H.E.: Space efficiency in finnish mid-rise timber apartment buildings. Buildings. **13**, 8 (2023). https://doi.org/10.3390/buildings13082094
16. Hasani, N., Riggio, M.: Achieving circular economy through adaptable design: a comparative analysis of literature and practice using mass timber as a case scenario. J. Build. Eng., 111802 (2025). https://doi.org/10.1016/j.jobe.2025.111802
17. Ottenhaus, L.-M., Yan, Z., Brandner, R., Leardini, P., Fink, G., Jockwer, R.: Design for adaptability, disassembly and reuse—a review of reversible timber connection systems. Construct. Build Mater. **400**, 132823 (2023). https://doi.org/10.1016/j.conbuildmat.2023.132823
18. Ljunge, J., Nerhed Silfverhjelm, H.: Reuse of structural CLT elements - assessing the impact of inter-element joint solutions on the reuse potential and environmental impact of a load-bearing wall panel. Master's dissertation, Chalmers University of Technology, Gothenburg (2022)
19. Svatoš-Ražnjević, H., Orozco, L., Menges, A.: Advanced timber construction industry: a review of 350 multi-storey timber projects from 2000–2021. Buildings. **12**(4), 404 (2022). https://doi.org/10.3390/buildings12040404
20. Nesheim, S.Ø.: Competitive Timber Floors: Optimisation of Hollow Section Timber Floor Elements for Adaptable Buildings. Norwegian University of Science and Technology (2021)
21. McFarland, D., Ross, B., Naser, M.Z., Blok, R., Teuffel, P.: Quantitative evaluation of the relationship between physical parameters and building demolition or adaptation outcomes. Archit. Struct. Constr. **3** (2021). https://doi.org/10.1007/s44150-021-00014-9
22. Li, J., Andersen, L.V., Hudert, M.M.: The potential contribution of modular volumetric timber buildings to circular construction: a state-of-the-art review based on literature and 60 case studies. Sustainability. **15**(23), 16203 (2023). https://doi.org/10.3390/su152316203
23. Birk, S.: Convertible timber hybrid for differentiated expansion stages - Wandelbarer Holzhybrid für differenzierte Ausbaustufen. In: Presented at the Süddeutscher Holzbau Kongress SHK 2023 (2023)

An Investigation of Digital Infrastructure's Role in Repair and Reuse of Timber-Based Components in Taller Timber Buildings: An Interdisciplinary Case study of Digital Product Passport

Katja Rodionova[1(✉)] and Veikko Eeva[2]

[1] University of Tampere, Tampere, Finland
katja.rodionova@tuni.fi
[2] Lumoin Oy, Tampere, Finland

Abstract. This report represents an extended abstract of a multidisciplinary study (Rodionova and Eeva, Wood Sci Eng, 2025) exploring the role of digital infrastructure in the assessment, repair, and reuse of timber-based components in multistory timber buildings.

Sustainability in the built environment is closely tied to the service life of material-intensive assets. In multi-story timber buildings, this is challenged by evolving design guidance, inconsistent product declarations, and incomplete documentation of engineering decisions. This study focuses on digital infrastructure to support the assessment, repair, and reuse of structural components in light timber frame (LTF) buildings. Drawing on Finnish guidance RIL 277-2024, it presents a framework that integrates Digital Product Passports (DPPs) DPPs [2], Digital Building Logbooks (DBLs) [3, 4]. Additionally, the DBL facilitates access to information on operational safety, removability, and the monetary value of individual components [5], and Architectural Decision Records (ADRs) [6, 7] for traceable and context-aware documentation.

The framework is evaluated through four subcases based on a representative three-story Finnish LTF building completed in 2016. Reverse-engineered and comparative structural calculations reveal how changes in engineering guidance and product data from 2007 to 2024 affect performance assessments. These subcases, structured as decision points in asset data management, demonstrate the impact of information gaps and variability on structural reliability and reuse potential. The study establishes a foundation for applying the Value of Information (VoI) methodology to digital infrastructure, with ADRs proposed as a cost-effective enhancement to existing DPP workflows.

1 Introduction

"*Extending the lifespan of existing structural assets is a critical challenge for structure owners worldwide*", states the EU structural code for assessment and retrofit of existing assets CEN/TS 17440:2020. Accurate assessment of structural resistance can yield substantial environmental, economic, and socio-political benefits. Unlike the conservative

G. Fink et al. (Eds.): *Holistic Design of Taller Timber Buildings*, SPRTRCIENG, pp. 335–340, 2025.
https://doi.org/10.1007/978-3-032-02098-7_26

approaches used in new structure design, detailed assessment of existing structures can reveal excess capacity through more refined models of structural systems and material properties [8].

The management and processing of information plays a key role in the transition of societal functions towards the development of sustainable economic models [9]. Precision and availability of information on structural components are crucial for enabling reuse of components and leveraging innovative materials in both new construction and the repair of existing structures. However, this necessitates a robust and verifiable data infrastructure that is often lacking, causing instability in data integration and utilization.

Our study addresses the issue by exploring practicalities of data infrastructure in line with emerging EU regulations and existing ISO standards. The complementing integration with financial decision-making can be facilitated by Value of Information (VoI) concept [10–13], applied in e.g. development of the new Eurocode parts focussing on condition assessment and repair design of existing timber structures but also enabling value chain integrations, for instance with forestry management [12].

2 Literature Review Summary

The literature review provides a summary of research, standards, and technical frameworks shaping lifecycle management and reuse of timber structures. It highlights a shift from static documentation and conservative design toward data-driven, context-aware approaches. Key elements of digital infrastructure necessary for supporting lifecycle management and reuse in timber construction are assessed. The review also identifies barriers to adoption. Existing DPP and BIM tools often fail to support dynamic updates or decision traceability, limiting their value for reuse. Timber-specific challenges include inconsistent product data, undocumented engineering logic, and fragmented design guidance. A review of guidance and product declarations for Light Timber Frame (LTF) multistory systems highlights inconsistent practices in wall bracing, load distribution, and anchoring design. Evolving standards and undocumented assumptions create risks for both structural safety and regulatory compliance. Emerging standards such as ESPR[2] and W3C's [14–16] promise to improve data interoperability. Still, adoption is limited by fragmented regulation, lack of shared schemas, lack of formal mechanisms for recording design rationale, and multidisciplinary skill gaps. The literature supports structured, stakeholder-aware digital infrastructure as essential for advancing circular timber construction.

3 Methodology Summary

The full account of the case study includes assessment based on the historical design records of the timber multistory building, completed in Finland in 2016 and representing typical design workflow and solutions used in Finnish light timber framing (LTF) multistory buildings' design through 2010's. The documentation of the case study project is obtained from the municipality archives, which reflects a typical source of information for renovation design. The investigation is concerned with interpreting the information available from the documentation, and professional literature as well as documentation gaps. Selected findings related to material property declarations and calculation results are provided.

The investigation of digital infrastructure is organized around four subcases that address real-life documentation and design issues in timber multistory construction. These subcases represent key lifecycle decision points where changing assumptions or missing documentation can impact safety, cost, and reuse feasibility. Each highlights dependencies between product data, design logic, and structural performance, illustrating how decisions could be better captured through a system combining Digital Product Passports (DPPs), Digital Building Logbooks (DBLs), and Architectural Decision Records (ADRs). The engineering findings inform core requirements for DPP-based systems to support lifecycle design and reuse in multistory timber buildings.

4 Results

The findings of the case study highlight the divergence in evaluation of structural components' design load and strength estimates, obtained from calculations based on assumptions used in design practice and informed in industry guidance at the time and after the completion of the projects. The four illustrative subcases highlighted in the case study are concerned with:

- Product information related to strength and stiffness properties of product combinations
- Definition of horizontal loads and their distribution on the loadbearing wall
- Calculation of design shear resistance and stiffness of the racking wall elements
- Evaluation of tension anchoring demand at the leading edge of the racking wall

With some of the alternative calculations suggesting failure of the building components or their connections, the findings illustrate the role and mechanisms of documented decision-making required to transparently ensure viability of the building through its lifecycle. Apart from demonstrated subcases, data gaps were observed that could hinder full-scale appraisal work. For instance, evaluations related to compression perpendicular to the grain cannot be carried out in absence of element drawings.

5 Discussion

The discussion synthesizes the four structural subcases, highlighting their interdependencies and shared vulnerabilities through the lens of digital infrastructure. Each subcase reveals increasing complexity, progressing from missing product property records (Subcase 1) to combined evaluations of component and system performance (Subcases 2–3), and finally to challenges in aligning with national and international design standards (Subcase 4). The cases are reiterated as requirements towards digital infrastructure, including:

- Product Information: tracking of structural property changes
- Documenting of complex decision-making processes (such as diaphragm assumptions and load distribution)

- Documenting of design rationale, including sensitivity of assumptions
- Managing lifecycle revisions emerging due to innovation and design guidance updates

Key decision points supported by ADR-enabled digital framework are [1]:

- Decision Point (DP) 1: Record changes in component quantities, ensuring traceability and supply-chain accountability when material batches or construction elements are modified.
- Decision Point (DP) 2: Captures variations in component qualities, such as manufacturer updates to material properties, observed performance deviations, or changes in certification requirements.
- Decision Point (DP) 3: Documents re-estimation of structural system parameters, ensuring updated methodologies, new standards, or real-time monitoring data lead to informed reassessments.
- Decision Point (DP) 4: Logs re-estimation of real asset parameters, ensuring that lifecycle performance assessments, financial valuations, and compliance requirements are continuously updated.

At each decision point, we identify key interactions between components of the digital infrastructure. Table 1 summarizes key digital infrastructure components and their functions.

Table 1. Summary and functions of digital solutions [1].

Component	Problem addressed	Functional role in the infrastructure
Ecodesign for Sustainable Products Regulation (ESPR) (DPP)	Need to make reuse enforceable and measurable	Establishes binding requirements for traceability, repairability, and data sharing
Architectural Decision Records (ADRs)	Lack of decision transparency and contextual reasoning	Embed structured decision rationale to support traceability and explainability
Digital Product Passport	Fragmented, unverifiable, siloed data	Regulatory anchor and minimal structured unit for lifecycle data exchange
Digital Building Logbook (DBL)	No persistent memory of lifecycle events across actors	A conceptual that connects decisions, events, and material data over time
Decentralized Identifiers (DIDs) and Verifiable Credentials (VCs)	No trusted, verifiable identity or data provenance in value-chain	Enable secure, auditable, decentralized identity and data access infrastructure

6 Conclusions

This study aimed to clarify the benefits and costs of using advanced digital infrastructure in light timber frame (LTF) multi-story buildings. The goal was to prepare the ground for applying the Value of Information (VoI) approach in a structured and scalable way. The research focused on practical challenges, such as missing documentation, changes in product data, and updates in regulations.

The analysis was based on a case study of a three-story Finnish LTF building, completed in 2016. Four technical subcases were examined, each showing a different source of uncertainty that can affect how a building performs over time. The study further identified, what kind of digital infrastructure is needed to support better decisions over the lifecycle of a building. ADRs were introduced as a way to record the thinking behind design choices in a simple but structured format. This fills a major gap in current documentation systems, which often include the results of decisions but not the reasons behind them.

This supports structured dialogue and scenario-based exploration among stakeholders, improving the basis for applying Value of Information (VoI) analysis to decisions on infrastructure adoption. By extending the guidance of RIL 277-2024, which primarily addresses construction clients, we also offer insights relevant to owners of existing buildings with long-term performance concerns. The proposed framework provides a verifiable, interoperable foundation for lifecycle-informed decisions and invites broader collaboration, particularly among asset owners and regulatory bodies.

References

1. Rodionova, K., Eeva, V.: An investigation of digital infrastructure's role in repair and reuse of timber-based components in multistory timber buildings: an interdisciplinary case study of digital product passport. Wood Sci. Eng. (2025)
2. EU: Ecodesign for Sustainable Products Regulation 2024/1781 (2024)
3. Alonso, R., Olivadese, R., Ibba, A., Reforgiato Recupero, D.: Towards the definition of a European Digital Building Logbook: a survey. Heliyon. **9**(9), e19285 (2023). https://doi.org/10.1016/j.heliyon.2023.e19285
4. Volt, J., et al.: Definition of the Digital Building Logbook (2020). https://doi.org/10.2826/480977
5. EPEA: Healthy buildings - sustainable & economical – EPEA. https://epea.com/en/services/buildings
6. van Heesch, U., Avgeriou, P., Hilliard, R.: A documentation framework for architecture decisions. J. Syst. Softw. **85**(4), 795–820 (2012). https://doi.org/10.1016/j.jss.2011.10.017
7. Buchgeher, G., Schöberl, S., Geist, V., Dorninger, B., Haindl, P., Weinreich, R.: Using Architecture Decision Records in Open Source Projects—An MSR Study on GitHub. IEEE Access. **11**, 63725–63740 (2023). https://doi.org/10.1109/ACCESS.2023.3287654
8. CEN/TS 17440:2020: CEN/TS 17440:2020 Assessment and Retrofitting of Existing Structures, p. 17216 (2020)
9. T. Kauppila et al., Handbook for a Data-Driven Circular Economy in Finland (2022). https://cris.vtt.fi/en/publications/handbook-for-a-data-driven-circular-economy-in-finland-data-sourc

10. Keisler, J.M., Collier, Z.A., Chu, E., Sinatra, N., Linkov, I.: Value of information analysis: the state of application. Environ. Syst. Decis. **34**(1), 3–23 (2014). https://doi.org/10.1007/s10669-013-9439-4
11. Long, L., Döhler, M., Thöns, S.: Determination of structural and damage detection system influencing parameters on the value of information. Struct. Health Monit. **21**(1), 19–36 (2022). https://doi.org/10.1177/1475921719900918
12. Kangas, A.S., Horne, P., Leskinen, P.: Measuring the value of information in multicriteria decisionmaking. For. Sci. **56**(6), 558–566 (2010)
13. Stepinac, M., Rajčić, V., Honfi, D.: Decision analysis and scenarios for the assessment of existing timber structures. In: IABSE Symposium, Guimaraes 2019: Towards a Resilient Built Environment Risk and Asset Management - Report, no. January, pp. 479–486 (2019). https://doi.org/10.2749/guimaraes.2019.0479
14. W3C: W3C DID Services. https://www.w3.org/TR/did-1.0/#services
15. W3C: Decentralized Identifiers. https://www.w3.org/TR/did-1.0/. Accessed 5 Jun 2025
16. W3C: Verifiable Credentials Data Model v2.0. https://www.w3.org/TR/vc-data-model-2.0/. Accessed 5 Jun 2025

Disassembly and Reuse in Tall Timber Buildings: Advancing Circular Construction Practices

José Manuel Cabrero[1(✉)], Lisa-Mareike Ottenhaus[2], Pedro Palma[3], and Shady Attia[4]

[1] University of Navarra, Pamplona, Navarra, Spain
jcabrero@unav.es

[2] The University of Queensland, St Lucia, QLD, Australia
l.ottenhaus@uq.edu.au

[3] Empa – Swiss Federal Laboratories for Materials Science and Technology, Structural Engineering Research Laboratory, Dübendorf, Switzerland
pedro.palma@empa.ch

[4] Sustainable Building Design Lab, Department of UEE, Faculty of Applied Sciences, Université de Liège, Liège, Belgium
shady.attia@uliege.be

Abstract. The construction industry is a major contributor to global CO_2 emissions, necessitating innovative solutions to reduce its environmental footprint. Tall Timber Buildings (TTBs) represent a sustainable alternative to traditional construction methods by leveraging engineered wood products like Cross-Laminated Timber (CLT) and Glued-Laminated Timber (GLT). However, conventional linear design processes limit the sustainability potential of TTBs. This paper explores the role of Design for Disassembly (DfD) in enhancing the reuse and recycling of timber components, aligning with circular economy principles. Key strategies such as modular design, reversible connections, and the adoption of digital tools like Building Information Modeling (BIM) are discussed. Challenges, including regulatory barriers and material degradation, are addressed alongside case studies highlighting successful TTB projects. By embracing DfD principles, the industry can extend material lifecycles, reduce waste, and transition toward a regenerative built environment. This work underscores the importance of collaboration across disciplines to achieve sustainable construction goals.

Keywords: Tall Timber Buildings · Design for Disassembly · Reversible Connections · dismantle · Demountability

1 Introduction

1.1 Relevance of Taller Timber Buildings

The construction industry is increasingly focused on decarbonization, since it is responsible for 40% of global CO_2 emissions [1], with cement production contributing a significant share. This, combined with urban densification trends that demand sustainable housing and infrastructure solutions, has driven the development of Taller Timber Buildings

G. Fink et al. (Eds.): *Holistic Design of Taller Timber Buildings*, SPRTRCIENG, pp. 341–349, 2025.
https://doi.org/10.1007/978-3-032-02098-7_27

(TTB). TTBs leverage the sustainability advantages of timber and the performance of Engineered Wood Products such as Cross-Laminated Timber (CLT), Glued-Laminated Timber (GLT), and Laminated Veneer Lumber (LVL) to meet current engineering and construction requirements.

However, a "linear" design, construction, demolition, and disposal process for TTBs is not sufficient to mitigate the current environmental impacts of the built environment. Instead, adopting approaches based on circular economy concepts offers an opportunity to reduce the demand for raw materials and further decarbonize the built environment by extending the lifespan of material resources through reuse and recycling [2]. By embracing these principles, timber buildings can shift from the traditional "take-make-dispose" model to a "reuse-recycle-retain value" framework (Fig. 1).

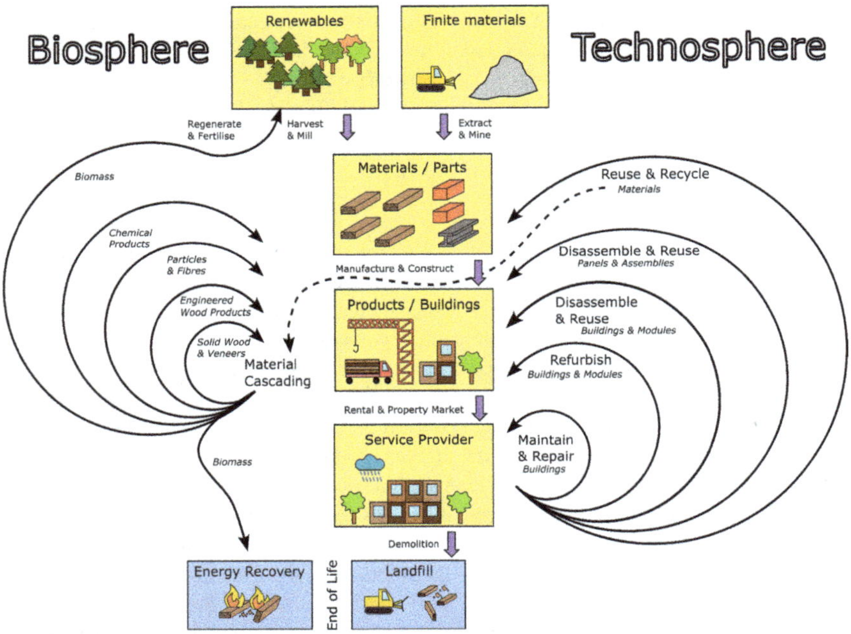

Fig. 1. Product and material streams of circular timber buildings [3].

1.2 Design for Disassembly in Taller Timber Buildings

In developing circular concepts and designs for TTBs, Design for Disassembly (DfD) is essential not only to maximize the reuse potential of construction components but also to enhance the adaptability and repairability of buildings, thereby extending their service life [4, 5]. DfD involves designing buildings, or parts of buildings, to be easily replaced or dismantled and reassembled, aiming at components being reclaimed with minimal processing.

Key principles guiding DfD include: i) Modularity – the building is designed so that certain parts or components can be replaced without compromising the integrity of the whole structure (this requires a clear separation between the primary load-carrying structure and other components); ii) Demountability of connections – using standardized parts and reducing complexity to allow easy separation without damaging components, thus enabling reuse.

It is important to acknowledge Brand's concept of a building as "shearing layers of change" [6], which recognizes that building components have different lifespans. Thus, modification and adaptation of the different building parts to new functions and requirements must be simplified. Moreover, DfD aligns with the concept of "buildings as material banks," where materials are viewed as valuable resources rather than waste [5].

1.3 Tools and Technologies Enabling DfD

The adoption of digital tools such as Building Information Modeling (BIM) enhances DfD by providing accurate documentation of components, connections, and their assembly sequence. Material passports and digital twins further facilitate the tracking of components, ensuring their reuse potential is preserved [7]. Prefabrication plays a pivotal role in DfD by allowing components to be manufactured offsite in optimal conditions and with minimal waste and then assembled onsite. If done correctly, this may later allow for a clean removal of those components. However, the prefabricated components themselves might not be optimized for disassembly, like a prefabricated shear wall or floor segments with nailed or stapled wood-based sheathing.

2 Rationale for DfD in Tall Timber Buildings

2.1 Advantages of DfD

Buchanan [8] identified the major challenges in TTBs as fire, wind, earthquakes and moisture; challenges which also exist for low- and mid-rise mass-timber buildings that become 'more severe as the buildings get taller' [8]. Successful 'holistic' design for all requirements (serviceability and load-carrying capacity, durability, acoustics, etc.) requires significant collaboration across teams and disciplines [9]. As timber buildings get taller, the consequence of failure increases and retrofit and repair become more costly. Provision of insurance depends on the risk profile of buildings to estimate the potential for loss and on the ease and cost of repairs [10]. Tall timber structures, such as multi-story timber temples and pagodas, have a long history in Asian vernacular architecture [11]. Their existence through time is in part owed to regular maintenance and repair, including 'repair by disassembly' (解体修理 *kaitai shūri* in Japanese), with evidence that some temples were (partially) disassembled and repaired as early as 1596 [12]. It thus seems obvious to design modern timber structures such that they can be regularly inspected, maintained and repaired by selectively replacing components. In addition to longevity, DfD provides several more advantages at the end of the life of a timber building:

In terms of environmental benefits, DfD enables the reuse and recycling of materials, which minimizes waste generation and conserves natural resources [13]. In the timber industry, where reliance on forests is significant, DfD ensures harvested wood is utilized efficiently across multiple lifecycles [14].

DfD also offers economic advantages by fostering a secondary market for reclaimed timber components, creating opportunities for material resale. Additionally, integrating DfD principles enhances the financial viability of timber buildings by reducing long-term lifecycle costs, such as through design for structural adaptation [15].

2.2 Challenges of DfD

Several challenges must be addressed for the effective implementation of DfD in TTBs. Structurally, TTBs must meet load-carrying, serviceability, fire safety, durability, and other standard requirements while also accommodating the additional demands of demountability and, if applicable, reusability. Achieving disassemblable connections that remain functional over multiple cycles requires initial overdesign, and components must retain their dimensional stability and structural integrity for reuse [4, 16, 17]. Consumer perception of reclaimed timber can affect its uptake, while on the policy front, DfD faces barriers due to the lack of regulatory frameworks that promote circular construction [18]. Policies that incentivize sustainable design and material reuse are essential, as highlighted by the inclusion of reuse maximization in the new EU Construction Products Regulation [19]. Addressing these technological, social, and regulatory challenges is critical for advancing DfD in the timber industry.

3 Types of Connections

From a design perspective, reversible connections are a key component in developing a successful strategy for DfD. Reversible connections, defined as those that combine ease of disassembly with the reuse potential of both the connection and the joined members, are relevant in this context. Ottenhaus et al. [4] provide a detailed description of the several existing types of structural timber connections (i.e., carpentry, with mechanical fastener brackets, proprietary systems, glued, etc.) and their potential for DfD.

Historic examples show the potential of carpentry connections [13]. However, shrinkage, swelling and visco-elastic deformation in timber can pose challenges for disassembly, particularly in parts produced using CNC machinery and very low tolerances. Connections with common dowel-type fasteners like nails and screws exhibit limited potential for reuse. Nails are difficult and labor-intensive to remove, while screws are sometimes impossible to remove, either because of rupture or because their head breaks when attempting to remove them [20]. Yet, experiments have shown that new screws inserted in holes from previous screws loaded in shear in the elastic range exhibited the same stiffness and load-carrying capacity [21]. Connections with dowels and slotted-in steel plates have been used in reusable sports buildings in Switzerland due to the ease of assembly and disassembly of such connections [22], however moisture fluctuations and shrinkage, as well as overloading, can affect their reversibility [17]. Proprietary brackets have been developed as "plug-and-play" assemblies and may offer reversibility in the elastic range. However, it must be considered how moisture variations, creep, and friction may affect or even hinder disassembly [16].

4 Tall Timber Building Case Studies

The global adoption of TTB underscores the potential of wood products as potentially more sustainable alternatives to traditional construction materials, particularly concrete. Recent projects, such as HAUT in Amsterdam and Mjøstårnet in Norway (Table 1), exemplify the viability of TTBs by demonstrating their ability to compete with steel and concrete in terms of performance and architectural design.

Table 1. Tall timber buildings inventory based on their year of completion.

Building name	Location	Height, Stories	Key features	Disassembly potential
Treet	Bergen, Norway	49 m, 14, 2015	Prefabricated modules, entirely timber structure	High - Prefabricated modules allowing full disassembly
Brock Commons Tallwood	Vancouver, Canada	53 m, 18, 2017	Hybrid structure with CLT panels & concrete cores	Low - Hybrid structure with concrete cores
25 King	Brisbane, Australia	45 m, 10, 2018	Australia's tallest timber building, featuring CLT and glulam	High - Modular prefab. Construction CLT & glulam
Mjøstårnet	Brumunddal, Norway	85 m, 18, 2019	Tallest timber building at completion, glulam and CLT construction	Medium – Prefab. components w. limited modularity
HoHo Wien	Vienna, Austria	84 m, 24, 2019	Mixed-use, hybrid timber-concrete design	Medium - Hybrid design with some prefab. Components
Sara Kulturhus	Skellefteå, Sweden	75 m, 20, 2021	Multi-functional building, CLT structure	High - Modular CLT components designed for sustainability
HAUT	Amsterdam, NL	73 m, 21, 2022	Hybrid construction with timber and concrete, focus on sustainability	High - Modular components with a hybrid design
25 King	Brisbane, Australia	45 m, 10, 2018	Australia's tallest timber building, featuring CLT and glulam	High - Modular prefab. Construction with CLT and glulam

5 Learnt Lessons and Challenges in Disassembly

As building heights exceed 30 m (approximately 8–10 stories), significant structural challenges arise [23]. At this scale, timber-only construction often becomes impractical, necessitating hybrid systems with steel or concrete to ensure structural stability [5]. These materials are typically used in critical components like reinforced cores and connections to enhance resistance to wind, seismic forces, and vertical loads [24]. However, the trade-off between connection efficiency and demountability complicates disassembly.

As timber buildings grow taller, integrating DfD principles becomes more complex. DfD emphasizes adaptability, modularity, and material reuse, aligning with circular construction principles. Yet, hybridization, e.g. with concrete casted on site, may reduce disassembly potential and limit circularity. Addressing these trade-offs requires innovative approaches to balance structural demands and sustainability, ensuring timber remains a viable low-carbon alternative in high-rise construction.

6 Learnt Lessons and Challenges in Reuse

A key challenge limiting the structural reuse of timber members has been the absence of grading standards for the recertification of salvaged timber [25]. The recent introduction of FWPA Standard G01 [26] and prNS 3691 [27] represents a significant step toward enabling the reuse of recycled timber. However, these standards do not address the (re)certification of engineered wood products, which constitute the majority of structural timber in TTBs. Additionally, grading salvaged timber often results in significant downgrading due to unknown load histories, undermining the financial viability of reuse [18]. Demolition remains more economically attractive than deconstruction due to high labour costs and limited residual timber value. Recent research into reversible timber connections shows promise for facilitating disassembly and reuse [4], but challenges persist with both proprietary and bespoke systems, particularly regarding ease of disassembly [16, 17]:

- Non-Reversible Deformations: Wood fibres may swell, shrink or deform during the service life of the building, making it difficult to demount connections [16]. Even when connections are demountable, residual deformations may reduce the fit and performance of reused connections. Controlled pre-compression during initial assembly, as well as innovative designs that provide sufficient tolerances, can help mitigate these effects.
- Corrosion of Metal Components: Metal fasteners may degrade over time, complicating disassembly and reuse. Advances in corrosion-resistant alloys and surface treatments are helping address this challenge.
- Assessment of Structural Integrity: Components must be inspected and regraded to ensure their suitability for reuse, particularly in load-bearing applications. Non-destructive testing methods, such as ultrasound and X-ray imaging, are increasingly used to assess the condition of timber components and timber joints.
- Wear and Fatigue: Repeated cycles of assembly and disassembly can lead to fatigue in connectors and increased tolerances, particularly in high-stress applications. Solutions like sacrificial layers or replaceable connector components are emerging to address these issues.

7 Conclusion

The disassembly and reuse of timber connections are central to achieving sustainable construction practices, particularly in Tall Timber Buildings (TTBs). By integrating Design for Disassembly (DfD) principles, the industry can extend the lifecycle of timber components, reduce waste, and enhance resource efficiency.

While technological advancements have improved the feasibility of DfD, challenges remain in addressing material degradation, connection durability, and regulatory barriers. Collaborative efforts between designers, engineers, policymakers, and manufacturers are essential to overcoming these obstacles and promoting circular construction practices.

Timber buildings present unique challenges in regard to disassembly, in comparison to other structural materials (i.e. steel). Furthermore, hybrid structures can complicate disassembly workflows and should be carefully addressed in the design process; conversely, the use of precast concrete panels and beams may offer advantages by facilitating DfD.

The future of TTBs lies in balancing performance, sustainability, and adaptability. By prioritizing disassembly and reuse, the timber industry can lead the transition toward a more resilient and regenerative built environment.

References

1. Androsevic, R., Brocato, M., Bijedic, D., Rustempasic, N.: Life cycle assessment as a tool for low waste building systems. Case study building external wall. IOP Conf. Ser. Mater. Sci. Eng. **603**(4), 042067 (2019)
2. Zhu, S., Feng, H.: Enhancing circularity of wood waste through deconstruction in building sector. J. Clean. Prod., 144382 (2024)
3. Ottenhaus, L.-M.: Butterfly Diagram of Circular Buildings. figshare (2022). https://doi.org/10.6084/m9.figshare.21249573.v1
4. Ottenhaus, L.M., Yan, Z., Brandner, R., Leardini, P., Fink, G., Jockwer, R.: Design for adaptability, disassembly and reuse – a review of reversible timber connection systems. Construct. Build Mater. **400**, 132823 (2023)
5. Attia, S., Al-Obaidy, M., Mori, M., Campain, C., Giannasi, E., van Vliet, M., Gasparri, E.: Disassembly calculation criteria and methods for circular construction. Autom. Constr. **165**, 105521 (2024)
6. Brand, S.: How buildings learn: what happens after they're built. Penguin Books (1994)
7. Byers, B.S., Cheriyamulla, S., Ewason, J., Hall, D.M., De Wolf, C.: Using engraved QR codes to connect building components to materials passports for circular construction. In: EC3 Conference, vol. 3, (2022)
8. Buchanan, A.H.: The challenges for designers of tall timber buildings. In: Proc. WCTE 2016 - World Conference on Timber Engineering (2016)
9. Fink, G., et al.: Holistic Design of Taller Timber Buildings - Cost Action Helen (Ca20139). In: Proc. COST Action HELEN, pp. 1001–1008 (2023). https://doi.org/10.52202/069179-0137
10. P. Callow, J. Glocking, Mass Timber Risk Consulting, Mass Timber Insurance Playbook, Alliance for Sustainable Building Products, London, UK, 2023
11. Wu, Y., Song, X., Luo, L.: Seismic energy analysis of a multi-story traditional timber pagoda. In: Proc. WCTE 2018—World Conference on Timber Engineering (2018)

12. Fukuda, M.: Repair by Disassembly (Jap. Kaitai Shūri) in Japan. In: Authenticity in Architectural Heritage Conservation, pp. 247–260 (2017)
13. Crowther, P.: Design for Disassembly, vol. 6, pp. 1–2. Royal Australian Institute of Architects (1999). https://doi.org/10.4324/9781315270326-56
14. Walker, J., Norman, J.: Designing timber buildings for longevity. Struct. Eng. **99**(11), 56–59 (2021)
15. Öberg, V., Jockwer, R., Goto, Y.: Design for Structural Adaptation in timber buildings: Industry perspectives and implementation roadmap for Sweden and Australia. J. Build. Eng. **98** (2024). https://doi.org/10.1016/j.jobe.2024.111413
16. Yan, Z., Ottenhaus, L.M., Leardini, P., Jockwer, R.: Performance of reversible timber connections in Australian light timber framed panelised construction. J. Build. Eng. **61**, 105244 (2022). https://doi.org/10.1016/j.jobe.2022.105244
17. Kontra, S., et al.: Modular deconstruction and reuse: a mass timber shake-table test building case study. In: Proceedings of World Conference on Earthquake Engineering (2024)
18. Brandner, R., Ottenhaus, L.-M.: Properties of salvaged structural timber components: how to account for long-term loading effects given unknown load histories? In: COST Action CA 20139, State-of-the-art report (2022)
19. European Union: New Regulation on Construction Products, Brussels, Belgium (2024). https://www.europarl.europa.eu/doceo/document/TA-9-2024-0188_EN.pdf
20. Graf, J., Poteschkin, V., Viezens, V.: Wandelbarer holzhybrid für differenzierte ausbaustufen. In: TU-Munich, Munich, Germany (2022) https://shorturl.at/mcy9i
21. Bompa, D.V., Ungureanu, V., Elghazouli, A.Y., Afsal, A.: Disassembly and structural reuse potential of steel-timber shear connections with screws. In: International Conference Coordinating Engineering for Sustainability and Resilience, pp. 231–242. Springer, Berlin (2024)
22. Prix Lignum: Wiederverwendbare Sportbauten (2024). https://prixlignum.ch/de/279/projektarchiv/projektarchiv-prixlignum/2889/wiederverwendbare-sportbauten/. Accessed 25 Jan 2025
23. van den Oever, M., et al.: Biobased grondstoffen voor hoogbouw: Geïndustrialiseerde modulaire en lage emissie hoogbouw in de G4. Wageningen University and Research (2024)
24. Ahn, N., et al.: Envisioning mass timber buildings for circularity: Life cycle assessment of a mass timber building with different end-of-life (EOL) and post-EOL options. In: Proc. WCTE (2023)
25. Crews, K., MacKenzie, C.: Development of grading rules for recycled timber used in structural applications. In: Proc. 10th World Conf. on Timber Engineering, vol. 1, pp. 231–238 (2008)
26. Forest & Wood Products Australia, FWPA Standard G01: Recycled Timber – Visually Stress Graded Recycled Timber for Structural Purposes, Draft (2024). https://shorturl.at/JTBEm
27. Standard Norge, prNS 3691: Evaluation of Recycled Wood (2024)

Evaluating the Disassembly Potential of Timber Buildings: Development of Calculation Tool and Proof of Concept

Lisa-Mareike Ottenhaus[1](✉), Marta Hernández-Alda[2], Anna Davies[3], and José Manuel Cabrero[4]

[1] Advanced Engineering Building, The University of Queensland, St Lucia, QLD, Australia
l.ottenhaus@uq.edu.au
[2] University of Navarra, Pamplona, Navarra, Spain
mhernandeza.1@alumni.unav.es
[3] The University of Queensland, St Lucia, QLD, Australia
[4] University of Navarra, Pamplona, Navarra, Spain
jcabrero@unav.es

Abstract. The traditional linear model in construction contributes significantly to waste, with over a third of all waste in the European Union stemming from this sector. Design for Disassembly (DfD) offers a path towards a circular construction model by facilitating the reuse of building components. This study specifically develops and validates a novel framework and an accompanying calculation tool to assess the disassembly potential of timber buildings, with a particular emphasis on their connections.

Keywords: Design for Disassembly · Disassembly Potential · Timber connections

1 Introduction

The construction industry's linear "take-make-dispose" approach is unsustainable, leading to vast amounts of waste as buildings often become obsolete before their stipulated 50-year service life [1]. Design for Disassembly (DfD) is emerging as a critical strategy to foster more sustainable and circular construction practices, enabling the sequential or parallel dismantling of buildings for component salvage and reuse [2]. Timber buildings, lauded for their environmental benefits like carbon storage, can further enhance their sustainability profile through DfD. By applying DfD principles the lifespan of timber components through reuse and cascading can be extended, thereby retaining value and maximizing their environmental advantages [3].

This is an extended abstract of a paper published in the Special Issue of the COST Action HELEN in the journal *Wood Material Science & Engineering*. The full open-access article is available at: https://doi.org/10.1080/17480272.2025.2510562

G. Fink et al. (Eds.): *Holistic Design of Taller Timber Buildings*, SPRTRCIENG, pp. 350–354, 2025.
https://doi.org/10.1007/978-3-032-02098-7_28

Existing tools for quantifying disassembly potential often lack clarity in criteria weighting. This research aimed to bridge this gap by developing a practical calculation tool grounded in a multi-stakeholder perspective, incorporating industry survey results and user feedback to define and weight key disassembly criteria.

2 Review of Existing Frameworks and Development of a New Tool

Several frameworks exist for assessing disassembly, including the Disassembly Effort Index (DEI) by Das et al. [4] focusing on product disassembly costs, Durmisevic's [5] model for building adaptability, Akanbi et al.'s [6] BIM-integrated D-DAS system, Pozzi's [7] framework for timber connections, Van Vliet et al.'s [8] method considering reuse, adaptability, and maintenance, and Laasonen and Pajunen's [9] criteria for timber joints.

Building upon these, the current research developed a new tool by:

- Identifying empirically quantifiable criteria for timber building joints.
- Incorporating industry professional perspectives through a stakeholder survey (53 responses) and practical user feedback from the disassembly of a 10-storey mass-timber building.
- Calibrating weightings for these criteria to compute an overall disassembly potential score.

2.1 Key Disassembly Criteria and Weightings

The developed tool evaluates connections based on several key criteria, scored from 0 (worst) to 5 (best). Considered criteria and a brief description of them are given in Table 1. The tool also includes criteria for the building as a whole, which are described in Table 2.

Weightings for these criteria were informed by stakeholder surveys and user feedback, as shown in Table 1. For example, the survey highlighted 'Access' and 'Hazard' as highly important across different geographical regions. Feedback from real-world disassembly emphasized 'Hazard' and 'Connection Type', and also pointed to the need for better calibration of 'Disassembly Time'. The tool allows users to adjust weightings to suit specific project contexts.

3 Application and Benchmarking

The tool's utility was demonstrated by applying it to different case studies, such as a five-storey rental apartment building constructed with glulam beams, steel columns, and CLT walls/floors. The most common beam-to-column connection, utilizing a flitch steel plate with dowels to join timber beams to a steel column, scored well due to its mechanical nature, moderate fastener count, and high standardization within the building. The overall disassembly potential for the case study building was computed as high, mostly due to the high number of similar connections used, resulting in a high standardization.

Table 1. Connection related disassembly criteria.

Criterion	Description (Reflecting worst to best conditions)	Priority
Hazard	Degree of danger posed during disassembly. Ranges from negligible risk (standard PPE) to life-threatening conditions requiring complex safeguards	Critical
Accessibility	Ease of reaching the connection during disassembly. Full visibility and reachability are ideal	Critical
Disassembly damage	Inevitable damage introduced during disassembly	Critical
Connection Type	Reversibility of the joining system, from irreversible (glued) to fully reversible (bolted)	High
Disassembly Time	Estimated time required to fully disassemble the joint. Shorter, predictable durations are favorable	High
Labour Required	Number of workers and auxiliary means required	High
Tools Required	Complexity of tools needed. Ranges from basic hand tools to demolition equipment	Medium
Number of Fasteners	Total number of fasteners required to be removed. Fewer fasteners reduce effort and time	Medium
Number of connected Elements	Number of structural parts joined at a node. Simpler joints with fewer elements score higher	Medium
Degrees of Freedom (DoF)	Directions available to extract the component. Greater freedom eases disassembly	Low
Training Required	Skill level needed for disassembly. From specialist knowledge to general-site operability	Low
Environmental exposure	Exposure to moisture and related timber swelling and fastener corrosion which can affect disassembly	Low

Table 2. Building structure related disassembly criteria.

Criterion	Description (Reflecting worst to best conditions)
Disassembly Plan	Reflects whether the structure is designed with disassembly in mind. Comprehensive planning scores highest
Prefabrication Level	Degree of prefabrication. Fully prefabricated assemblies facilitate modular disassembly
Standardisation	Uniformity of connection types across the project. High standardisation simplifies labor, tooling, and sequencing

4 Conclusion

The developed tool (which can be downloaded at https://espace.library.uq.edu.au/view/UQ:2312b0b) offers an intuitive method for assessing the disassembly potential of timber buildings. However, certain limitations are acknowledged. Analyzing connections within the full building context is crucial, as non-structural layers significantly impact criteria like 'Access' and 'Tools Required'. Criteria such as 'Disassembly Time' and 'Labour' carry uncertainty without real-life data, as site constraints can greatly influence scores. Furthermore, theoretical reversibility can differ from practical disassembly due to issues like tight tolerances.

The tool's primary strength lies in its application during the early design stages, enabling designers to optimize connection details for disassembly before construction. While it can provide insights for existing buildings, its prospective use is key to advancing sustainable practices. The flexibility in adjusting criteria weightings allows for adaptation to regional practices and project scales, although capturing all variations remains a challenge.

Further feedback on the tool is encouraged and can be provided in an online survey (https://forms.gle/RjKVfLeggjCQoNVR7).

References

1. Huuhka, S., Lahdensivu, J.: Statistical and geographical study on demolished buildings. Build. Res. Inf. **44**(1), 73–96 (2016). https://doi.org/10.1080/09613218.2014.980101
2. Walsh, S., Shotton, E.: Adding value to timber components through consideration of demolition and disassembly. In: WCTE 2021 - World Conference on Timber Engineering (2021)
3. Ottenhaus, L.M., Yan, Z., Brandner, R., Leardini, P., Fink, G., Jockwer, R.: Design for adaptability, disassembly and reuse—a review of reversible timber connection systems. Construct. Build Mater. **400**(July), 132823 (2023). https://doi.org/10.1016/j.conbuildmat.2023.132823
4. Das, S.K., Yedlarajiah, P., Narendra, R.: An approach for estimating the end-of-life product disassembly effort and cost. Int. J. Prod. Res. (2000)
5. Elma, D.: Transformable building structures. Design for Dissassembly as a Way to Introduce Sustainable Engineering to Building Design & Construction. PhD Thesis. TU Delft (2006)
6. Akanbi, L.A., Oyedele, L.O., Omoteso, K., Bilal, M., Akinade, O.O., Ajayi, A.O., Davila Delgado, J.M., Owolabi, H.A.: Disassembly and deconstruction analytics system (D-DAS) for construction in a circular economy. J. Clean. Prod. **223**, 386–396 (2019). https://doi.org/10.1016/j.jclepro.2019.03.172

7. Pozzi, L.E.: Design for disassembly with structural timber connections. Analysis and assessment of different connection systems used in engineered timber building structures to maximise demountability and reuse of the elements. TU Delft (2019)
8. Van Vliet, M., van Grinsven, J., Teunizen, J.: Circular Buildings Disassembly Potential Measurement Method, pp. 1–32. Dutch Green Buildings Council (2021)
9. Laasonen, S., Pajunen, S.: Assessment of load-bearing timber elements for the design for disassembly. Buildings. **13**, 1878 (2023). https://doi.org/10.3390/buildings13071878

Design with Reclaimed Components in Timber Construction

Kristina Kröll[1(✉)], Alec Singh[2], Aída Santana Sosa[3], Esther Vandamme[4], and Felipe Riola-Parada[5]

[1] Architectural Design and Engineering, Eindhoven University of Technology, Eindhoven, Netherlands
k.v.kroell@tue.nl

[2] Faculty of Architecture and Civil Engineering, University of Wuppertal, Wuppertal, Germany

[3] Department for Building and Design, University of Applied Sciences Campus Wien, Vienna, Austria

[4] Faculty of Design Sciences, University of Antwerp, Antwerp, Belgium

[5] Future-Oriented Construction: Mechanics/Timber Construction/Sustainability, Bremen City University of Applied Sciences, Bremen, Germany

Abstract. In the context of a circular economy, material reuse and design with reclaimed components are increasingly viewed as essential strategies for reducing waste and extending product life cycles. Circular economy principles aim to minimize waste through the direct reuse of products, components, or materials, thereby prolonging their life before final recycling is required. This approach is particularly significant in the construction industry, one of the largest consumers of raw materials and producers of waste worldwide. Timber construction, with its high cascade utilization potential, offers significant opportunities for reuse. However, despite the advantages of reclaiming timber for new uses, traditional downcycling practices, such as converting old timber into wood-based products or using it for energy production, remain prevalent. This chapter explores the challenges and strategies associated with timber reuse, focusing on its role in circular design. It outlines the key economic, environmental, technological, regulatory, and organizational challenges when implementing timber reuse, followed by strategies for overcoming these obstacles through technological, geometric, and organizational approaches. Practical examples of timber reuse are highlighted through three case studies: the structural reuse of glulam at Brussels' Recypark, the temporary office building for the Austrian Parliament, and the PopUp Dorms in Vienna. The chapter concludes by identifying potential future research areas and emphasizing the continued evolution of timber reuse within the construction sector.

Keywords: Circular Economy · Timber Reuse · Reclaimed Components · Technological Strategies · Geometric Optimization · Organizational Strategies

G. Fink et al. (Eds.): *Holistic Design of Taller Timber Buildings*, SPRTRCIENG, pp. 355–371, 2025.
https://doi.org/10.1007/978-3-032-02098-7_29

1 Introduction

1.1 Circular Design and Reuse Principles

Strategies for material reuse and design with reclaimed components are changing from being something desirable to a priority and necessity in the frame of a circular economy. Circular economy is a model of production and consumption with the objective of extending as much as possible the life cycle of products and reduce waste to a minimum. Direct reuse of products, or at least their components or materials, is one of its key strategies, allowing extended life cycles before a final recycling is needed [1].

The design with reclaimed components can be seen as the upper link of cascade utilization concepts and therefore should be prioritized. Direct reuse allows extended life cycles without loss of quality and thus prevents premature downcycling.

The benefits of reuse strategies are double: on one side waste itself, its treatment, disposal and/or recycling costs are avoided. On the other side, the quantity of primary raw materials needed for new uses is diminished. Both sides of the equation, the existence of waste itself and the pressure put on primary resources, are responsible of present and very relevant environmental problems and direct or indirect costs.

The construction industry heavily relies on raw materials. At the same time, the demolition of buildings—whether partial or complete—creates a significant amount of waste worldwide. Accordingly, the implementation of circular economy principles in the construction sector is a priority for addressing both kind of problems [2].

1.2 Specifics of Timber Construction in Circular Design

In traditional timber construction, the fact that timber elements were usually mechanically joined and could be disassembled, or the material itself could be easily cut out from an assembly, allowed an easy reclamation of timber elements for new uses. Rebuilding with old components was a common strategy when building timber frames [3].

Nowadays timber as a construction material has a high cascade utilization potential. Sawn timber or glulam products can be used in its original form, be recovered and used as a source for producing derived timber products and these can be recycled again or burnt for energy gaining purposes at their end of life.

Surprisingly, this positive and intuitive cascade utilization potential can be seen also a challenge: using old timber components for producing new wood-based boards or burning old timber components for producing energy, both examples of downcycling, are so established that comparatively too few efforts are made for reclaiming timber components and reuse them in their original form without loss of quality.

The existing of easy and immediate downcycling options has prevented a wider settling of possible design alternatives with reclaimed components which should be prioritized with the final objective of extending the life cycle and the function of timber as carbon storage as much as possible.

These facts explain that the "burning timber" option as end-of-life scenario is taken nowadays as standard and, according to EN 15804 Sustainability of construction works [4], compulsory to declare for construction products that contain biogenic carbon. Considering this, the following chapter addresses the challenges and strategies related to the reuse of timber components.

The chapter begins by outlining the key challenges of timber reuse, categorized into economic, environmental, technological, regulatory and organizational aspects. Following this, various strategies to promote timber reuse are presented, divided into three main categories: technological, geometric, and organizational approaches. To illustrate the practical implementation of these strategies, three case studies are examined: the structural reuse of glulam at Brussels' Recypark, the temporary office building for the Austrian Parliament, and the PopUp Dorms in Vienna. The chapter concludes with an outlook on potential future research areas, emphasizing the ongoing development of timber reuse in construction.

2 Challenges with Reusing Timber

Reusing timber elements for structural purposes is an innovative and sustainable practice that offers significant potential for reducing environmental impacts in the construction industry. However, its implementation comes with a variety of challenges, which can be grouped into economic, environmental, technological, and regulatory categories. The following section will provide an overview of the key challenges associated with each of these aspects.

2.1 Economic Challenges

The reuse of timber often complicates the design and construction process, potentially increasing overall project costs despite savings on raw material expenses [5]. The process of reclaiming timber, classifying and properly storing it are sources of costs and, at least in specific markets, like traditional post-and-beam construction, prices of reclaimed components can be several times more expensive than new ones. It can be argued that comparable kind of costs (sourcing, classifying and storage) are also present in the case of new timber components. Therefore, the difference in final prices suggests that the main drivers are the standardization, scale and continuity of supply and not the processes themselves.

Additional costs arise from the need to adapt reclaimed timber to meet structural requirements. Moreover, the market demand for reused timber is predominantly driven by regulatory and policy mandates rather than business incentives, which suggests a pressing need for economic models that enhance the financial attractiveness of reuse practices [6].

2.2 Environmental Challenges

From an environmental standpoint, reusing timber significantly reduces the ecological footprint of construction by extending the life cycle of materials. However, the variability in the quality and size of reclaimed timber can limit its effective reuse, challenging its sustainability potential [7, 8]. Furthermore, while the practice helps delay the release of stored carbon, the environmental benefits depend on optimizing the reuse process to ensure maximum efficiency and impact [9].

Chemically treated timber components can be seen also as an environmental challenge and their reuse is not allowed and desirable due to health reasons. Processes for detecting if timber was originally treated with toxic products are nowadays already possible but their use is not widespread and practical at deconstruction sites [10]. The result is that, if suspicion about chemical treatments appears, those timber components are usually burnt and not even tested for toxicity and evaluating their reuse-potential. The presence of old metal components (nails or screws) is usually seen as a similar barrier, although their detection and removal would be comparatively easier [11].

2.3 Technological Challenges

Technological barriers are among the most significant hurdles in reusing timber for structural purposes. The variability in the quality of reclaimed elements poses difficulties in ensuring structural integrity. These challenges often necessitate the use of oversized components or additional reinforcements, which can diminish economic feasibility [12]. A further complication is the lack of standardized criteria for assessing the structural characteristics of reused timber, making its application inconsistent and complex [6, 8]. The irregular sizes and shapes of reclaimed timber elements require innovative design strategies and optimization algorithms to minimize waste while maximizing usability [7, 13].

2.4 Regulatory Challenges

Regulatory frameworks further complicate the reuse of timber in construction. Current visual grading standards fail to account for the previous life cycles of timber, leading to inefficiencies and missed opportunities for effective reuse [8]. To establish reuse as a common practice in the building industry, new regulatory approaches and industry collaborations are necessary. These would facilitate the development of clear guidelines and best practices tailored to the characteristics of reclaimed timber [5].

2.5 Organizational Challenges

Organizational barriers significantly affect stakeholders involved in timber construction. The limited availability of companies with expertise in deconstruction and material reuse [14, 15] requires additional effort to identify suitable partners and coordinate processes effectively. A lack of specialized skills and knowledge in material reuse complicates project planning, often necessitating extra training or consultation to ensure reused components are properly implemented [15–19]. Logistical challenges, such as the need for additional storage space and increased transport costs, add to project complexity and require flexible budgeting [14, 20, 21].

3 Strategies to Enhance Timber Reuse

The successful reuse of timber in construction projects requires a multifaceted approach that addresses the challenges of material recovery, adaptation, and reintegration. This section explores three key areas: **Technological Strategies**, which focus on methods for efficient disassembly and processing; **Geometric Optimization Strategies**, which tackle the complexities of adapting reclaimed materials to new forms and applications; and **Organizational Strategies**, which emphasize the importance of planning, collaboration, and resource management to maximize the potential of timber reuse.

3.1 Technological Strategies for Timber Reuse

Advancing the reuse of timber in construction relies on technological innovations that address the challenges of disassembly and integration into new designs. In the following section, the key topics of connection and joint systems as well as system-oriented assembly and disassembly and their significance for working with reused components are explained.

3.1.1 Connection and Joint Systems

One of the main challenges of implementing circular economy principles in the construction sector in general is that the existent building stock was not designed and built for being disassembled and reused. Monolithic and rigid joints common in masonry and concrete construction make concepts like urban mining and design with reused components difficult to implement in practice [22]. In the case of timber and timber-based construction this barrier is less present due to the predominance of mechanical joints, which are comparatively easier to disassemble.

Usual fasteners and connectors for timber construction are nowadays being tested for evaluating their assembly and disassembly potential and assessing how many construction cycles they could support [23]. The possibility of a damage-free disassembly, supporting modularity and flexibility in application, must be a key aspect for the development of innovative joint solutions.

3.1.2 Levels of Reuse - System-Oriented Assembly and Disassembly

Extending the life cycle of construction and its components can be linked to diverse levels of reuse. The easiest and most effective strategy for reuse would be simply the further use of a building in its original state and location, with as few changes as possible. This is often not possible due to new programmatic, spatial, or technical requirements and therefore the definition of additional levels of reuse is desirable, planning for allowing the reuse of material, linear components, panels and spaces.

In timber-based construction distinct kinds of construction systems (frame systems, panel systems and volumetric modules) are well-established and offer different potential for reusing their components [24]. Three-dimensional modules can be assembled and disassembled as such, meanwhile panel or frame systems should be designed for making components recoverable.

A key aspect of this approach is the definition of the construction level that should be primarily planned for being reused (the material, the linear element, the panel or the volumetric module) and plan their dimensions and joints accordingly for allowing easy and realistic handling, transport, assembly, and disassembly. A best-case scenario should allow for a cascading of reuse levels.

3.2 Geometric Optimization Strategies for Timber Reuse

The concept of reuse aims to minimize the adaptation of components to meet the requirements of their new "task," thereby reducing labor, fabrication time, testing, and material waste. This approach necessitates greater flexibility in other parameters, such as accepting variations in the final form of the structure or building. While this may challenge traditional architectural design processes, it should be embraced as an opportunity. By leveraging digital process chains, architects can gain greater control over production and fabrication, ultimately influencing the outcome with precision and creativity.

The following section delves into strategies for addressing geometric complexity in timber reuse, focusing on reducing material waste and simplifying adaptation processes. It examines specific approaches tailored to different types of components: starting with strategies for linear elements, followed by those for panels, and finally addressing volumetric aggregations.

3.2.1 Strategies for Linear Elements

To minimize the impact on the final form, selecting suitable structural systems is crucial. Reciprocal frame systems (Figs. 1 and 2) present a promising solution, as demonstrated in different research projects [25, 26]. These systems allow short elements to be assembled in ways that span greater distances, and the typical T-shaped connections enable overhangs for larger components. Additionally, reciprocal frames function as redundant structural systems, reducing hierarchical dependencies. This can be beneficial when the material properties of reclaimed components are not fully known. Reciprocal frameworks can serve as standalone structural systems or act as bracing structures for smaller prefabricated components.

Fig. 1. Assembly of a reciprocal framework by Alec Singh.

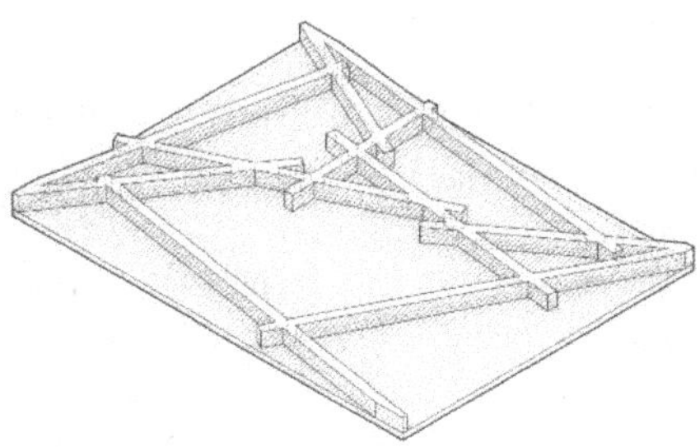

Fig. 2. Rectangular plate reinforced by ribs in form of a reciprocal framework by Alec Singh.

There are also innovative approaches to integrate reclaimed timber into conventional structural elements. For example, the project "Wood ReFramed" [27] showcases a built research demonstrator, which features trusses made from reclaimed timber. The parametric approach involved here does not produce a highly complex geometry but is in fact used to distribute material according to its structural requirements. Computational optimization is used to counter unknown material properties, a typical problem that arises with reclaimed timber.

3.2.2 Strategies for Panels

Reusing two-dimensional timber elements poses greater geometric challenges, particularly when aiming to minimize cut-off waste. A study that aims to address this issue is the DTC Timbershell project [28], which reuses cross-laminated timber (CLT) production waste. Rectangular cut-offs from door and window openings in CLT production, typically measuring 1 m x 1 m and 2 m x 1 m, are shaped into polygonal elements that form a wood-only shell structure. While the material properties of CLT are well-documented, the proper orientation of panels—specifically aligning the main direction of the wood fibers—is crucial and demands a carefully coordinated digital fabrication workflow.

When reducing cut-off waste and machining time is a priority, effective stock material management through matchmaking strategies becomes essential. Moreover, the increased degrees of freedom associated with 2D elements introduce higher computational demands and require extended optimization time. As highlighted by Singh (2023), geometric compromise solutions often emerge because exact matches are rarely achievable—only approximations. While it may not always be possible to adhere precisely to the initially intended global geometry, adjusting the proportion of reused elements allows for significant influence on the final design outcome.

3.2.3 Strategies for Volumetric Aggregations

Glulam timber is a well-known structural component, which basically follows the idea of a controlled assembly of a large object from smaller timber elements. By categorizing and testing the quality of the timber, glulam beams achieve superior structural performance and are capable of spanning greater distances and accommodating larger dimensions. A detailed study by Bergsagel et al. (2022) shows, that this approach is also feasible for structural glulam manufactured from reclaimed timber.

Similarly, cross-laminated timber (CLT) panels produced from secondary timber, known as CLST, offer potential for reuse. However, as highlighted by Llana et al. (2022), the mechanical properties of CLST do not fully match those of panels made from new timber. To address this, newly developed methods, such as non-destructive testing, have been introduced to evaluate the performance of recovered timber, as conventional visual grading standards are insufficient for this purpose.

Advancements in computational methods, such as topology optimization, further enhance the reuse of timber by enabling the design of components assembled from small-scale parts. The Re-VoxLam Truss [31] recombines numerous small blocks of timber into a volumetric structure. Each "voxel" is classified based on its density, which serves as an indicator of its mechanical properties. These voxels are then strategically positioned and oriented to meet local structural requirements, showcasing the potential of computational tools in optimizing reclaimed timber usage.

3.3 Organizational Strategies for Timber Reuse

Additionally, when dealing with technological and geometrical challenges, effective reuse of timber relies on two key organizational strategies: location-based reuse and early collaboration.

3.3.1 Location-Based Reuse

This approach encompasses on-site and off-site reclamation strategies. On-site reuse minimizes transportation costs and environmental impact by directly reclaiming timber from the project site. It is particularly effective when demolition and construction are closely linked, requiring flexibility, rapid decision-making, and thorough planning to integrate reclaimed materials into new designs. Off-site reuse, on the other hand, involves collecting timber from various locations and processing it through centralized facilities. This allows for higher-quality preparation, such as grading and treatment, but introduces logistical challenges, such as transportation and inventory management. Both methods highlight the importance of integrating reuse considerations early in project planning to maximize benefits.

3.3.2 Early Collaborations

The integration of timber reuse into construction projects is strengthened by early collaboration among architects, engineers, contractors, and material suppliers. Aligning goals and processes from the outset ensures that reusable materials are identified, assessed, and incorporated efficiently. Steps such as pre-demolition audits and material testing should be embedded into the timeline to address the unique characteristics of reclaimed timber. These proactive measures enhance adaptability, reduce delays, and support the successful recovery and reuse of materials.

4 Case Studies

This section delves into the potential and challenges of material reuse in construction, examining three diverse case studies: the structural reuse of glulam at Brussels' Recypark, the temporary office building for the Austrian Parliament and the PopUp Dorms in Vienna. Each case highlights innovative approaches to reusing materials, offering valuable insights into the practical, ecological, economic, and regulatory aspects.

4.1 Structural Reuse of Glulam at Brussels' Recypark

The Recypark project (Fig. 3), commissioned by Bruxelles Propreté, exemplifies the process of incorporating reclaimed structural timber into a new construction, highlighting the importance of rigorous testing and collaborative problem-solving in reuse projects. The project in Brussels, designed by 51N4E architects and engineered by Witteveen and Bos and Greisch, spanned from 2016 to 2022. It involved the reuse of glulam (Fig. 4) from a decommissioned riding arena, integrating it into the new recycling facility.

Fig. 3. Recypark overview by Esther Vandamme.

Fig. 4. Reused glulam in forefront, new glulam arches in the back by Esther Vandamme.

In 2016, as part of the Recypark competition preparation, four demolition-slated halls were identified as potential timber sources. Rotor, the reuse consultants, evaluated various scenarios and ultimately selected the Liège Hall—an old riding arena located 95 km from the site—for its structural integrity and availability. This choice underscores the importance of thorough preliminary assessments in reuse projects.

4.1.1 Ecological Challenges

A comprehensive analysis of the Liège Hall revealed varying conditions of the 22 glulam arches: five were deemed irrecoverable, two required local repairs, and the remainder were salvageable under specific conditions such as drying, insect protection, and base repairs. This early evaluation including plans, photographs, and moisture readings, was crucial in determining the feasibility of the reuse strategy and informed subsequent design decisions. The main ecological challenge was upkeeping the highest possible amount of recoverable material for structural reuse by securing storage and limited destructive testing.

4.1.2 Technological Challenges

The design team implemented several risk mitigation strategies. They cut arches by 1 meter at the base and placed them on concrete plinths to prevent water infiltration. Additionally, they doubled arches for enhanced safety and load-bearing capacity while employing conservative assumptions in structural calculations. The new arches were designed to mimic the shape of reclaimed ones to address geometric complexity and potential losses, while glued-in rods reinforced overstressed areas to counter wind-induced stresses. These design strategies demonstrate the need for adaptability and innovation in reuse projects as the prior technological challenges for projects where standard solutions may not suffice.

4.1.3 Economic and Organizational Challenges

In 2018, the contractor negotiated with the landowner for careful dismantling rather than destructive demolition. The contractor stored the arches in a dry, ventilated warehouse for over two years to ensure proper drying and transported the large structural elements from the original site to the storage facility and then to the construction site. These primarily organizational challenges result in a higher cost for the recovered arches, and thus generate economic challenges which are hard to overcome without a sustainability driven client.

4.1.4 Regulatory Challenges

The project team proactively addressed potential regulatory challenges, by engaging early with the general contractor's insurer and regulatory bodies, like the technical control office SECO. In 2021, the project team conducted flexural, shear and aging tests on declassified arches as well as density measurements to correlate degradation with mass loss. Construction began in Autumn 2021, and engineering office Greisch confirmed the test results met the required specifications, allowing the project to proceed.

4.1.5 Strategies

The Recypark project highlights the importance of early collaboration with reuse and timber experts, thorough testing and certification processes, and the need for redundancy in design to accommodate uncertainties. Acquiring materials before finalizing designs can enhance project flexibility and maintaining similar use conditions as the original application reduces technical challenges. Furthermore, early identification of available materials and project goals is vital for success. By adhering to these principles, the project team successfully navigated the complexities of integrating reclaimed structural timber into a new construction, setting a precedent for future sustainable building practices.

4.2 Post-Beam: Temporary Office Building for the Austrian Parliament

The Lukas Lang Building Technologies system offers a modular construction approach centered on reusability, adaptability, and circularity. The system uses a skeleton structure with standardized components that facilitate easy assembly (Fig. 5), disassembly (Fig. 6), and reuse across various projects. A defined grid with steel-to-steel and steel-to-wood connections provide strength and flexibility, allowing the system to adapt to different building requirements. The components can be dismantled and reused without compromising structural integrity, and the universal grid design enables integration across projects, supporting future additions. In projects like the Kindergarten Spillern, the system's adaptability was demonstrated. A temporary kindergarten was built during the construction of a new elementary school. Once the permanent site was ready, the temporary building was dismantled and repurposed into a new permanent structure in just two months. The modular system allowed for efficient reuse and integration of new components to accommodate size modifications.

Similarly, the 3 to 4-story temporary buildings used for the Austrian Parliament offices tested the system's scalability. Originally designed for a 5-year lifespan, the

Fig. 5. Assembly process (Lukas Lang Building Technologies).

Fig. 6. Disassembly process (Lukas Lang Building Technologies).

project lasted 7 years. Lukas Lang's buyback concept was planned, but rising raw material prices and depreciation due to extended use led to complications. Additionally, Lukas Lang's role as a subcontractor, rather than the main contractor, created further challenges during the dismantling process. A third-party company caused significant damage to the first floor, resulting in a 10% loss of component value. However, lessons learned improved the dismantling of subsequent floors, and today, 10,000 square meters of components are stored, with the first project—using 3000 square meters—already scheduled. This experience highlighted the importance of detailed deconstruction manuals, skilled dismantling teams, and efficient storage solutions to preserve the value of components over time.

4.2.1 Economic Challenges

Storage costs immobilize capital, preventing its productive use in generating rental income. In this case, the building elements have been stored for two years, further tiring up resources and escalating associated costs. Material price inflation, particularly for wood—and the depreciation of components during extended storage and reuse further erode profitability. The dismantling process added another layer of expense, driven by labor, storage, transport, and further reassembly costs. While reused components offer material cost savings, the labor and reassembly expenses will remain substantial, calculated in this project for up to 75% of the cost of new components. The Lukas Lang system started to get developed almost 30 years ago and despite its innovative design, continues to struggle with high development costs and insufficient returns, raising concerns among investors. Furthermore, the absence of an established and profitable market for reselling dismantled components exacerbates these financial hurdles, underscoring the need for more sustainable and economically viable solutions.

4.2.2 Ecological Challenges

The Lukas Lang Parliament building, composed of 70% timber and 30% steel, offers a favorable balance, where the environmental benefits of timber compensate for the negative impacts associated with steel and other components, particularly in the façade. However, the varied materials used in the façade have different lifespans, complicating the maintenance processes, including refurbishment, repair, and eventual replacement.

These activities, combined with the use of unsustainable insulation materials driven by cost pressures, reduce the overall environmental sustainability of the system and limit its lifecycle benefits. Furthermore, processes related to refurbishment, dismantling, transportation, processing, and reassembly generate considerable carbon emissions, which could offset the carbon savings from reusing components. This highlights the need to carefully assess the CO_2 impact of deconstruction and reassembly when evaluating sustainability.

4.2.3 Technological Challenges

Current joinery solutions often fail to support repeated dismantling and reassembly effectively. While advanced connection points, such as those developed by Lukas Lang with detachable steel-steel and permanent steel-wood joints, offer effective disassembly, they are both complex and costly to produce. The proprietary nature of these systems limits their broader applicability, reducing interoperability and creating inefficiencies in scaling or adapting solutions across different projects. Additionally, the dismantling process requires specialized expertise and can lead to material loss, as observed in the first store of the Lukas Lang Pavilion. Efficient tracking of the dismantling, transport, and reassembly of components is essential to maximize reuse potential. However, the testing and certification of reclaimed components remain significant barriers, as these processes can be both time-consuming and expensive, complicating the integration of reused materials into new projects.

4.2.4 Regulatory and Organizational Challenges

A key challenge is the lack of clear standards for reintroducing reused components into the market. The absence of definitive guidelines, such as those for CE markings and product declarations, alongside evolving safety and compliance issues—particularly related to fire safety and structural stability—complicates the reuse of reclaimed timber components. Reused materials often lack standardized documentation or proven fire performance, making certification difficult and increasing liability concerns for stakeholders. This regulatory uncertainty creates significant barriers in certifying reused materials, often limiting their market viability to buy-back models or certifications provided exclusively by the original producer, as is the case with Lukas Lang. Furthermore, legal constraints, such as federal tendering requirements, may hinder optimal resource allocation. For example, the Lukas Lang Pavilion faced a two-year storage period before its first implementation project this year, underscoring the delays caused by such legal and regulatory challenges. Additionally, storage limitations, including insufficient or costly warehousing options, delay or prevent reuse projects from moving forward.

4.3 Volumetric - Pop-Up Dorms

The PopUp Dorms by OeAD Housing (Fig. 8) is a pioneering solution for sustainable and affordable student housing, designed under Passive House standards. The concept was first introduced in 1991 to address the high costs of student accommodation caused by rising land prices. Intended for temporary use on unbuilt land, the modular dormitories are designed for relocation every five years and have an estimated 40-year lifespan. The project emerged from a 2014 competition, with 45 participants aiming to develop modular and reusable designs. The winning proposal was a collaboration between F2 Architekten, Obermayr Holzkonstruktionen, and Passive House consultant Günter Lang. Prefabrication took three months, followed by transportation (230 km), one week for assembly (housing: 4 days; atrium: 1 day), and an additional three weeks for finishing. This streamlined process enables completion within 6–9 months. Each apartment module measures 5.5 x 16.8 x 3.5 meters, accommodating four students across 75 m^2. Constructed with timber frames, mineral wool insulation, and a fir façade, the modules sit on point foundations, reducing environmental impact. Common spaces cover 200 m^2, enhancing community life. Temporary land leases eliminate ground costs, resulting in affordable rent. First built in 2015, the dorms won the 2019 Gold Affordable Housing Award. In 2021, the first relocation demonstrated the system's mobility, with 30-ton modules transported overnight by truck (Fig. 7). The PopUp Dorms, 24% cheaper than conventional housing, set a benchmark for circular, nomadic student accommodation.

Fig. 7. Dismantling process (© Seestadt TV/Roland Thurner).

Fig. 8. Finished building in new location (© Obermayr).

4.3.1 Economic Challenges

Contrary to expectations, the cost of building components does not decrease with repeated use. Rising material prices have made constructing a similar building today more expensive. During the COVID-19 crisis, timber scarcity and high prices made reuse economically viable. The modules were transported directly to the new location without requiring additional processing, which significantly reduced transportation and processing costs. However, storing the modules at the new location incurred substantial costs, including those for weather protection, particularly for the first-floor modules. Furthermore, damage sustained during transport and handling, such as cracks in drywall and broken tiles, necessitated repairs. Although these damages were superficial and accounted for in the original competition bid, they still added to the overall expenses. The reuse of foundations also emerged as a major challenge. Reusing foundations was time-consuming and expensive, involving excavation, layering, transport, and reinstallation. While this method was adopted for ecological reasons, it proved economically inefficient.

4.3.2 Technological Challenges

Technological challenges included issues with module design and the potential for damage during transport. Despite being specifically designed for relocation, the modules still suffered damage during transport and handling, such as cracks in drywall and broken tiles. This risk of damage highlighted the need for further technological refinement in the module design. Another challenge was the autonomy of the modules themselves. Each module was designed to be self-sufficient, which made relocation easier. However, this autonomy also required precise coordination to ensure everything functioned correctly once reassembled.

4.3.3 Regulatory and Organizational Challenges

When relocating to a new site after the first five years, new building permits were required due to changes in building regulations. This could have significantly impacted the module configuration, increasing costs. Fortunately, no modifications were necessary, and only formal paperwork was required. From an organizational standpoint, the relocation process faced numerous challenges due to tight deadlines. The dismantling, transportation, and reassembly of the modules had to be completed within a very narrow time frame. The sequential nature of the work, with dismantling taking place before reassembly (including the foundations), further extended the timeline, as parallel workflows were not possible. The on-site storage at the new location and logistics were initially underestimated, leading to unexpected challenges. This time pressure also placed considerable strain on the team, who had to work nights and weekends to meet the deadlines, increasing both the labor intensity and pressure on workers. Additionally, the division of ownership and operational responsibilities added complexity. The modules were owned by the housing association, while the construction company was responsible for relocation, requiring careful coordination to manage expectations and ensure smooth execution. Furthermore, the modules are not standardized, requiring specialized expertise for their assembly and disassembly. This makes it difficult for third-party companies to handle the relocation, as they lack the specific knowledge needed, making the process a closed solution.

5 Potential and Future Research

As shown by the previous sections, the continued advancement of timber reuse in construction requires innovative research and practical solutions across several key areas. These directions should aim to address financial, technical, regulatory, and design-related challenges. The following directions could be identified for future research:

1. Economic Viability and Financial Models:

   Future research should explore innovative financial models to support the adoption of circular practices. Buy-back systems for timber components, tax incentives for reclaimed materials, and improved storage strategies to minimize logistical expenses represent innovative approaches. Expanding modularity in design could enhance scalability and reusability, improving long-term financial returns. Furthermore, the introduction of CO_2 credits for successful material reclamation could incentivize reuse practices on a larger scale.

2. Lifecycle Assessment and Material Innovation:
   Developing and refining lifecycle impact assessment tools is critical for accurately evaluating the environmental benefits of timber reuse.
3. Standardization, Certification, and Data Management:
   The lack of universal standards remains a significant barrier to timber reuse. Future research could focus on establishing universal joinery standards and creating open-source systems for easy integration across projects. Streamlined certification processes are essential to ensuring reclaimed materials meet safety and quality requirements Additionally, efficient data management systems to track dismantling, transportation, and reassembly processes could improve logistics. Manuals for disassembly, dedicated dismantling teams, and training programs for workers would professionalize and standardize the reuse process.
4. Regulatory Frameworks and Collaborative Models:
   Research should support the development of robust regulatory frameworks and roadmaps to guide the certification and reuse of reclaimed materials. Fostering public-private partnerships could help establish centralized storage facilities and digital marketplaces for reused components, increasing accessibility and adoption rates.
5. Expansion of Structural Integrity in Reclaimed Timber:
   Investigating advanced methods to assess and enhance the structural integrity of reclaimed engineered timber products is essential. Developing repair and reinforcement techniques for damaged or aged timber components would extend their usability and ensure their safety in new applications.
6. Innovative Design Strategies:
   Adaptable and modular design approaches that integrate reclaimed components are critical to maximizing reuse potential. Future research should explore digital tools and technologies, such as parametric design, computational optimization, and machine learning, to streamline the use of non-standard reclaimed materials in architectural and structural designs.
7. Case Studies and Best Practices:
   Documenting and analyzing successful reuse projects are essential for understanding the practical challenges and solutions. A repository of best practices and lessons learned would provide valuable guidance for future projects, helping to scale timber reuse across the construction industry.

These potential research areas could collectively aim to address the existing barriers to timber reuse while creating a foundation for its broader adoption. By advancing financial, technical, regulatory, and design innovations, the construction sector can move closer to incorporate reused components in practice.

References

1. European Commission: A new Circular Economy Action Plan for a Cleaner and More Competitive Europe, Brussels (2020). https://eur-lex.europa.eu/resource.html?uri=cellar:9903b325-6388-11ea-b735-01aa75ed71a1.0017.02/DOC_1&format=PDF. Accessed 1 Jan 2025
2. Heisel, F., Hebel, D.E., Webster, K.: Building Better-Less-Different: Circular Construction and Circular Economy: Fundamentals, Case Studies, Strategies. Birkhäuser, Basel (2022)

3. Meier, H.-R.: Spolien: Phänomene der Wiederverwendung in der Architektur. De Gruyter, Berlin (2021). https://doi.org/10.1515/9783868599442
4. EN15804: EN 15804:2012 + A2:2019 + AC:2021, Sustainability of construction works - Environmental product declarations - Core rules for the product category of construction products (2012)
5. Pronk, A., Brancart, S., Sanders, F.: Reusing timber formwork in building construction: testing, redesign, and socio-economic reflection. Urban Plan. **7**(2), 81–96 (2022). https://doi.org/10.17645/up.v7i2.5117
6. Niu, Y., Rasi, K., Hughes, M., Halme, M., Fink, G.: Prolonging life cycles of construction materials and combating climate change by cascading: the case of reusing timber in Finland. Resour. Conserv. Recycl. **170** (2021). https://doi.org/10.1016/j.resconrec.2021.105555
7. Parigi, D.: Minimal-waste design of timber layouts from non-standard reclaimed elements: a combinatorial approach based on structural reciprocity. Int. J. Space Struct. **36**(4), 270–280 (2021). https://doi.org/10.1177/09560599211064091
8. De Gregorio, S.: Reuse process for timber elements to optimise residual performances in subsequent life cycles. VITRUVIO. **7**(2), 88–99 (2022). https://doi.org/10.4995/vitruvioijats.2022.18809
9. Grüter, C., et al.: Design for and from disassembly with timber elements: strategies based on two case studies from Switzerland. Front. Built Environ. **9** (2023). https://doi.org/10.3389/fbuil.2023.1307632
10. Fraunhofer, W.K.I.: Current project: Bewertung und Aufbereitung von Altholz zur Rückführung in den Kreislauf im Rahmen eines Wiederverwendungskonzeptes (2024–2026). https://www.wki.fraunhofer.de/de/forschungsprojekte/2024/refore-bewertung-und-aufbereitung-von-bau-altholz-zur-wiederverwendung-konstruktiven-holzbauteilen.html#1797858745. Accessed 1 Jan 2025
11. Klinge, A., Roswag-Klinge, E., Paganoni, S., Radeljic, L., Lehmann, M.: Design concept for prefabricated elements from CDW timber for a circurlar building. IOP Conf. Ser. Earth Environ. Sci. **323**(1) (2019). https://doi.org/10.1088/1755-1315/323/1/012022
12. Gao, Y., Shao, Y., Akbarzadeh, M.: Application of graphic statics and strut-and-tie models optimization algorithm in innovative timber structure design. Buildings. **13**(12) (2023). https://doi.org/10.3390/buildings13122946
13. Giordano, L., Derikvand, M., Fink, G.: Bending properties and vibration characteristics of dowel-laminated timber panels made with short salvaged timber elements. Buildings. **13**(1) (2023). https://doi.org/10.3390/buildings13010199
14. Dunant, C.F., Drewniok, M.P., Sansom, M., Corbey, S., Allwood, J.M., Cullen, J.M.: Real and perceived barriers to steel reuse across the UK construction value chain. Resour. Conserv. Recycl. **126**, 118–131 (2017). https://doi.org/10.1016/j.resconrec.2017.07.036
15. Chileshe, N., Rameezdeen, R., Hosseini, M.R., Lehmann, S., Udeaja, C.: Analysis of reverse logistics implementation practices by South Australian construction organisations. Int. J. Oper. Prod. Manag. **36**(3), 332–356 (2016). https://doi.org/10.1108/IJOPM-01-2014-0024
16. Chileshe, N., Rameezdeen, R., Hosseini, M.R.: Barriers to implementing reverse logistics in South Australian construction organisations. Supply Chain Manag. **20**(2), 179–204 (2015). https://doi.org/10.1108/SCM-10-2014-0325
17. Gorgolewski, M., Morettin, L.: The process of designing with reused building components. In: Lifecycle Design of Buildings, Systems and Materials (2009)
18. Shaurette, M.: Demolition contractors perceptions of impediments to salvage and reuse of wood structural components. J. Green Build. (2006). https://doi.org/10.3992/jgb.1.2.145
19. Yeung, J., Walbridge, S., Haas, C.: The role of geometric characterization in supporting structural steel reuse decisions. Resour. Conserv. Recycl. **104**, 120–130 (2015). https://doi.org/10.1016/j.resconrec.2015.08.017

20. Chinda, T., Ammarapala, V.: Decision-making on reverse logistics in the construction industry. Songklanakarin J. Sci. Technol. **38**(1), 7–14 (2016) http://www.sjst.psu.ac.th
21. Dunant, C.F., Drewniok, M.P., Sansom, M., Corbey, S., Cullen, J.M., Allwood, J.M.: Options to make steel reuse profitable: An analysis of cost and risk distribution across the UK construction value chain. J. Clean. Prod. **183**, 102–111 (2018). https://doi.org/10.1016/j.jclepro.2018.02.141
22. Hebel, D., Wappner, L.: Sortenrein bauen - Material, Konstruktion. Methodik. DETAIL. (2023). https://doi.org/10.11129/9783955536145
23. Carvahlo Cardoso, E.: Re-Use auf dem Prüfstand: Performance zerlegbarer Holzfachwerke mit unterschiedlichen Verbindungen. Master Thesis, Universität Stuttgart (2025)
24. Kolb, J., Kolb, H., Müller, A.: Holzbau mit System: Tragkonstruktion und Schichtaufbau, 4th edn. Birkhäuser, Basel (2024)
25. Singh, A.: Formfindungsmethoden für das Upcycling von geometrisch irregulären Holzbauteilen. Dissertation (PhD), University of Wuppertal (2023) https://doi.org/10.25926/BUW/0-99
26. Browne, X., Larsn, O.P., Castriotto, C.: Utilising waste wood through reciprocal frame systems. In: Proceedings of IASS Annual Symposia, vol. 2020, pp. 1–10. IASS (2020)
27. Browne, X., Larsen, O.P.: Prototyping new scenarios for fabricating structural components from reclaimed timber. In: Fabricate 2024: Creating Resourceful Futures, pp. 114–121. UCL Press (2024). https://doi.org/10.2307/jj.11374766.18
28. Robeller, C., Von Haaren, N.: Recycleshell: Wood-only Shell Structures Made From Cross-Laminated Timber (CLT) Production Waste. J. Int. Assoc. Shell Spatial Struct. **61**(2), 125–139 (2020). https://doi.org/10.20898/j.iass.2020.204.045
29. Bergsagel, D., Isaac, P., Koeck, A.: Reusing Timber in Construction - Feasibility Study Exploring Turning Reclaimed Timber into Structural Glulam (2022). https://doi.org/10.13140/RG.2.2.22225.54888
30. Llana, D.F., González-Alegre, V., Portela, M., Íñiguez-González, G.: Cross Laminated Timber (CLT) manufactured with European oak recovered from demolition: Structural properties and non-destructive evaluation. Construct. Build Mater. (2022). https://doi.org/10.1016/j.conbuildmat.2022.127635
31. Naboni, R., Kunic, A., Jensen, C., Sander, S.: Re-VoxLam Truss. Topology optimisation and reclaimed voxel lamination of horizontal wood structures. In: Proceedings of the 27th SIGraDi Conference 2023: Accelerated Landscapes, pp. 1298–1309 (2024). https://doi.org/10.5151/sigradi2023-510

Challenges and Potentials for Circulating Structural Timber and Engineered Wood Products

Henri Ranttila[1(✉)], Gerhard Fink[1], Reinhard Brandner[2], and Daniel Ridley-Ellis[3]

[1] Aalto University, Espoo, Finland
Henri.Ranttila@aalto.fi
[2] Graz University of Technology, Graz, Austria
[3] Edinburgh Napier University, Edinburgh, UK

Abstract. There are several major barriers in maintaining the load-bearing function of timber products in a further service life beyond extending the life of the building. The main barriers include missing standards and regulations for re-grading, re-classifying and re-certifying salvaged timber for load-bearing structural purposes. Current grading standards, which are formulated for new timber, are limited in this way due to several reasons. The same applies to current product and design standards which are also explicitly formulated only for new timber and new timber products. As a result, there are legal and insurance issues which currently act as barriers for circulating structural timber and timber construction products. This chapter presents a short summary of these issues.

Keywords: salvaged timber · circulation · strength grading · engineered wood products

1 Re-Grading Salvaged Solid Timber

Current grading standards, such as the standard EN 14081–1 [1] and visual strength grading rules that comply with it, are not formulated for salvaged timber or previously graded new timber. In formulating grading standards for salvaged timber, several aspects need to be considered:

- The timber species and corresponding growth region are usually unknown, unless apparent from stamps and records. Useful species identification from inspection is restricted by aging, weathering, coatings, dirt etc., and ultimately by what is possible to differentiate by wood microstructure [2–4].
- Salvaged structural timber might be treated or penetrated by (potentially) harmful chemicals which prevent any further circulation. Other impurities, such as (non-removeable) fasteners and fittings, concrete residues and coatings, might restrict circulation due to processing problems which interfere with strength grading methods [5, 6]. Furthermore, the current standards for strength grading are restricted to treatments for biological durability only, which exclude other kinds of treatments (e.g. fire protection) or chemical or thermal modification.

G. Fink et al. (Eds.): *Holistic Design of Taller Timber Buildings*, SPRTRCIENG, pp. 372–376, 2025.
https://doi.org/10.1007/978-3-032-02098-7_30

- Biological damage (due to insects, bacteria, fungi, etc.), modifications from first construction becoming damage for reuse (holes for fasteners, notches, slots, etc.), and damage arising from use and deconstruction might limit or even prevent circularity. Local reductions in cross sections due to fastener holes, notches or slots need to be evaluated in respect to their dimensions, combined effect, and position within the cross section [2, 3, 6–12].
- Splitting of timber, such as cracks, checks and fissures from drying, processing and moisture content changes in use, as well as geometric deformations, such as twist, bow, spring and cup, might prevent (or at least limit) any further use significantly [3, 13–17].
- Unknown history of the exposure (e.g. climate, weather) and loads result in uncertainties regarding aging of timber and long-term loading effects.
- Studies on aging and long-term loading effects, in general, indicate decreasing strengths but sustained elastic properties which mean a consequential shift in prediction models for strength when based on visual characteristics and non-destructive testing (NDT) of stiffness [18].
- The current test standards can be difficult to apply to salvaged timber, due to e.g. typically shorter lengths. Also the requirement for large test data, is likely disproportionate to the scale of the recovered wood supply.

Considering this aspects, deriving a grading system for salvaged timbers is challenging. As a result, the grading system would need to transform from the current EN 14081–1 [1] population-based system to one that is free of this dependency on a reasonably stable and defined population. Grading salvaged timbers ideally requires separate indicators for the three grade determining properties (strength, stiffness and density). Thresholds for indicators need to adjust dynamically to the quality of the resource being graded, otherwise grading needs to be very conservative to account for the missing prior population characteristics.

2 Reusing and Recycling Glued Laminated Timber Construction Products

Compared to reusing structural sawn timber the physical size and consistency of products like cross-laminated timber (CLT) and glulam remove much of the complexity related to resource identification or previous grading. Nevertheless, to ensure that salvaged glued laminated timber construction products (GLTCPs) can be used widely again for structural, load bearing purposes a sufficient evaluation procedure is essential. It must be ensured that timber with environmental degradation is excluded and the mechanical properties still need to be quantified and re-classified. For the evaluation different kinds of available information can be used. Therefore, the development and establishment of widely applicable standardized procedures are crucial.

Depending on the investigation, different types of information can be collected. They can be grouped as direct and indirect information, and as equality type and inequality type information. A possible formal framework to consider different types of information is Bayesian updating. For Bayesian updating, quantifiable prior information needs to be available, which can be planned conditions (if available) or an expert opinion. However, it has to be considered that prior information is associated with uncertainties (see e.g [19]. for more information). Depending on the type of information, different updating procedures are available, see e.g. [19–22]. A framework for the estimation of the strength properties of existing timber structures using Bayesian updating is presented in [22]. Although the selected investigation methods might be different, the general principles are also valid for the estimation of mechanical properties of timber elements for the purpose of reuse. Furthermore, the approach can be also extended for the evaluation of timber connections or even entire structural systems. This could also be potentially used for the evaluation of existing buildings, for example for the sake of adaptations and renovations.

3 Conclusions

The quality and properties of salvaged solid timber are different to those of new timber in multiple ways. Thus, re-grading salvaged timber requires new forms of grading system in which these differences are considered. Salvaged GLTCPs are more consistent than salvaged solid timber both in material properties and in prior production processes, and as a result more easily assessed and evaluated for reuse. In all cases, structurally reliable reuse of timber requires some form of assessment, either re-grading of salvaged timber or an assessment-based adjustment of the initially declared material properties of structural timber products. A more detailed discussion about the issues can be found in [23].

References

1. Timber structures – strength graded structural timber with rectangular cross section – Part 1: General requirements. EN 14081–1:2016+A1:2019, CEN (2019)
2. Crews, K.: Development of grading rules for re-cycled timber used in structural applications. In: Presented at the International Council for Research and Innovation in Building and Construction, Working Commission W18 – Timber Structures, CIB-W18/40–5-1, Bled, Slovenia (2007)
3. Falk, R.H., Maul, D.G., Cramer, S.M., Evans, J., Herian, V.: Engineering properties of Douglas-fir lumber reclaimed from deconstructed buildings. In: Research Paper FPL-RP-650. U.S. Department of Agriculture, Forest Service, Forest Products Laboratory, Madison, WI (2008). https://doi.org/10.2737/FPL-RP-650
4. Sandberg, K., Sandin, Y., Harte, A.M., Shotton, E., Hughes, M., Ridley-Ellis, D., Turk, G., Íñiguez-González, G., Risse, M., Cristescu, C.: Summary report InFutUReWood – Innovative Design for the Future – Use and Reuse of Wood (Building) Components. In: InFutUReWood, RISE report 2022:08 (2022). https://doi.org/10.23699/p41e-ae46
5. Hafner, A., Ott, S., Winter, S.: Recycling and end-of-life scenarios for timber structures. In: SB13, Graz, Austria (2013)
6. Evaluation of Recycled Wood. NS 3691, parts 1 to 3, Standards Norway (2025)
7. Falk, R.H., DeVisser, D., Cook, S., Stansbury, D.: Effect of damage on the grade yield of recycled lumber. For. Prod. J. **49**(7/8), 71–79 (1999)
8. Chini, A.R., Acquaye, L., Rinker, M.E.: Grading and mechanical properties of salvaged lumber. In: Proceedings of the Task Group 39-Deconstruction Meeting, Wellington, NZ (2001)
9. Fridley, K.J., Williams, J.M., Cofer, W.F.: Reliability of reused sawn lumber: effect of fastener holes. J. Struct. Eng. **127**(4), 412–417 (2001)

10. Falk, R.H., DeVisser, D., Plume, G.R., Fridley, K.J.: Effect of drilled holes on the bending strength of large dimension Douglas-fir lumber. For. Prod. J. **53**(5), 55–60 (2003)
11. Recycled timber – visually stress graded for structural purposes, FWPA Standard G01, Draft. Australia (2024)
12. Pasca, D.P., Aloisio, A., De Santis, Y., Burkart, H., Øvrum, A.: Visual-based classification models for grading reclaimed structural timber for reuse: a theoretical, numerical and experimental investigation. Eng. Struct. **322**, 119218 (2025). https://doi.org/10.1016/j.engstruct.2024.119218
13. Falk, R.H.: The properties of lumber and timber recycled from deconstructed buildings. In: Presented at the Pacific Timber Engineering Conference (PTEC), Rotorua, New Zealand, Proceedings, vol. 2, (1999)
14. Rammer, D.R.: Evaluation of recycled timber members. In: Presented at Materials and Construction: Exploring the Connection: Proceedings of the Fifth ASCE Materials Engineering Congress, Cincinnati, Ohio. Reston, VA, pp. 46–51. American Society of Civil Engineers (1999)
15. Green, D.W., Falk, R.H., Lantz, S.: Effect of heart checks on flexural properties of reclaimed 6 by 8 Douglas-fir timbers. For. Prod. J. **51**(7/8), 82–88 (2001)
16. Llana, D.F., Íñiguez-González, G., Plos, M., Turk, G.: Grading of recovered Norway spruce (Picea abies) timber for structural purposes. Construct. Build Mater. **398**, 132440 (2023). https://doi.org/10.1016/j.conbuildmat.2023.132440
17. Llana, D.F., González-Alegre, V., Portela, M., García-Navarro, J., Íñiguez-González, G.: Engineered wood products manufactured from reclaimed hardwood timber. In: Presented at the World Conference on Timber Engineering (WCTE), Oslo, Norway, pp. 3594–3599 (2023). https://doi.org/10.52202/069179-0468
18. Brandner, R., Ottenhaus, L.M.: Properties of salvaged structural timber components: How to account for long-term loading effects given unknown load histories? In: COST Action CA 20139 "Holistic design of taller timber buildings", State-of-the-art report, pp. 71–75 (2022)
19. Rackwitz, R.: Predictive distribution of strength under control. Mater. Constr. **16**(4), 259–267 (1983)
20. Faber, M., Val, D., Stewart, M.: Proof load testing for bridge assessment and upgrading. Eng. Struct. **22**(12), 1677–1689 (2000). https://doi.org/10.1016/S0141-0296(99)00111-X
21. Faber, M.: Statistics and probability theory. In: Pursuit of Engineering Decision Support, vol. 18. Springer Verlag, Berlin (2012). https://doi.org/10.1007/978-94-007-4056-3
22. Fink, G., Köhler, J.: Risk based investigations of partly failed or damaged timber constructions. In: Aicher, S., Reinhardt, H.W., Garrecht, H. (eds.) Materials and Joints in Timber Structures. RILEM Bookseries, vol. 9. Springer, Dordrecht (2014). https://doi.org/10.1007/978-94-007-7811-5_6
23. Ranttila, H., Fink, G., Brandner, R., Ridley-Ellis, D.: Circulating structural timber and engineered wood products – challenges and potentials towards reliable evaluation of mechanical properties. Submitted to Wood Material Science & Engineering (2025). https://doi.org/10.1080/17480272.2025.2517186

Circulation of Structural Timber: A Review of Opportunities and Challenges

R. N. Passarelli[1](✉), T. O. Chiletto[1], R. Brandner[2], and L. M. Ottenhaus[3]

[1] Hasselt University, Hasselt, Belgium
rafael.novaispassarelli@uhasselt.be
[2] Graz University of Technology, Graz, Austria
[3] The University of Queensland, St Lucia, QLD, Australia

Abstract. There is a growing interest in transitioning to a circular economy as an alternative to linear approach. In this context, there has been a renewed interest in materials such as timber, which is part of a natural renewable cycle with lower embodied energy compared to conventional construction. Academics and practitioners are investigating methods to implement Design for Circularity strategies with the objective of extending the service life of building materials, thereby reducing the necessity for virgin material extraction and the environmental impact of construction. The circulation of structural timber is crucial for making the most effective and long-term use of timber. This article discusses prerequisites for the circulation of structural timber, considering aspects of architectural and structural design. The fundamental concepts discussed are combined with findings from a comprehensive literature review focusing on the circulation of structural timber. The objective of this article is to identify possibilities, challenges, and critical knowledge gaps, with the aim of creating new knowledge that contributes to an increased and facilitated circulation of structural timber.

Keywords: circularity · reuse · recycling · timber architecture · timber engineering

1 Introduction

The urgent need for sustainable construction has renewed interest in bio-based materials like timber. When sourced from sustainably managed forests, timber is a carbon-negative material, sequestering CO_2 during growth and storage, significantly lowering embodied energy compared to concrete [1]. However, short-term use of bio-based products limits benefits, emphasizing the need for extended service life through innovative design [2].

Circulating structural timber is key to advancing a circular economy, maximizing the grey energy from harvesting, transport, and processing while enhancing CO_2 storage. The literature offers growing recommendations to improve timber reclaimability and

This is an extended abstract of a paper published in the Special Issue of the COST Action HELEN in the journal *Wood Material Science & Engineering*. The full article is available at: https://doi.org/10.1080/17480272.2025.2493711

G. Fink et al. (Eds.): *Helistic Design of Taller Timber Buildings*, SPRTRCIENG, pp. 377–380, 2025.
https://doi.org/10.1007/978-3-032-02098-7_31

facilitate its circulation in a second lifecycle [3]. Despite the relevance of timber re-use and re-cycling in the much-needed transition to more circular practices, cohesion in the field remains lacking. Piccardo & Hughes [4] highlight limited research on circular timber strategies, while Munaro et al. [5] call for clearer conceptual definitions to aid implementation. Eberhardt et al. [6] stress the need for standardized terminology, as strategies are inconsistently interpreted. Moreover, achieving flexibility, easy dismantling, and reuse while maintaining architectural appeal presents challenges for architects and engineers.

This chapter is the extended abstract version of a study [7] that conducts a comprehensive literature review on the structural reuse of timber from both architectural and engineering perspectives. It critically examines key aspects within each discipline, synthesizing trends and identifying knowledge gaps. The study's original contribution lies in providing an up-to-date analysis to inform sustainable design, support circular construction, and lay the groundwork for a cohesive theoretical framework for timber reuse and recycling.

2 Method

This study presents a narrative literature review on the structural reuse and recycling of timber, integrating architectural design and timber engineering perspectives.

From an architectural perspective, the review builds on two systematic literature reviews (SLRs) by the first two authors, which included a bibliometric metadata analysis [8] and a meta-synthesis assessment [9] of papers from 2006 to 2022. This study extends those analyses to include publications up to October 2024, capturing recent developments and examining frequently discussed design strategies.

From a timber engineering perspective, findings build on prior research on salvaged timber circulation [10], which explored reuse potential, challenges, and factors such as erection, use, deconstruction, re-grading, and material aging. This study expands that work with new literature, particularly on aging and duration of load (DoL), as both significantly impact the structural viability of recycled timber.

3 Results and Discussion

From an architectural design perspective, strategies such as designing for disassembly, avoiding the use of toxic materials to impregnate or connect timber elements, modular prefabrication can facilitate the structural circulation of timber in the future. The lack of real-world validation of design strategies, variability and uncertainty of properties and sizes of available timber elements for reclamation, lack of standardized material databases, and limited market incentives pose significant challenges for its current and future circulation. Addressing these issues requires a holistic perspective, integrating research and development of new technologies, policy support, and the regional markets that promote reclaimed timber as a viable material for structural re-use and re-cycling. The findings suggest several areas for future exploration, that include 1) creating policy frameworks and economic incentives to foster local markets for salvaged timber, 2) expanding real-world case studies to validate upstream strategies like design for disassembly and adaptability, 3) investigating the economic and environmental trade-offs of innovative timber-to-timber connections and modular prefabrication systems, 4) developing standardized material databases that integrate regional inventories, mechanical properties, and life-cycle assessments, 5) refining computational algorithms and 3D modeling techniques to better assess and optimize salvaged timber's structural properties.

Considering environmental impact and circularity assessment methods, the literature identified several emerging approaches tackling different fronts and trying to incorporate the complex aspects of the circular potential of timber buildings. Some of the challenges identified are the subjectivity of the assessment methods to analyse the circularity indicators, the complexity of the assessment, which includes and combines several criteria, and the uncertainties related to the longevity of the materials as the future use and end of life conditions of a building and their components remains difficult to predict.

From a timber engineering perspective, several areas of uncertainty and research gaps remain critical to the broader adoption of timber circulation. Effects caused by aging and in particular by the loading history, involving long-term static and cyclic effects, affect the material performance and represent significant challenges in circulating as the knowledge basis is still insufficient or at least on a level at which only pragmatic solutions appear meaningful. The physical / mechanical properties of salvaged timber, even within the same species, can differ from new timber due to complex interactions of environmental exposure and load history. Current testing methods lack standardization, with approaches varying from small specimen testing to those on structural timber, complicating direct comparisons between salvaged and new timber. Addressing this issue will require internationally harmonized protocols for evaluating aging, including relationships between artificial and natural weathering, and DoL effects. Advanced methods like accelerated aging in controlled environments hold promise, yet further refinement is needed to ensure these tests accurately represent real-world conditions.

Follow-up research topics could address 1) expanding the literature, and re-evaluate existing data sets to identify potential counteracting effects from variations in timber quality between reference values and salvaged timber, 2) expanding knowledge on the long-term loading effects on timber and timber joints to establish a solid foundation for harmonised regulations and re-grading standards on salvaged timber, 3) conducting exemplary analyses of long-term loading effects using reliability methods, 4) developing proposals for regulating long-term loading effects on salvaged timber, considering unknown factors such as load history and the number and duration of previous service lives.

References

1. Gustavsson, L., Sathre, R.: Variability in energy and carbon dioxide balances of wood and concrete building materials. Build. Environ. **41**(7), 940–951 (2006). https://doi.org/10.1016/j.buildenv.2005.04.008
2. Brunet-Navarro, P., Jochheim, H., Muys, B.: The effect of increasing lifespan and recycling rate on carbon storage in wood products from theoretical model to application for the European wood sector. Mitig. Adapt. Strat. Glob. Chang. **22**(8), 1193–1205 (2017). https://doi.org/10.1007/s11027-016-9722-z

3. Crowther, P.: Design for Disassembly—themes and principles. Environ. Des. Guide. **31**, 1–7 (2005). https://doi.org/10.2307/26149108
4. Piccardo, C., Hughes, M.: Design strategies to increase the reuse of wood materials in buildings: Lessons from architectural practice. J. Clean. Prod. **368**, 133083 (2022). https://doi.org/10.1016/j.jclepro.2022.133083
5. Munaro, M.R., Tavares, S.F., Bragança, L.: The ecodesign methodologies to achieve buildings' deconstruction: a review and framework. In: Sustainable Production and Consumption, vol. 30, pp. 566–583. Elsevier B.V, Amsterdam (2022). https://doi.org/10.1016/j.spc.2021.12.032
6. Eberhardt, L.C.M., Birkved, M., Birgisdottir, H.: Building design and construction strategies for a circular economy. Archit. Eng. Des. Manag. **18**(2), 93–113 (2022). https://doi.org/10.1080/17452007.2020.1781588
7. Passarelli, R.N., Chiletto, T.O., Brandner, R., Ottenhaus, L.M.: Circulation of structural timber for structural purposes: a review of opportunities and challenges. Wood Mater. Sci. Eng., 1–22 (2025). https://doi.org/10.1080/17480272.2025.2493711
8. Passarelli, R.N.: Design for disassembly and reuse of timber in construction: identification of trends and knowledge gaps. In: Springer Tracts in Civil Engineering, vol. F1844, pp. 57–67. Springer, Cham (2024). https://doi.org/10.1007/978-3-031-45980-1_6
9. Chiletto, T.O., Knapen, E., Passarelli, R.N.: Design for disassembly and reuse: a synthesis for timber construction. World Sustainable Built Environment. IOP Conf. Ser. Earth Environ. Sci. **1363**, 012040 (2024)
10. Brandner, R., Ottenhaus, L.M.: Properties of salvaged structural timber components: how to account for long-term loading effects given unknown load histories? In: COST Action CA 20139 "Holistic Design of Taller Timber Buildings", State-of-the-art report, pp. 71–75 (2022)

Special Aspects of Taller Timber Buildings

Experimental and Numerical Investigations of Timber-Concrete Composite Slabs Subjected to Negative Bending Moments

Sebastian Krug(✉) and Jörg Schänzlin

Institute for Timber Design, Hochschule Biberach University of Applied Sciences, Karlstrasse 11, 88400 Biberach, Germany
krug@hochschule-bc.de

Abstract. Timber-concrete composite (TCC) floors offer an efficient solution for taller buildings by combining the advantages of timber and concrete. To optimise material use and reduce emissions, multi-span configurations could replace single-span beams. However, in practice, TCC slabs are still designed as simply supported. This study investigates the moment-rotation relation over the centre support of multi-span TCC slabs using experimental and numerical methods. Nine three-point bending tests were conducted, complemented by finite element modelling and a probabilistic component analysis. Results indicate that the structural configuration achieves significant rotational stiffness, which can be exploited to reduce material consumption and lower the carbon footprint of TCC construction.

This is an extended abstract of a paper published in the Special Issue of the COST Action HELEN in the journal Wood Material Science & Engineering. The full open-access article is available at: https://doi.org/10.1080/17480272.2025.2509083

1 Introduction

Timber-concrete composite (TCC) systems combine timber and concrete, making them efficient for floor spans of 6 to 8 m. However, most TCC floors are designed as single-span beams, despite the potential benefits of multi-span configurations. The recent introduction of the Technical Specification CEN/TS 19103 [1] permits continuous composite systems, provided both materials remain continuous over the centre support. Otherwise, stiffness and load-bearing capacity reductions due to joints must be considered.

While simply supported TCC slabs have been widely studied (e.g. [2,3]), research on multi-span systems remains limited. Experimental studies show that continuous configurations provide rotational stiffness at the centre support, reducing deflections and enhancing material efficiency [4,5]. However, discrepancies between analytical models and experimental results highlight the need for

G. Fink et al. (Eds.): *Holistic Design of Taller Timber Buildings*, SPRTRCIENG, pp. 383–388, 2025.
https://doi.org/10.1007/978-3-032-02098-7_32

further sensitivity studies. Numerical analyses [6] confirm that multi-span TCC floors exhibit lower deflections, but existing calculation models remain case-specific [7]. This study addresses this gap through experimental and numerical investigations, validating a probabilistic model for moment-rotation behaviour at the centre support. Further details on the experimental setup, modelling, and analysis are provided in the corresponding full publication [8].

2 Material and Methods

Continuous TCC systems are characterised by the presence of at least one internal support, enabling load redistribution and enhanced structural efficiency. In contrast to simply supported spans, continuous systems develop negative bending moments at intermediate supports, which poses specific challenges in design and detailing. These include the appropriate placement of reinforcement steel, the use of continuous concrete slabs without joints, and the selection of shear connectors that perform adequately under cracked concrete conditions. Despite these advantages, such as reduced mid-span deflections and increased load capacity, continuous TCC slabs are not yet common in practice due to limited design guidelines and experimental validation.

To address this gap, the present study investigates the load-bearing and deformation behaviour of continuous TCC floors using both experimental and numerical methods. The configuration shown in Fig. 1 combines jointed, simply supported glued-laminated timber elements with a continuous reinforced concrete slab. This design minimises timber transport lengths, enhances acoustic and fire performance, and reduces compression perpendicular to the grain at the centre support.

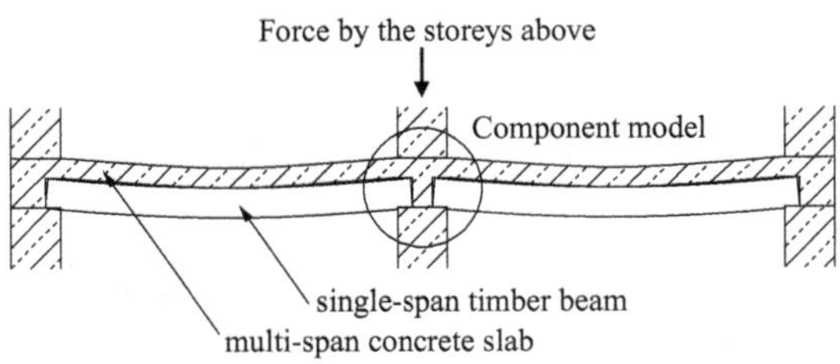

Fig. 1. Timber-concrete composite floor as continuous system [9].

Due to the lack of research on shear connectors in negative bending zones [5,10], only notches in regions with positive bending moments were considered. Experimental and numerical investigations were conducted to evaluate the moment-rotation behaviour at the centre support. The experimental setup comprised nine three-point bending tests with varying reinforcement amounts to assess rotational stiffness and failure mechanisms.

Finite element (FE) simulations, incorporating nonlinear material behaviour of timber, concrete, and reinforcement, were performed to replicate the experiments. A probabilistic component model [9] was used to predict the moment-rotation relation, considering material variability. While this model has been numerically validated, its experimental verification is part of this study.

The experimental investigations focused on the negative bending moment region of a continuous TCC slab. Nine three-point bending tests were carried out on specimens with a span of 3.20 m. Three test series, each consisting of three identical specimens, were examined, with the primary variable being the amount of embedded reinforcement. An overview of the test setup is presented in Fig. 2, where an inverted TCC slab was used to simulate the negative moment region of a two-span beam.

(a) Front view (b) Cross-sectional view

Fig. 2. Overview of the 3-point bending test configuration.

The specimens consisted of jointed timber beams with a continuous concrete slab, employing a concrete-to-timber height ratio of 1:2 (80 mm concrete, 160 mm timber).

A finite element (FE) model was developed to numerically replicate the experimental investigations and validate the results. A analysis was performed using Ansys [11], based on [9] and adapted to the tested specimens. Timber was modelled using an orthotropic material model, and concrete was defined with a Coupled Damage-Plasticity Microplane Model. Reinforcement was modelled with bilinear isotropic hardening plasticity, using reinforcement elements directly connected to the concrete. Material parameters and mesh density were optimised based on [9]. The model accurately captured strain localisation in the cracked concrete, timber joint, and notches.

To determine the moment-rotation relation over the middle support of continuous TCC slabs, a probabilistic calculation model was developed in [9] and validated through experimental investigations. The model is based on the component method, commonly used in steel construction [12], accounting for deformations within the joint. The negative bending moment region is represented by individual load-carrying components, each described by nonlinear springs. The overall moment-rotation relation is determined iteratively, considering the variation of the lever arm with rotation, the nonlinear cracking behaviour of concrete, and load distribution in the timber. Originally deterministic, the model was extended to a probabilistic component model incorporating material variability based on [13] and analysed using the Monte Carlo method. Results indicate that input parameter variability significantly affects joint stiffness, leading to notable scatter in the moment-rotation relation, particularly during concrete cracking. A probabilistic analysis is therefore essential for accurately predicting stiffness behaviour at the middle support. Further details are provided in [9].

3 Results

To validate the probabilistic component model, a comparison was conducted between the experimental results, finite element analysis (FEA), and the probabilistic model predictions. The analysis focused on the moment-rotation relation over the centre support of continuous TCC slabs and the variation in the compression contact height in the timber.

The comparison of moment-rotation relations demonstrated good agreement between the experimental results, the probabilistic component model, and FEA. It was evident that increasing reinforcement significantly enhanced the moment capacity of the connection. Additionally, for low reinforcement ratios, the cracking behaviour of the concrete had a notable influence, as individual cracks were distinctly visible in the evaluation. Since cracks were already present in the concrete before testing, both uncracked and pre-cracked conditions were considered in the analysis. Figure 3 illustrates the moment-rotation relation for test series 1 under pre-cracked conditions, which provided a good match with the experimental results.

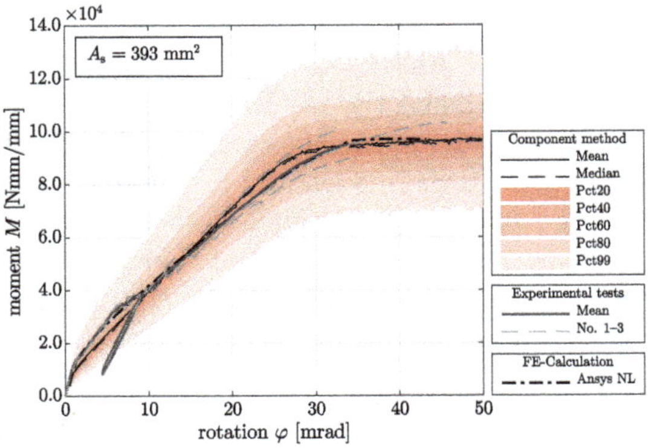

Fig. 3. Moment-rotation relation of series 1; Comparison between the experimental test results, the probabilistic component model with distribution percentiles (Pct), and the finite element analysis using Ansys.

Overall, the results confirm that the probabilistic component model provides an accurate representation of the moment-rotation behaviour at the centre support of continuous TCC slabs. The probabilistic approach effectively captures the influence of material variability, particularly in the cracking phase of the concrete.

4 Conclusions & Outlook

Experimental and numerical investigations confirmed that multi-span TCC floors achieve significant rotational stiffness, enabling material optimisation and enhanced sustainability. The moment-rotation relations obtained from experiments align well with numerical predictions, validating the probabilistic component model.

Pre-cracked concrete conditions were found to influence initial stiffness, particularly in prefabricated systems. The study highlights the importance of reinforcement ratios in controlling stiffness and load-bearing capacity.

Future work should address structural reliability, considering stiffness variability and long-term effects such as creep. The validated probabilistic model offers a foundation for optimising multi-span TCC design using rotational springs.

References

1. CEN/TS 19103. Technical specification: Berechnung und Konstruktion von Holzbauten — Bemessung und Berechnung von Holz-Beton-Verbundbauteilen — Allgemeine Regeln und Regeln für den Hochbau. DIN-Deutsches Institut für Normung (2022). https://doi.org/10.31030/3232343.
2. Schänzlin, J.: Zum Langzeitverhalten von Brettstapel-Beton-Verbunddecken. Doctoral thesis, Universität Stuttgart (2003)
3. Frohnmüller, J.: Geklebter Holz-Beton-Verbund: Verbundfestigkeit und Bauteilbemessung. Doctoral thesis, Universität Kassel (2024)
4. Kaluza, C., Krug, S., Schänzlin, J., Sieder, M.: Durchlaufwirkung von HBV-Decken - Durchlaufwirkung von Holz-Beton-Verbunddecken für die Anwendung im Mehrgeschoss- und Bürobau. Forschungsinitiative Zukunft Bau, unpublished (2024)
5. Adema, A., Chacón, M.F., María, H.S., Opazo-Vega, A., Casanova, E., Guindos, P.: Flexural performance of full-scale two-span nail-laminated timber concrete composite slabs (2024). https://doi.org/10.1016/j.conbuildmat.2024.136528
6. Marschall, J.: Untersuchung von Holz-Beton-Verbunddecken als Mehrfeldsystem. Master's thesis, Hochschule für Forstwirtschaft Rottenburg (2018)
7. Müller, K.: Timber-concrete composite slabs with micro-notches. Doctoral thesis, ETH Zürich (2020)
8. Krug, S., Schänzlin, J.: Experimental and numerical investigations of timber-concrete composite slabs subjected to negative bending moments. Special Issue of COST Action CA20139 - Holistic Design of Taller Timber Buildings (HELEN). Submitted to Wood Material Science & Engineering for publication (2025)

9. Krug, S., Jaaranen, J., Fink, G., Schänzlin, J.: Determination of the moment–rotation relation of continuous timber–concrete composite floors based on the component method. Structures **69**, 107365 (2024). https://doi.org/10.1016/j.istruc.2024.107365. ISSN 2352-0124
10. Sebastian, W., Lawrence, A., Smith, A.: Commentary: The potential for multi-span continuous timber-concrete composite floors (2018). https://doi.org/10.1680/jstbu.171.9.661
11. Ansys Inc. Ansys - Structural Analysis Guide - 2024 R1. Ansys Inc. (2024)
12. DIN EN 1993-1-8. Eurocode 3: Bemessung und Konstruktion von Stahlbauten - Teil 1-8: Bemessung von Anschlüssen. DIN-Deutsches Institut für Normung (2005)
13. JCSS. Probabilistic Model Code Part III - Resistance Models (2001)

Experimentally Validated Bending Capacity Models for Strip-Reinforced Timber Beams

Alex Sixie Cao[1,2](✉), René Steiger[2], and Andrea Frangi[1]

[1] Institute of Structural Engineering, ETH Zurich, Zurich, Switzerland
cao@ibk.baug.ethz.ch

[2] Structural Engineering Laboratory, Empa – Swiss Federal Laboratories for Materials Science and Technology, Dübendorf, Switzerland

Abstract. We present analytically derived and experimentally validated bending capacity models for strip-reinforced glued laminated timber beams using carbon-fiber reinforced polymer sheets. Motivated by sustainability and efficiency concerns, the study demonstrates substantial mechanical improvements, with bending strength increases up to 74%, flexural rigidity enhancements up to 65%, and timber volume reductions of up to 44% for comparable reliability. Reliability analyses further confirm significant timber savings while maintaining structural safety. Although reinforced beams showed higher cradle-to-gate environmental impacts, these are justified by mechanical performance and reduced timber usage.

1 Introduction

Increasing global demand for timber as a sustainable construction material requires improved efficiency in timber use. Significant portions of available timber currently remain underutilized structurally due to low-quality mechanical properties or competing uses in non-structural applications. Strengthening timber beams through reinforcement has emerged as an effective approach to maximize the structural potential of lower-grade timber. Among various reinforcement options, carbon-fiber reinforced polymers (CFRP) sheets, due to their high strength-to-weight ratio and ease of integration with timber, show great promise. This research aims to develop analytically robust bending capacity models for both symmetrically and asymmetrically strip-reinforced GLT beams, validate these models through full-scale experiments, and evaluate their effectiveness in enhancing structural performance and sustainability.

2 Model

The bending capacity models developed by Cao et al. [1] rely on assumptions of linear-elastic ideal-plastic behavior of timber and linear-elastic behavior in CFRP sheets. Timber is modeled to fail primarily through tensile stresses, representing the governing failure mode. For symmetrically reinforced beams, neutral axis locations and stress resultants are analytically determined, while asymmetrically reinforced beams are categorized as under-reinforced or over-reinforced depending on reinforcement ratios and failure

G. Fink et al. (Eds.): *Holistic Design of Taller Timber Buildings*, SPRTRCIENG, pp. 389–392, 2025.
https://doi.org/10.1007/978-3-032-02098-7_33

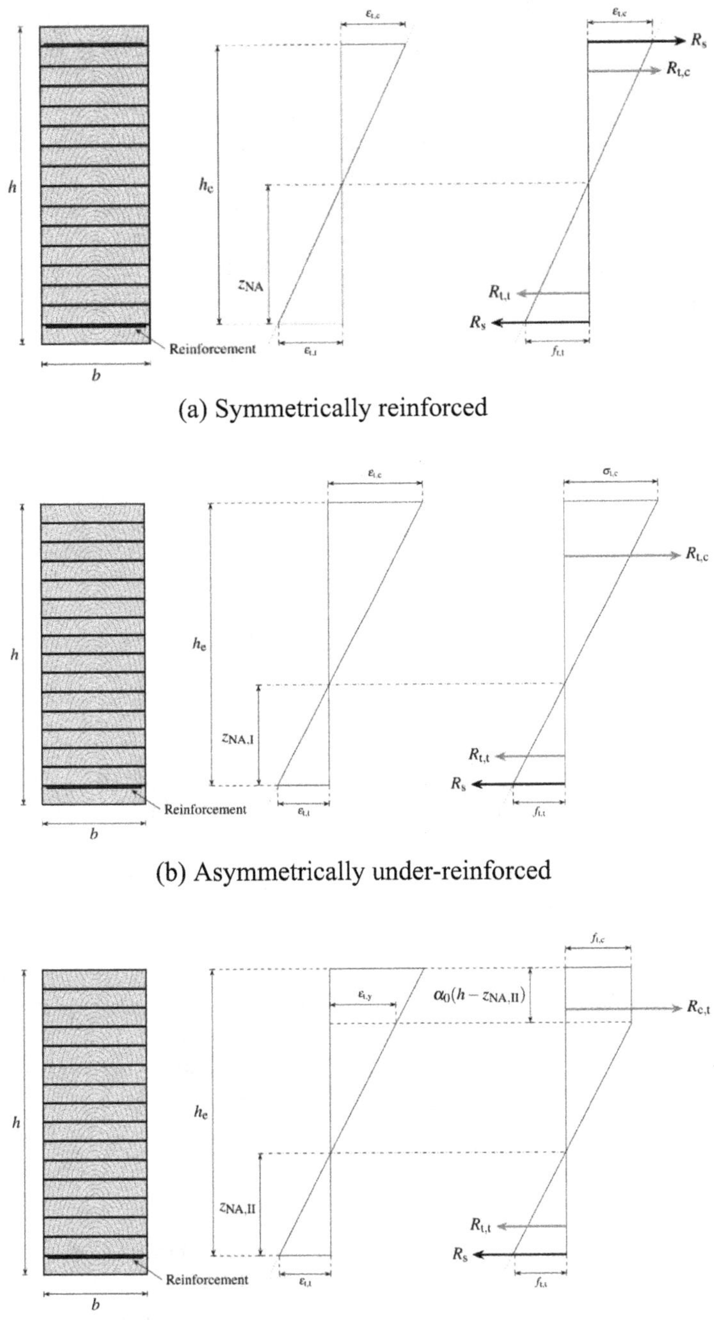

(a) Symmetrically reinforced

(b) Asymmetrically under-reinforced

(c) Asymmetrically over-reinforced

Fig. 1. Capacity models for the reinforced timber beams.

mechanisms. Equations are derived for bending resistance, neutral axis positioning, and resultant stresses based on material and geometric properties, allowing predictions of performance for various reinforcement layouts. The models are shown in Fig. 1.

3 Results and Discussion

Experiments were conducted on full-size GLT beams strip-reinforced with CFRP sheets in four-point bending. The experimental results validated the analytical models, highlighting notable enhancements in mechanical performance. Specifically, reinforced beams showed bending strength improvements of up to 74% and flexural rigidity enhancements reaching up to 65% compared to unreinforced beams.

Reliability analyses were performed to assess the structural safety and efficiency of the strip-reinforced beams. These analyses indicated significant timber volume reductions of up to 44% without compromising structural reliability. Asymmetrically reinforced beams, particularly over-reinforced layouts, exhibited the highest efficiency gains, demonstrating their potential for optimal material usage.

Sensitivity analyses further identified timber tensile strength as the primary parameter influencing variability in bending capacity predictions, emphasizing the need for precise characterization of timber properties. Interestingly, the plastification factor, although crucial for structural applications involving catenary actions, showed minimal influence under pure bending scenarios.

Environmental impacts assessed through a cradle-to-gate analysis indicated higher global warming potential for CFRP-reinforced beams compared to their unreinforced counterparts. However, these higher environmental costs may be justified by the improved mechanical properties, reduced timber volume requirements, and enhanced structural efficiencies. This shows that the use of CFRP may be justified in some specific structural scenarios. Other reinforcing materials may be more beneficial provided that they have comparable mechanical properties and reduced global warming potentials compared with CFRP.

4 Conclusion

The presented research delivers practical, experimentally validated analytical bending capacity models for CFRP-reinforced timber beams, demonstrating clear mechanical and sustainability advantages. While CFRP reinforcement incurs higher environmental impacts, the overall benefits, such as substantial mechanical performance improvements and significant timber usage reductions, may justify its application. Future research recommendations include detailed experimental validation of the plastification factor, strategies to mitigate potential shear failures, comprehensive evaluations addressing serviceability limits, and complete cradle-to-grave environmental analyses. Addressing these areas will facilitate broader adoption and integration of reinforced timber beams within the construction industry, contributing meaningfully to sustainable building practices.

Reference

Cao, A.S., Steiger, R., Frangi, A.: Experimentally validated bending capacity models for strip-reinforced timber beams. Wood Mater. Sci. Eng. (2025). https://doi.org/10.1080/17480272.2025.2509076

Deformation and Creep Recovery of Structural Timber Elements

Conan O'Ceallaigh[1,2(✉)], Patrick J. McGetrick[2,3], Thomas Walsh[3], Gary Moran[3], Karol Sikora[4], Daniel McPolin[5], and Annette M. Harte[2,3]

[1] Department of Building and Civil Engineering, Atlantic Technological University, Galway, Ireland
conan.oceallaigh@atu.ie
[2] Construct Innovate, Galway, Ireland
[3] College of Science and Engineering, University of Galway, Galway, Ireland
[4] Faculty of Engineering and Information Sciences, University of Wollongong in Dubai, Dubai, UAE
[5] School of Planning, Architecture and Civil Engineering, Queen's University Belfast, Belfast, UK

Abstract. The use of reinforcement in timber beams has been shown to improve the short-term flexural behaviour but the long-term behaviour is often more complex, particularly when subjected to a variable climate condition. In this study, the creep deflection behaviour of unreinforced and Basalt Fibre Reinforced Polymer (FRP) reinforced beams are subjected to creep tests at a common maximum compressive stress of 8 MPa over a 450-week ($\approx$ 8.5 years) period. This study builds upon data previously presented by the authors for a test period of 75 weeks. Furthermore, the tested elements are unloaded, and the creep recovery data is also presented. Results demonstrated a significant reduction in total creep deflection due to the FRP reinforcement in both a variable and constant climate. Once unloaded, the results indicate that a significant proportion of creep can be recovered when subjected to a constant climate, however, when subject to a variable climate, there appears to be a significant proportion of non-recoverable deformation.

Keywords: Basalt Fibre · Engineered Wood Products · Creep Behaviour · Creep Recovery; Reinforced Timber · Sitka Spruce

1 Introduction

This chapter is a summary of a study on the long-term performance of both unreinforced and reinforced structural timber elements. FRP (Fibre Reinforced Polymer) materials have been used to successfully strengthen and stiffen structural timber products, not only in new structures but in the upgrading and repair of existing structures [1, 2]. This can significantly improve the flexural performance of timber elements [1–13]. The long-term performance, or the creep behavior, of reinforced structural elements has received less attention in the literature and required further investigation, particularly as recent innovation in the timber construction industry has resulted in the increased use of mass-timber elements and a greater number of tall timber buildings [14–17].

G. Fink et al. (Eds.): *Holistic Design of Taller Timber Buildings*, SPRTRCIENG, pp. 393–398, 2025.
https://doi.org/10.1007/978-3-032-02098-7_34

When using timber as a structural material, particularly in bending, the behaviour is often not governed by its structural strength or ultimate limit state, but by the serviceability limit state, namely deformations and vibrations. As a result, careful consideration must be given to creep deformation. When timber structures are subjected to long-term structural loads, they are subject to an initial elastic response which may be followed by a creep response with time. This creep response may be further split into a time dependent or viscoelastic creep response, a mechano-sorptive creep response due to moisture changes, and a pseudo-creep response that is attributed to the swelling and shrinkage of the timber [18, 19].

The viscoelastic response is characterised as the deformation of an element with time at constant stress under constant environmental conditions, whereas the mechano-sorptive and pseudo-creep responses consider additional effects due to fluctuations in the relative humidity of the surrounding environment. Typically, for timber element carrying gravity loads e.g. a beam, there is an increase in the deflection during each drying phase and a decrease during the wetting phase of a relative humidity cycle. In a variable climate, the total strain/deformation comprises the elastic, viscoelastic, mechano-sorptive and swelling/shrinkage components. It can be difficult to characterise the different creep responses as they often occur simultaneously and as a result, when determining the overall global creep behaviour of a tall timber structure, the impact of the expected environmental conditions needs to be carefully assessed.

This study builds upon the work produced by the authors [19–22] whereby the unreinforced and reinforced glued laminated beams were previously subject to creep testing over a 75-week period in a controlled constant and variable climate. This paper presents the analysis of the continuation of the experimental creep data from 75 weeks up to a period of 450 weeks or 8.5 years. Furthermore, the experimental beams have also been unloaded and creep recovery of the structural elements is presented. The knowledge gathered in this study will contribute to the future development of design guidelines for the creep behaviour and creep recovery behaviour of structural timber elements subjected to combined loading and climatic changes during their in-service life in addition to examining the influence of reinforcement on the long-term behaviour of timber elements.

2 Experimental Programme Summary

The test programme developed by O'Ceallaigh et al. [19–22] was designed to observe the elastic, viscoelastic, mechano-sorptive creep components and swelling/shrinkage behaviour of unreinforced and reinforced beams in constant and variable climates. The 98 mm x 125 mm cross-section of the reinforced beams is shown in Fig. 1a; all beams had a span of 2300 mm. The reinforced beams were reinforced with two, 12 mm diameter basalt fibre-reinforced polymer (BFRP) rods inserted into routed grooves along the bottom lamination.

Four matched groups (two unreinforced and two reinforced groups comprising nine beams each), equal in terms of stiffness, were subjected to long-term creep testing at a common stress of 8 MPa. The reader is directed to the original articles [19–22] for full in-depth details on the experimental programme, formation of matched groups, and loading procedures. The matched groups are herein referred to as:

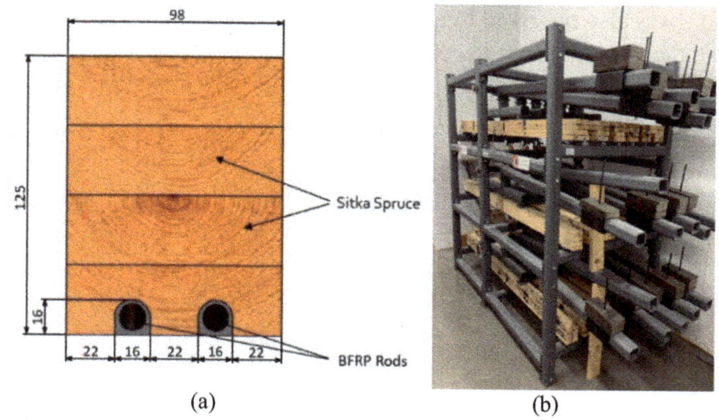

Fig. 1. Creep Testing, (a) FRP reinforced glued laminated beam [22] and (b) the creep testing frame.

- Group UC (UC = **U**nreinforced **C**onstant Climate)
- Group RC (RC = **R**einforced **C**onstant Climate)
- Group UV (UV = **U**nreinforced **V**ariable Climate)
- Group RV (RV = **R**einforced **V**ariable Climate)

Groups UC and RC were subjected to four-point bending in a constant climate (Fig. 1b) with a relative humidity of 65% ± 5% and a temperature of 20 °C ± 2 °C and experiences an initial elastic deflection when loaded, followed by viscoelastic creep with time. Groups UV and RV were also subjected to four-point bending but in a variable climate. The variable climate cycled between a relative humidity of 65% and 90% ± 5% with a cycle length of 8 weeks; this was repeated until a duration of 280 weeks had elapsed, after which the relative humidity was set to 65% and the temperature remained at approximately 20 °C. After 450 weeks had passed, the beams were also unloaded and the initial elastic deflection recovered was monitored followed by the creep recovery behaviour of the unloaded beams.

3 Experimental Results

The relative creep behaviour of the unreinforced and reinforced groups over a 450-week period for the constant climate and variable climate are presented in Fig. 2. This creep behaviour generally can be seen to increase with time, while the differences between Groups UC and RC in Fig. 2 also increase with time and become statistically significant after around 103 weeks (≈2 years), indicating that the use of reinforcement has a positive effect on the creep deflection behaviour of the reinforced beam. A similar trend between Groups UV and RV can be seen for the variable climate however, the difference was statistically significant after only 35 weeks. Following unloading, there were no significant differences in the elastic recovery between unreinforced and reinforced beams, but the climate had a bigger impact, with significantly higher mean unrecovered elastic deformation for beams in the variable climate. Unrecovered creep deformation after 150 days was also observed to follow a similar trend, and this may be attributed to irrecoverable mechanosorptive creep.

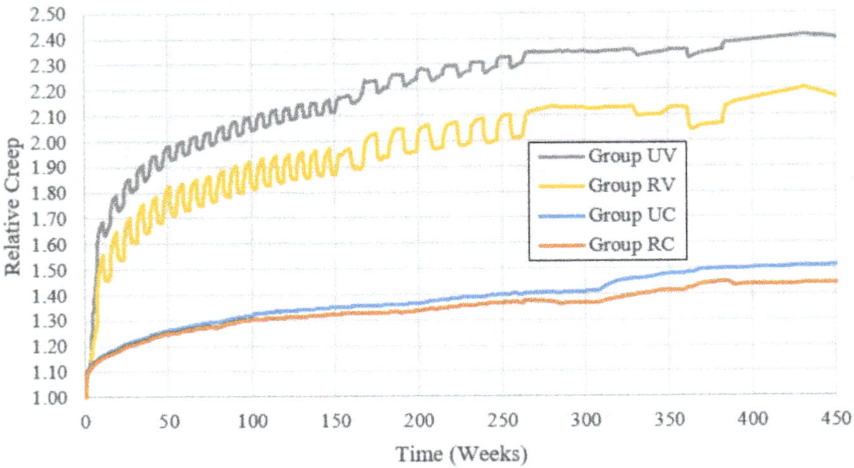

Fig. 2. Comparison between the unreinforced (Group UC) and reinforced (Group RC) beam groups in a constant climate condition and the unreinforced (Group UV) and reinforced (Group RV) beam groups in a variable climate condition over a period of 450 weeks.

4 Conclusions

The creep deflection and creep recovery results presented in the full study provide important test data for examining the future long-term behaviour of timber structures subjected to ever-increasing structural requirements. It shows that reinforcing timber with a suitable FRP material has a positive effect on the creep behaviour of timber elements and may also aid creep recovery. The findings suggest that current creep modification factors used in design could potentially be reduced for reinforced structural timber elements. It should be noted that only one type and reinforcement ratio was utilised in this study and various other parameters, such as stress level, timber species and timber grade will need to be examined before changes to creep modification factors can be proposed. Furthermore, the data can be used to validate a numerical model to predict the creep behaviour of structural timber elements in taller timber buildings.

Acknowledgements. This work was carried out as part of the project 'Innovation in Irish timber Usage' (Project Ref. 11/C/207) funded by the Department of Agriculture, Food and the Marine of the Republic of Ireland under the FIRM/RSF/COFORD scheme. The authors would also like to thank ECC Ltd. (Earrai Coillte Chonnacht Teoranta) for supplying all the timber used in this project. The contribution of the technical staff of the College of Science and Engineering at University of Galway, in particular, Peter Fahy, Colm Walsh and Gerard Hynes, is gratefully acknowledged. This article has also benefited from collaboration and the work from COST Action HELEN CA20139, supported by COST (European Cooperation in Science and Technology).

References

1. Schober, K.-U., Harte, A.M., Kliger, R., Jockwer, R., Xu, Q., Chen, J.-F.: FRP reinforcement of timber structures. Construct. Build Mater. **97**, 106–118 (2015). https://doi.org/10.1016/j.conbuildmat.2015.06.020
2. Kliger, I.R., Haghani, R., Brunner, M., Harte, A.M., Schober, K.-U.: Wood-based beams strengthened with FRP laminates: improved performance with pre-stressed systems. Eur. J. Wood Wood Prod. **74**, 319–330 (2016). https://doi.org/10.1007/s00107-015-0970-5
3. Gilfillan, J.R., Gilbert, S.G., Patrick, G.R.H.: The improved performance of home grown timber glulam beams using fibre reinforcement. J. Inst. Wood Sci. **15**, 307–317 (2001)
4. Raftery, G., Harte, A.: Low-grade glued laminated timber reinforced with FRP plate. Compos. Part B Eng. **42**, 724–735 (2011). https://doi.org/10.1016/j.compositesb.2011.01.029
5. Franke, S., Franke, B., Harte, A.M.: Failure modes and reinforcement techniques for timber beams—State of the art. Construct. Build Mater. **97**, 2–13 (2015). https://doi.org/10.1016/j.conbuildmat.2015.06.021
6. O'Neill, C., McPolin, D., Taylor, S.E., Harte, A.M., O'Ceallaigh, C., Sikora, K.S.: Timber moment connections using glued-in basalt FRP rods. Construct. Build Mater. **145**, 226–235 (2017). https://doi.org/10.1016/j.conbuildmat.2017.03.241
7. Borri, A., Corradi, M., Grazini, A.: A method for flexural reinforcement of old wood beams with CFRP materials. Compos. Part B Eng. **36**, 143–153 (2005). https://doi.org/10.1016/j.compositesb.2004.04.013
8. De la Rosa, G.P., Escamilla, A.C., Nieves González García, M.: Bending reinforcement of timber beams with composite carbon fiber and basalt fiber materials. Compos. Part B Eng. **55**, 528–536 (2013). https://doi.org/10.1016/j.compositesb.2013.07.016
9. Raftery, G., Harte, A.: Nonlinear numerical modelling of FRP reinforced glued laminated timber. Compos. Part B Eng. **52**, 40–50 (2013). https://doi.org/10.1016/j.compositesb.2013.03.038
10. Thorhallsson, E.R., Hinriksson, G.I., Snæbjörnsson, J.T.: Strength and stiffness of glulam beams reinforced with glass and basalt fibres. Compos. Part B Eng. **115**, 300–307 (2017). https://doi.org/10.1016/j.compositesb.2016.09.074
11. Harte, A.M., Dietsch, P.: Reinforcement of timber structures: a state-of-the-art report. Shaker Verlag GmbH, Berlin (2015)
12. Brunetti, M., Christovasilis, I.P., Micheloni, M., Nocetti, M., Pizzo, B.: Production feasibility and performance of carbon fibre reinforced glulam beams manufactured with polyurethane adhesive. Compos. Part B Eng. **156**, 212–219 (2018). https://doi.org/10.1016/j.compositesb.2018.08.075
13. De la Rosa, G.P., Cobo Escamilla, A., González García, M.N.: Analysis of the flexural stiffness of timber beams reinforced with carbon and basalt composite materials. Compos. Part B Eng. **86**, 152–159 (2016). https://doi.org/10.1016/j.compositesb.2015.10.003
14. Ramage, M., Foster, R., Smith, S., Flanagan, K., Bakker, R.: Super Tall Timber: design research for the next generation of natural structure. J. Architect. **22**, 104–122 (2017). https://doi.org/10.1080/13602365.2016.1276094
15. Harley, T., White, G., Dowdall, A., Bawcombe, J., Mcrobie, A., Steinke, R.: Dalston Lane—the world's tallest CLT building. In: Proceedings of the World Conference on Timber Engineering (WCTE 2016) (2016)
16. Abrahamsen, R.: Mjøstårnet—Construction of an 81 m tall timber building. Internationales Holzbau-Forum IHF (2017)
17. Jockwer, R., Grönquist, P., Frangi, A.: Long-term deformation behaviour of timber columns: monitoring of a tall timber building in Switzerland. Eng. Struct. **234**, 111855 (2021). https://doi.org/10.1016/j.engstruct.2021.111855

18. Hunt, D.G.: A unified approach to creep of wood. Proc. R. Soc. Lond. Ser. A Math. Phys. Eng. Sci. **455**, 4077–4095 (1999). https://doi.org/10.1098/rspa.1999.0491
19. O'Ceallaigh, C., Sikora, K., McPolin, D., Harte, A.M.: The mechano-sorptive creep behaviour of basalt FRP reinforced timber elements in a variable climate. Eng. Struct. **200**, 109702 (2019). https://doi.org/10.1016/j.engstruct.2019.109702
20. O'Ceallaigh, C.: An Investigation of the Viscoelastic and Mechano-sorptive Creep Behaviour of Reinforced Timber Elements. PhD Thesis. National University of Ireland, Galway (2016)
21. O'Ceallaigh, C., Sikora, K., McPolin, D., Harte, A.M.: Modelling the hygro-mechanical creep behaviour of FRP reinforced timber elements. Construct. Build Mater. **259**, 119899 (2020). https://doi.org/10.1016/j.conbuildmat.2020.119899
22. O'Ceallaigh, C., Sikora, K., McPolin, D., Harte, A.M.: An investigation of the viscoelastic creep behaviour of basalt fibre reinforced timber elements. Construct. Build Mater. **187**, 220–230 (2018). https://doi.org/10.1016/j.conbuildmat.2018.07.193

Creep Behaviour and Creep Recovery of FRP Reinforced Timber Elements

Conan O'Ceallaigh[1,2](✉), Patrick J. McGetrick[2,3], Thomas Walsh[3], Gary Moran[3], Karol Sikora[4], Daniel McPolin[5], and Annette M. Harte[2,3]

[1] Department of Building and Civil Engineering, Atlantic Technological University, Galway, Ireland
conan.oceallaigh@atu.ie
[2] Construct Innovate, Galway, Ireland
[3] College of Science and Engineering, University of Galway, Galway, Ireland
[4] Faculty of Engineering and Information Sciences, University of Wollongong in Dubai, Dubai, UAE
[5] School of Planning, Architecture and Civil Engineering, Queen's University Belfast, Belfast, UK

Abstract. The use of FRP (Fibre Reinforced Polymer) materials in strategic locations has been shown to successfully strengthen and stiffen structural timber products when required, not only in new structures but also in the upgrading and repair of existing structures (Kliger et al., Eur. J. Wood Wood Prod 74:319–330, 2016; Schober et al., Constr Build Mater 97:106–118, 2015). The addition of such materials which typically have superior properties to that of timber has been shown to significantly improve the flexural performance of timber (Franke et al., Constr Build Mater 97:2–13, 2015; Harte and Dietsch, Reinforcement of timber structures: A state-of-the-art report, Shaker Verlag GmbH, Germany, 2015; Kliger et al., Eur J Wood Wood Prod 74:319–330, 2016; O'Neill et al., Constr Build Mater 145:226–235, 2017; Raftery and Harte, Compos Part B Eng 52:40–50, 2013; Schober et al., Constr Build Mater 97:106–118, 2015; Thorhallsson et al., Compos B Eng 115:300–307, 2017)) however, the long-term performance or the creep behaviour of reinforced structural elements has received less attention in the literature and requires further investigation, particularly as recent innovations in the timber construction industry have resulted in the increased use of mass-timber elements and a greater number of tall timber buildings (Abrahamsen, Mjøstårnet—Construction of an 81 m tall timber building. Internationales Holzbau-Forum IHF, 2017; Harley et al., Proceedings of the World Conference on Timber Engineering (WCTE 2016), 2016; Jockwer et al., Eng Struct 234:111855, 2021; Ramage et al., J Architect 22:104–122, 2017). The long-term behaviour of timber is often complex, particularly when subjected to a variable climate condition and in this study, the creep deflection behaviour of unreinforced and Basalt Fibre Reinforced Polymer reinforced beams are subjected to creep testing at a common maximum compressive stress of 8 MPa in both constant and variable climate conditions. This

This is an extended abstract of a paper published in the Special Issue of the COST Action HELEN in the journal *Wood Material Science & Engineering*. The full article is available at: https://doi.org/10.1080/17480272.2025.2480248

G. Fink et al. (Eds.): *Holistic Design of Taller Timber Buildings*, SPRTRCIENG, pp. 399–402, 2025.
https://doi.org/10.1007/978-3-032-02098-7_35

study builds upon the work produced by the authors (O'Ceallaigh, An Investigation of the Viscoelastic and Mechano-sorptive Creep Behaviour of Reinforced Timber Elements. National University of Ireland, Galway, 2016; O'Ceallaigh et al., Constr Build Mater 259:119899, 2020, 2019, 2018) whereby the creep behaviour of unreinforced and reinforced glued laminated beams over a 75-week period in a controlled constant and variable climate was presented. This study herein presents a significant extension of the experimental creep data from 75 weeks up to a period of 450 weeks or 8.5 years. Over this period, the results have shown that reinforcing timber with an FRP material of superior properties has a positive effect on the creep behaviour of timber elements. In the constant climate, the percentage difference between the unreinforced (Group UC) and reinforced (Group RC) is statistically significant after approximately 103 weeks demonstrating a positive effect of FRP reinforcement on the creep deflection behaviour of reinforced beams. In a variable climate, the mean relative creep results demonstrated a statistically significant difference between the unreinforced (Group UV) and reinforced (Group RV) after just the first relative humidity cycle. This trend has continued for the duration of the creep testing (450 weeks) and further demonstrates that the FRP reinforcement has a statistically significant impact on the creep deflection of structural timber elements.

Furthermore, the tested elements were unloaded, and the creep recovery data was also presented. A significant reduction in total creep deflection due to the FRP reinforcement was observed in the results obtained in both a variable and constant climate and once unloaded, the results indicate that a significant proportion of creep can be recovered in beams subjected to a constant climate, however, when subject to a variable climate, there appears to be a significant proportion of non-recoverable deformation. In conclusion, the use of FRP reinforcement can reduce the creep behaviour of structural timber elements but the influence of the climatic conditions is significant, and the magnitude of the creep behaviour is a product of the coupled load and environmental history of the beams.

Keywords: Basalt Fibre · Engineered Wood Products · Creep Behaviour · Creep Recovery · Reinforced Timber · Sitka Spruce

Acknowledgements This work was carried out as part of the project 'Innovation in Irish timber Usage' (Project Ref. 11/C/207) funded by the Department of Agriculture, Food and the Marine of the Republic of Ireland under the FIRM/RSF/COFORD scheme. The authors would also like to thank ECC Ltd. (Earrai Coillte Chonnacht Teoranta) for supplying all the timber used in this project. The contribution of the technical staff of the College of Science and Engineering at University of Galway, in particular, Peter Fahy, Colm Walsh and Gerard Hynes, is gratefully acknowledged. This article has also benefited from collaboration and the work from COST Action HELEN CA20139, supported by COST (European Cooperation in Science and Technology).

References

Abrahamsen, R.: Mjøstårnet—Construction of an 81 m tall timber building. Internationales Holzbau-Forum IHF (2017)

Franke, S., Franke, B., Harte, A.M.: Failure modes and reinforcement techniques for timber beams—State of the art. Construct. Build Mater. **97**, 2–13 (2015). https://doi.org/10.1016/j.conbuildmat.2015.06.021

Harley, T., White, G., Dowdall, A., Bawcombe, J., Mcrobie, A., Steinke, R.: Dalston Lane—The world's tallest CLT building. In: Proceedings of the World Conference on Timber Engineering (WCTE 2016) (2016)

Harte, A.M., Dietsch, P.: Reinforcement of timber structures: a state-of-the-art report. Shaker Verlag GmbH, Berlin (2015)

Jockwer, R., Grönquist, P., Frangi, A.: Long-term deformation behaviour of timber columns: monitoring of a tall timber building in Switzerland. Eng. Struct. **234**, 111855 (2021). https://doi.org/10.1016/j.engstruct.2021.111855

Kliger, I.R., Haghani, R., Brunner, M., Harte, A.M., Schober, K.-U.: Wood-based beams strengthened with FRP laminates: improved performance with pre-stressed systems. Eur. J. Wood Wood Prod. **74**, 319–330 (2016). https://doi.org/10.1007/s00107-015-0970-5

O'Ceallaigh, C.: An Investigation of the Viscoelastic and Mechano-sorptive Creep Behaviour of Reinforced Timber Elements. PhD Thesis, National University of Ireland, Galway (2016)

O'Ceallaigh, C., Sikora, K., McPolin, D., Harte, A.M.: Modelling the hygro-mechanical creep behaviour of FRP reinforced timber elements. Construct. Build Mater. **259**, 119899 (2020). https://doi.org/10.1016/j.conbuildmat.2020.119899

O'Ceallaigh, C., Sikora, K., McPolin, D., Harte, A.M.: The mechano-sorptive creep behaviour of basalt FRP reinforced timber elements in a variable climate. Eng. Struct. **200**, 109702 (2019). https://doi.org/10.1016/j.engstruct.2019.109702

O'Ceallaigh, C., Sikora, K., McPolin, D., Harte, A.M.: An investigation of the viscoelastic creep behaviour of basalt fibre reinforced timber elements. Construct. Build Mater. **187**, 220–230 (2018). https://doi.org/10.1016/j.conbuildmat.2018.07.193

O'Neill, C., McPolin, D., Taylor, S.E., Harte, A.M., O'Ceallaigh, C., Sikora, K.S.: Timber moment connections using glued-in basalt FRP rods. Construct. Build Mater. **145**, 226–235 (2017). https://doi.org/10.1016/j.conbuildmat.2017.03.241

Raftery, G., Harte, A.: Nonlinear numerical modelling of FRP reinforced glued laminated timber. Compos. Part B Eng. **52**, 40–50 (2013). https://doi.org/10.1016/j.compositesb.2013.03.038

Ramage, M., Foster, R., Smith, S., Flanagan, K., Bakker, R.: Super Tall Timber: design research for the next generation of natural structure. J. Architect. **22**, 104–122 (2017). https://doi.org/10.1080/13602365.2016.1276094

Schober, K.-U., Harte, A.M., Kliger, R., Jockwer, R., Xu, Q., Chen, J.-F.: FRP reinforcement of timber structures. Construct. Build Mater. **97**, 106–118 (2015). https://doi.org/10.1016/j.conbuildmat.2015.06.020

Thorhallsson, E.R., Hinriksson, G.I., Snæbjörnsson, J.T.: Strength and stiffness of glulam beams reinforced with glass and basalt fibres. Compos. Part B Eng. **115**, 300–307 (2017). https://doi.org/10.1016/j.compositesb.2016.09.074

Rate-Dependency and Hysteresis of Timber Connections with a Single Laterally-Loaded Dowel and a Slotted-in Steel Plate

Katharina Sroka[1,2](✉), Alex Sixie Cao[1,2], Pedro Palma[1], and Andrea Frangi[2]

[1] Empa - Swiss Federal Laboratories for Materials Science and Technology, 8600 Dübendorf, Switzerland
{katharina.sroka,alex.cao,pedro.palma}@empa.ch
[2] ETH Zurich, 8093 Zurich, Switzerland
frangi@ibk.baug.ethz.ch

Abstract. In timber structures, connections are vital for plasticity and ductile failure modes but knowledge on their behaviour under high loading rates or cyclic loading is limited. This chapter summarizes an experimental campaign on laterally-loaded timber connections with a single dowel and a slotted-in steel plate. Quasi-static cyclic experiments showed severe degradation phenomena, while high loading rates in monotonic experiments led to increased yield force and, at loading parallel to the grain, to reduced ductility. The effects on ductility, maximum force, and failure modes varied with the load-to-grain angle.

1 Introduction

Wind, earthquake, traffic, blast, or impact can induce cyclic or high-rate loading into a building. In these conditions, ductile failure modes are required to avoid sudden collapse without prior warning. In timber structures, the connections are critical to provide this ductility [7]. Connections with laterally-loaded dowels and one or more slotted-in steel plates are commonly used and can exhibit three basic failure modes of increasing ductility [6]: embedment, development of one plastic hinge in the dowel, or development of three plastic hinges in a slender dowel. Understanding the behaviour of these connections under high loading rates and cyclic loading is critical to designing safe, robust, and efficient timber structures.

High loading rate effects in timber connections with metal fasteners depend on the load-to-grain angle and failure mode [5,13]. Strain rates above 10^{-3} s^{-1} increase the yield and tensile strength of steel [9], facilitating brittle failure in these connections. A logarithmic dependency on the loading rate is typically suggested for timber connections [5,8]. Embedment and bolted connection experiments in LVL demonstrated an increase in the embedment strength and elastic stiffness, but a reduction in the ultimate displacement [2]. The rate-dependency

G. Fink et al. (Eds.): *Holistic Design of Taller Timber Buildings*, SPRTRCIENG, pp. 403–406, 2025.
https://doi.org/10.1007/978-3-032-02098-7_36

of the maximum force of nailed timber joints was found to be stronger for loading perpendicular to the grain, while the elastic stiffness did not depend on the loading rate [5].

Experiments on laterally-loaded dowel-type connections under cyclic loading consistently showed pinching, strength degradation, and stiffness degradation [3,11]. Larger dowel spacing or reinforcement with self-tapping screws increase the load-carrying capacity, ductility, and energy dissipation and can ensure ductile failure modes [1,10]. The steel properties and detailing of the steel plates significantly influence the ductility and failure modes [4].

The objectives of this study were to (i) investigate rate-dependent effects and (ii) provide experimental data for modelling hysteresis in timber connections with a single laterally-loaded 10 mm S235 steel dowel and a slotted-in steel plate. Approximately 100 single-fastener connections in Norway spruce LVL were tested in displacement control under (i) monotonic loading with displacement rates of 0.05, 15, and 150 $\mathrm{mm \cdot s^{-1}}$ and (ii) quasi-static reversed cyclic loading, both at load-to-grain angles 0°, 30°, 60°, and 90° [12].

2 Experimental Research

The experimental setup is shown in Figs. 1a and 2a. Ductile failure with three plastic hinges in the dowel was observed under monotonic loading with ultimate splitting failure at 0° load-to-grain angle. The rate-dependent effects were not the same across different load-to-grain angles. At 0° load-to-grain angle, a significant loss of ductility was observed at high displacement rates, while the main rate-dependent effect at 90° was an increase in maximum force (Fig. 1b). The cyclic experiments showed stiffness and strength degradation, pinching, zero-stiffness regions, and force plateaus during load reversal at large displacement cycles (Fig. 2b). For more information, see Sroka et al. [12].

(a) (b)

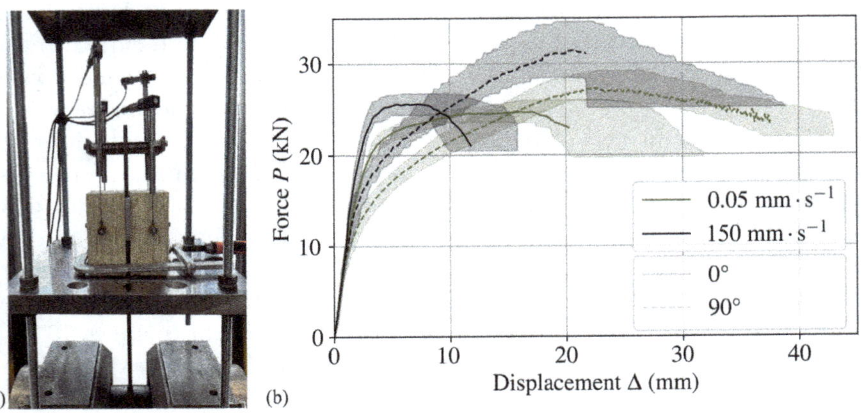

Fig. 1. Experimental setup for monotonic loading parallel to the grain (a) and mean force-displacement curves with upper and lower bound for quasi-static and high-speed loading at load-to-grain angles 0° and 90° (b).

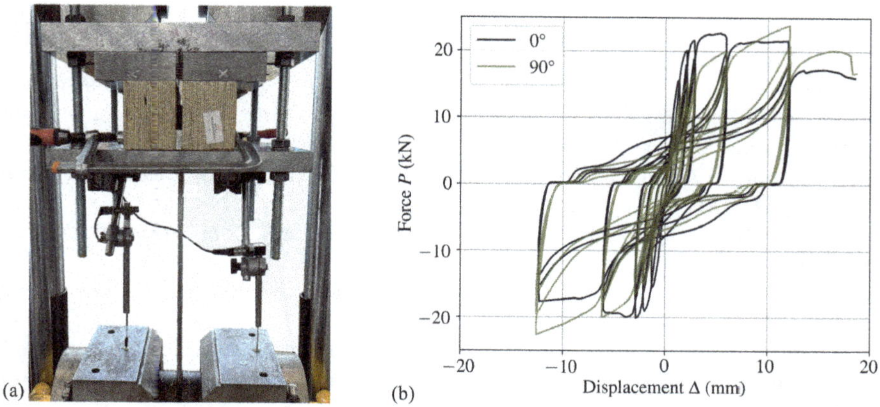

Fig. 2. Experimental setup for cyclic loading (a) and mean hysteresis curves for load-to-grain angles 0° and 90° (b).

3 Conclusions

This experimental campaign studied timber connections with a single laterally-loaded dowel and a slotted-in steel plate under quasi-static cyclic loading and monotonic loading at various displacement rates. Higher displacement rates delayed yielding, but rate-dependent effects on the maximum force and plastic hardening varied by load-to-grain angle. At loading parallel to the grain, the connections showed reduced ductility at high displacement rates due to earlier splitting. Cyclic experiments revealed pinching, strength and stiffness degradation, and low-cycle fatigue failure of the dowel. These findings are relevant for designing connections under high displacement rates and cyclic loading, with applications in impact, blast loading, seismic analysis, and progressive collapse.

Acknowledgements. The authors thank Kilian Paul and Maksim Borovlev for their help in the experimental investigation and initial data analysis, as well as the team of the Bauhalle at ETH Zurich.

Competing Interests. The authors gratefully acknowledge the financial support provided by SNSF [grant no. 213364] and SNSF and Formas [grant no. SNSF 220112, Formas 2023-01603]. The authors have no conflicts of interest to declare that are relevant to the content of this chapter.

References

1. Baird, Z., Woods, J.E., Viau, C., Doudak, G.: Cyclic behavior of bolted glued-laminated timber brace connections with slotted-in steel plates. J. Struct. Eng. **150** (2024)
2. Cheng, X., Gilbert, B.P., Lyu, C., Guan, H.: Strain rate effect on the embedment mechanical properties and fracture behavior of softwood laminated veneer lumbers: implications to timber connections. J. Struct. Eng. **149** (2023)

3. Chui, Y.H., Ni, C.: Load-embedment response of timber to reversed cyclic load. Wood Fiber Sci. **29**, 148–160 (1997)
4. Geiser, M., Furrer, L., Kramer, L., Blumer, S., Follesa, M.: Investigations of connection detailing and steel properties for high ductility doweled timber connections. Constr. Build. Mater. **324** (2022)
5. Girhammar, U.A., Andersson, H.: Effect of loading rate on nailed timber joint capacity. J. Struct. Eng. **114**, 2439–2456 (1988)
6. Johansen, K.W.: Theory of timber connections. IABSE Publications **9**, 249–262 (1949)
7. Jorissen, A., Fragiacomo, M.: General notes on ductility in timber structures. Eng. Struct. **33**, 2987–2997 (2011)
8. Lacroix, D.N., Doudak, G.: Investigation of dynamic increase factors in light-frame wood stud walls subjected to out-of-plane blast loading. J. Struct. Eng. **141** (2015)
9. Malvar, L.J., Crawford, J.E.: Dynamic increase factors for steel reinforcing bars. In: DDESB Seminar, DoD Explosive Safety Board, Orlando, Florida (1998)
10. Ottenhaus, L.M., Li, M., Smith, T.: Structural performance of large-scale dowelled CLT connections under monotonic and cyclic loading. Eng. Struct. **176**, 41–48 (2018)
11. Shu, Z., Li, Z., Yu, X., Zhang, J., He, M.: Rotational performance of glulam bolted joints: experimental investigation and analytical approach. Constr. Build. Mater. **213**, 675–695 (2019)
12. Sroka, K., Cao, A.S., Palma, P., Frangi, A.: Rate-dependency and hysteresis of timber connections with a single laterally-loaded dowel and a slotted-in steel plate. Wood Mater Sci Eng (2025)
13. Viau, C., Doudak, G.: Effect of high strain-rates on heavy timber connections. In: CSCE Annual Conference, Laval, Canada (2019)

Fire Risk Perception of Timber Buildings: Implications for the Responsibility of Designers

Leo Willem Menzemer[1,2](✉)

[1] Division of Fire Safety Engineering, Lund University, Lund, Sweden
leo_willem.menzemer@brand.lth.se
[2] The Danish Institute of Fire and Security Technology, Hvidovre, Denmark

Abstract. Timber's sustainability benefits make it an attractive construction material, but public concerns about fire safety remain a challenge. Many associate timber buildings with higher fire risks, particularly in taller structures, despite trusting regulations and professionals to ensure safety. This highlights a gap between public perception and technical knowledge. Designers have a professional responsibility to address these concerns through clear communication and transparent, appropriate risk mitigation strategies to ensure the safety of building occupants. Disappointing this responsibility can negatively impact people's safety and lead to stigmatization of wooden building materials.

Keywords: Fire safety · Wood construction · Responsible Practice · Building design · Fire Engineering

1 Introduction

Building in biogenic materials like timber is a potential driver to reduce the impact of construction on the environment [1] and the market demand for engineered wood products (e.g. cross-laminated timber, CLT) is continuously increasing in recent years [2]. As the material is inherently combustible it does introduce challenges to its safe use in construction.

When timber is being involved in a fire, it reduces the time to flashover [3], thereby decreasing the available time during evacuations to reach a place of safety. It also leads to increased heat release rates and external flaming [4, 5]. Additionally, char fall-off or delamination can prevent auto-extinction after burnout of moveable fuels resulting in continued burning of the structure [6, 7]. The extent of such effects depends on the amount of exposed timber involved in a compartment fire, where a meta-analysis identified that with more than 20% of the timber surface being exposed, the probability of continued burning and no auto-extinction rises significantly [8]. While researchers in fire science in recent years outlined the potential hazards that need to be mitigated in timber building

This is an extended abstract of a paper published in the Special Issue of the COST Action HELEN in the journal *Wood Material Science & Engineering*. The full open-access article is available at: https://www.tandfonline.com/doi/full/10.1080/17480272.2025.2459350

G. Fink et al. (Eds.): *Holistic Design of Taller Timber Buildings*, SPRTRCIENG, pp. 407–411, 2025.
https://doi.org/10.1007/978-3-032-02098-7_37

design [9], advocates for the use of timber criticized an "irrational approach to fire risk" [10].

Surveying designers' perception and barriers to the use of wooden materials in practice, studies showed that the combustibility of timber raises concerns about its fire safety performance in construction [11, 12]. As a result, its use is often restricted due to fire safety considerations and building regulations [13, 14].

Recognizing differing professional opinions and practitioner uncertainties about timber's fire safety, this chapter examines how laypeople perceive timber as a construction material in terms of fire safety - an area of both professional debate and consumer interest. Understanding public risk perception in timber buildings informs on emergency behavior, intervention planning, mitigation strategies, and aids to improving communication about timber fire risks to public stakeholders, including policymakers [15].

2 Public Perception of Timber Buildings

Consumers generally view timber positively for its aesthetics, comfort, naturalness, and eco-friendliness, however technical properties such as stability, combustibility, and durability are often rather seen as potential disadvantages of the material. As such, consumer perceptions align with those of practitioners like engineers and architects, but beliefs may vary across different target groups. Figure 1 shows how timber is perceived with regard to a selection of characteristics in comparison to other building materials, based on results from research on public fire risk perception of timber buildings [15].

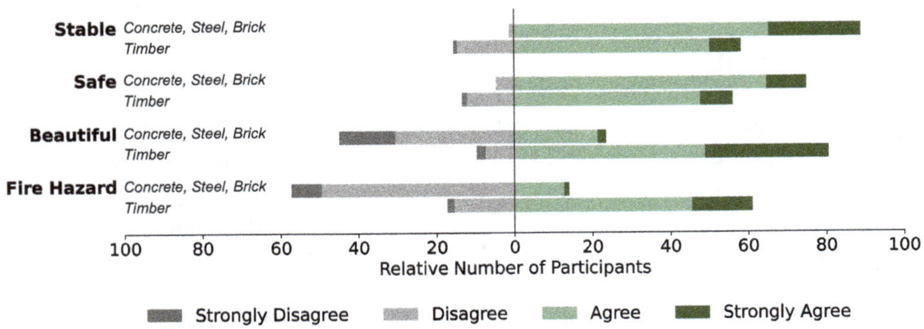

Fig. 1. Perception of timber in comparison to other building materials, based on [15]. Neutral opinions were indicated where percentages do not sum up to 100%.

2.1 Fire Risk Perception of Timber Buildings

Investigating fire risk perceptions of timber buildings among the public with a mixed-method study [15], combining an online survey with semi-structured interviews, respondents expressed less comfort living on higher floors in timber buildings, and were more likely to evacuate rather than adopt stay-put policies in fire emergencies. Fire incidents were perceived as severe across varied building types, but concerns about risk were more divided for timber buildings in the survey data.

Interviews with laypeople from Denmark and Sweden corroborate these findings, although underlying many accounts was a common theme on lack of knowledge about construction materials, with exposed or covered timber affecting perceptions. Many people tend to associate a diffuse risk with the inherent combustibility of timber when employed as a building material, while others argue that fires also occur in non-combustible buildings. Regardless of stance, most participants expressed uncertainty about fire safety in timber buildings. Interestingly, recognizing risks did not necessarily translate to feelings of unsafety; many trusted architects, engineers, and building regulations to ensure adequate fire safety. Greater concerns were expressed for taller buildings supporting trends in survey data that show people being less comfortable living on higher floors in timber buildings. In context of taller timber buildings, means and time for safe escape in case of a fire become an increased concern for some, with some interviewees stating they would research fire safety before moving in.

Despite fire concerns, timber's aesthetic and environmental appeal remained important, and many were open to living in timber buildings if safety could be ensured. However, practical factors like affordability and availability were constraints to opting for a timber building in the first place, aligning with findings from literature [16]. Greater exposure to timber buildings and clearer safety communication could enhance public confidence in their use.

3 Implications for Design Practice

Timber buildings are increasingly recognized for their sustainability, yet fire safety concerns can remain a barrier to its broad adoption. Public perception of timber buildings differs from that of non-combustible structures, with timber buildings being perceived as a greater fire hazard. These perceptions stem from experiences, cultural traditions, and intuitive associations with wood as a combustible material. Interestingly, despite recognizing these risks, the public largely places trust in professional expertise and regulatory systems to ensure safety, indicating a disconnect between technical understanding and public confidence. This aligns with the concept of the "certainty trough" [17], which suggests that those most distant from specialized knowledge—such as the general public—experience the greatest uncertainty, while professionals adjacent to fire science may hold overconfident simplifications of fire safety issues. In contrast, professionals and researchers within the field remain most aware of its complexities and limitations.

For designers, this highlights the need for responsible communication, risk mitigation strategies, and understanding limits to existing applications and knowledge. Public concern is particularly heightened in taller timber buildings. Rather than dismissing fire concerns as "prejudices," designers should adequately engage with the potential hazards associated to any building material and transparently communicate with the public, demonstrating the identification and management of hazards at a systemic level. A shift in communication towards the public—from debating oversimplified material properties to explaining conceptually how fire safety is ensured—can help bridge the certainty gap, maintain societal trust, and support the broader acceptance of timber construction.

Neglecting this responsibility not only undermines public trust but also risks fostering misconceptions that can lead to a misguided stigmatization of biobased building materials like timber for the future, while ultimately putting people at risk.

References

1. Churkina, G., Organschi, A., Reyer, C.P.O., Ruff, A., Vinke, K., Liu, Z., Reck, B.K., Graedel, T.E., Schellnhuber, H.J.: Buildings as a global carbon sink. Nat. Sustain. **3**, 269–276 (2020). https://doi.org/10.1038/s41893-019-0462-4
2. UNECE/FAO: Forest Products Annual Market Review 2020-2021, United Nations Economic Commission for Europe, Geneva, Switzerland (2021). https://unece.org/sites/default/files/2021-11/2114516E_Inside_Final_web.pdf. Accessed 28 May 2024
3. Su, J., Pier-Simon, L., Hoehler, M., Bundy, M.: Data from Fire Safety Challenges of Tall Wood Buildings - Phase 2: Task 3 - Cross Laminated Timber Compartment Fire Tests. National Institute of Standards and Technology (2018) https://data.nist.gov/od/id/5CF421B31BE8030 4E0531A57068197E41874. Accessed 31 May 2024
4. Gorska, C., Hidalgo, J.P., Torero, J.L.: Fire dynamics in mass timber compartments. Fire Saf. J. **120**, 103098 (2021). https://doi.org/10.1016/j.firesaf.2020.103098
5. Hadden, R.M., Bartlett, A.I., Hidalgo, J.P., Santamaria, S., Wiesner, F., Bisby, L.A., Deeny, S., Lane, B.: Effects of exposed cross laminated timber on compartment fire dynamics. Fire Saf. J. **91**, 480–489 (2017). https://doi.org/10.1016/J.FIRESAF.2017.03.074
6. McNamee, R., Zehfuss, J., Bartlett, A.I., Heidari, M., Robert, F., Bisby, L.A.: Enclosure fire dynamics with a cross-laminated timber ceiling. Fire Mater. **45**, 847–857 (2021). https://doi.org/10.1002/fam.2904
7. Pope, I., Xu, H., Gupta, V., Carrascal, J., Lange, D., McLaggan, M., Mendez, J., Solarte, A., Soriguer, D., Torero, J.L., Wiesner, F., Hidalgo, J.P.: Fire dynamics in under-ventilated mass timber room compartments. In: Proceedings of the 12th Asia-Oceania Symposium on Fire Science and Technology (AOSFST 2021). The University of Queensland (2021). https://doi.org/10.14264/7e19a8f
8. Ali Awadallah, A.A., Hadden, R.M., Law, A.: Meta-analysis of temperature, heat release rate, delamination and auto-extinction of timber compartments. Fire Saf. J. **146**, 104164 (2024). https://doi.org/10.1016/j.firesaf.2024.104164
9. Law, A., Hadden, R.: We need to talk about timber: fire safety design in tall buildings. Struct. Eng. (2020). https://www.istructe.org/journal/volumes/volume-98-(2020)/issue-3/we-need-to-talk-about-timber-fire-safety-design-in/
10. Wimmers, G.: Wood: a construction material for tall buildings. Nat. Rev. Mater. **2**, 17051 (2017). https://doi.org/10.1038/natrevmats.2017.51
11. Xia, B., O'Neill, T., Zuo, J., Skitmore, M., Chen, Q.: Perceived obstacles to multi-storey timber-frame construction: an Australian study. Archit. Sci. Rev. **57**, 169–176 (2014). https://doi.org/10.1080/00038628.2014.912198
12. Laguarda Mallo, M.F., Espinoza, O.: Awareness, perceptions and willingness to adopt cross-laminated timber by the architecture community in the United States. J. Clean. Prod. **94**, 198–210 (2015). https://doi.org/10.1016/j.jclepro.2015.01.090
13. Giorgio, B., Barlet, A., Blanchet, P., Cabral, M.R., Perez, C., Gaudelas, A.: Using mass timber in multi-storey and non-residential construction projects: Motivations and barriers for professionals in Quebec. BioRes. **20**, 1931–1970 (2025). https://doi.org/10.15376/biores.20.1.1931-1970
14. Viļuma, A., Bratuškins, U.: Barriers for use of wood in architecture: the Latvian case. Architect. Urban Plann. **13**, 43–47 (2017). https://doi.org/10.1515/aup-2017-0006

15. Menzemer, L.W., Vad Karsten, M.M., Gwynne, S., Dragsted, A., Ronchi, E.: Public perception of fire safety and risk of timber buildings. Wood Mater. Sci. Eng., 1–14 (2025). https://doi.org/10.1080/17480272.2025.2459350
16. Jussila, J., Franzini, F., Häyrinen, L., Lähtinen, K., Nagy, E., Mark-Herbert, C., Roos, A., Toppinen, A., Toivonen, R.: Consumer housing choices among residents living in wooden multi-storey buildings. Hous. Stud. **39**, 2654–2679 (2024). https://doi.org/10.1080/02673037.2023.2217765
17. MacKenzie, D.: The certainty trough. In: Williams, R., Faulkner, W., Fleck, J. (eds.) Exploring Expertise, pp. 325–329. Palgrave Macmillan UK, London (1998). https://doi.org/10.1007/978-1-349-13693-3_15

Author Index

G. Fink et al. (Eds.): *Holistic Design of Taller Timber Buildings*, SPRTRCIENG, pp. 413–414, 2025.
https://doi.org/10.1007/978-3-032-02098-7

www.ingramcontent.com/pod-product-compliance
Lightning Source LLC
Chambersburg PA
CBHW071739271025
34597CB00003B/125
* 9 7 8 3 0 3 2 0 2 0 9 7 0 *